DISCRETE MATHEMATICS AND ITS APPLICATIONS
Series Editor KENNETH H. ROSEN

DIOPHANTINE ANALYSIS

T0179170

DISCRETE MATHEMATICS AND ITS APPLICATIONS

Series Editor

Kenneth H. Rosen, Ph.D.

Continued Titles

DISCRETE MATHEMATICS AND ITS APPLICATIONS
Series Editor KENNETH H. ROSEN

DIOPHANTINE ANALYSIS

JÖRN STEUDING

CRC Press
Taylor & Francis Group
Boca Raton London New York

CRC Press is an imprint of the
Taylor & Francis Group, an **informa** business
A CHAPMAN & HALL BOOK

CRC Press
Taylor & Francis Group
6000 Broken Sound Parkway NW, Suite 300
Boca Raton, FL 33487-2742

First issued in paperback 2019

© 2005 by Taylor & Francis Group, LLC
CRC Press is an imprint of Taylor & Francis Group, an Informa business

No claim to original U.S. Government works

ISBN-13: 978-1-58488-482-8 (hbk)
ISBN-13: 978-0-367-39285-7 (pbk)

Library of Congress Cataloging-in-Publication Data

Catalog record is available from the Library of Congress

Visit the Taylor & Francis Web site at
http://www.taylorandfrancis.com

and the CRC Press Web site at
http://www.crcpress.com

Contents

Preface

Preface

The book is devoted to the theory of diophantine approximations and the theory of diophantine equations with emphasis on interactions between these subjects. Many diophantine problems have simple formulations but are extremely hard to attack. For instance, consider the rather simple looking equation

$$X^n + Y^n = Z^n,$$

where n is a positive integer and which has to be solved in integers x, y, z. Integer solutions x, y, z which satisfy $xyz = 0$ are — for obvious reasons — called *trivial*. In the case $n = 2$ there are many other solutions, e.g.,

$$3^2 + 4^2 = 5^2, \quad 5^2 + 12^2 = 13^2,$$

already known by the ancient Greeks. However, the situation becomes totally different when $n \geq 3$. For such exponents, Pierre de Fermat, a French lawyer and mathematician in the seventeenth century, claimed to be able to prove that there are no solutions other than trivial ones; however, he never published his proof. In Fermat's time publishing proofs was not very common; usually, mathematicians wrote letters to other mathematicians announcing what they could prove and asked whether the other could do the same. Fermat is famous for this method of doing math. But for more than three centuries no one succeeded in proving Fermat's statement, known as Fermat's last theorem (since it was the last of his statements to be proved).

In the twentieth century, diophantine analysis emerged from its very beginnings to an extraordinary modern and powerful theory (with three Fields medalists: Roth, Baker, Faltings); one example for this success story is Wiles' solution of Fermat's last theorem, the final proof that there are no solutions other than the trivial ones. However, there are still a lot of open problems and conjectures which turn diophantine analysis into an active and attractive field of interest for researchers and students.

The book's motivation is contained in its introduction. Here we present some basic principles of diophantine analysis. The second chapter deals with classical approximation theorems and the third chapter is devoted to the theory of continued fractions. Then we give a more detailed account on certain topics of the classic theory (Chapters 5 and 6) and present some of its applications (Chapters 4 and 7). Chapter 8 gives a short introduction to the geometry of numbers and its applications. The following two chapters deal with transcendental numbers and Roth's theorem with applications to Thue equations. In Chapter 11 we present the recent *abc*-conjecture and discuss its importance. We conclude with a short introduction to p-adic numbers and their applications to diophantine equations (Chapters 12–14).

Certain topics like elliptic curves or the metric theory of continued fractions are touched but not entirely included because an appropriate presentation would be beyond the scope of this book. However, there exist many excellent books describing these areas in detail and we refer to some of them for further reading.

Modern topics (which were only considered in a few earlier textbooks) are Apéry's celebrated proof of the irrationality of $\zeta(3)$ (Chapter 4), the polynomial Pell equation (parts of Chapters 6 and 11), and the abc-conjecture with its plenty of applications (Chapter 11). Furthermore, we touch several topics which are of interest in related fields of discrete mathematics: factoring methods for large integers with continued fractions (Chapter 7), the LLL-lattice reduction algorithm (Chapter 8), error-free computing with p-adic numbers (Chapter 12), and factoring polynomials with Hensel lifting (Chapter 13).

This book primarily aims at advanced undergraduate students or graduates who may want to learn the fundamentals of this subject. In some sense, diophantine analysis may be regarded as algebraic geometry over the spectrum of a Dedekind domain. This modern point of view is much beyond the scope of this introduction but we shall keep in mind the idea of using the geometry of the objects under observation to learn something about their arithmetic nature. Our approach is elementary; with exception of the proof of Roth's theorem, only a small background in algebra and number theory is needed. Many of the results are presented with respect to their historical development. I believe that the best way to learn math is to look at how it was developed. However, only practice makes perfect. For this purpose, at the end of each chapter, several exercises and problems of different degrees of difficulty are posed, the advanced ones indicated by an asterisk *. Furthermore, I recommend using a computer algebra package like Mathematica or Maple; excellent introductions to experimental mathematics are Bressoud & Wagon [29] and Vivaldi [165], respectively.

The concept of the book is based upon a course which I gave at Frankfurt University in the winter term 2002/2003. I am very grateful to the audience, in particular, Iqbal Lahseb, Thomas Müller, and Matthias Völz. I want to express my appreciation to my colleagues at the departments of the universities in Frankfurt, Vilnius, Šiauliai, and Madrid; and the many people from whom I learned this subject, in particular, my academic teachers Georg Johann Rieger, Wolfgang Schwarz, and Jürgen Wolfart. I am very much indebted to Jürgen Sander and Hessel Posthuma for inspiring conversations and careful reading of the manuscript. My thanks also go to Christian Beck, Artūras Dubickas, Carsten Elsner, Ernesto Girondo, and Harald König for their interest, help, and valuable remarks. Further, I want to thank Helena Redshaw, Bob Stern, and Kenneth Rosen from CRC Press for their encouragement, and the editorial staff at CRC Press for their support during the preparation of this book.

Finally, and most importantly, I want to thank Rasa.

Jörn Steuding, Madrid, December 2004

CHAPTER 1

Introduction: basic principles

Diophantine analysis is devoted to diophantine approximations and diophantine equations. Here the word *diophantine* means that we are concerned about integral or rational solutions. The exact meaning will become clear in the sequel. In this introductory chapter we shall learn some of the basic principles of diophantine analysis. These are special methods (as Fermat's method of infinite descent) as well as building bridges to other fields (e.g., the use of real analysis or geometry for understanding underlying arithmetic structures). In order to present these principles we prove some fundamental historical results.

1.1. Who was Diophantus?

Diophantus of Alexandria was a mysterious mathematician who lived around 250 A.D.; it is not known whether Diophantus was Greek or Egyptian and there are even some rumors that his name stands for a collective of authors (like Bourbaki). Virtually nothing more about his life is known than the following conundrum:

> *God granted him to be a boy for the sixth part of his life, and adding a twelfth part to this, He clothed his cheeks with down. He lit him the light of wedlock after a seventh part, and five years after his marriage He granted him a son. Alas! late-born wretched child; after attaining the measure of half his father's life, chill Fate took him. After consoling his grief by this science of numbers for four years he ended his life.* (cf. Singh [153])

How old was Diophantus when he died? This riddle is an example for the kind of problems in which Diophantus was interested. Diophantus wrote the influential monography *Arithmetica*, but, unfortunately, only 10 of the 13 books of it survived. He was the first writer who made a systematic study of the solutions of polynomial equations in integers or rationals. The *Arithmetica* is a collection of isolated problems for each of which Diophantus gave a special solution, but in general, not all solutions. To some extent he was aware of some general methods; however, there was no algebraic formalism (as denoting unknowns by symbols) in Diophantus' time. With the rise of algebra around the tenth century in Arabia Diophantus' monography was translated into algebraic language and mathematicians asked for generalizations of diophantine problems. Thus Diophantus' epoch-making

Arithmetica might be viewed as the birth to number theory. The first translation in Europe, by Regiomantus, appeared only in the fifteenth century. Maybe the most important one is that of Bachet from 1621 which was edited with Pierre de Fermat's remarks by his son Samuel de Fermat in 1679 as part of Fermat's collected works. For more details we refer to Schappacher [**140**] and Weil [**172**].*

1.2. Pythagorean triples

We shall take a closer look at one of the problems discussed in Diophantus' book. Consider the so-called **Pythagorean equation**

$$(1.1) \qquad\qquad X^2 + Y^2 = Z^2.$$

We are interested in solving this equation in integers and, of course, it suffices to consider non-negative integers. Such solutions (x, y, z) are called **Pythagorean triples** in honor of the contributions of Pythagoras (572–492 B.C.). With an integer solution (x, y, z) of (1.1) the triple (ax, ay, az) is also an integer solution provided $a \in \mathbb{Z}$. If we want to get an overview over all Pythagorean triples, it thus makes sense to consider only integer solutions (x, y, z) which are coprime; such solutions are called **primitive**. Given a solution (x, y, z) of (1.1), it is easily seen that the greatest common divisor of x and y, denoted by $\gcd(x, y)$, must divide z. Clearly, the same holds if we interchange x or y and z. Hence, pairwise coprimality of a Pythagorean triple is necessary and sufficient for having a primitive one.

Any solution of the Pythagorean equation in positive real numbers corresponds to a right angular triangle; this is Pythagoras' famous theorem in geometry. Integer solutions of (1.1) were known for quite a long time before Pythagoras. The ancient Babylonians, four millennia ago, were aware of the solutions

$$3^2 + 4^2 = 5^2, \quad 5^2 + 12^2 = 13^2, \quad 8^2 + 15^2 = 17^2$$

and many more. It is assumed that the Babylonians used Pythagorean triples for constructing right angles.

Pythagoras not only gave a mathematical proof for what the Babylonians knew in practice, but also constructed an infinitude of primitive Pythagorean triples by the identity

$$(2n + 1)^2 + (2n^2 + 2n)^2 = (2n^2 + 2n + 1)^2.$$

In the third century B.C. Euclid solved the problem of finding all solutions.

Theorem 1.1. *If a and b are positive coprime integers of opposite parity (i.e., a is even and b is odd, or vice versa) such that a > b, then the triple (x,y,z), given by*

$$(1.2) \qquad\qquad x = a^2 - b^2, \quad y = 2ab, \quad z = a^2 + b^2,$$

---

*Of course, MacTutor's Web page http://www-groups.dcs.st-and.ac.uk/∼history/ also gives plenty of information on the history of mathematics.

is a primitive solution of (1.1). This establishes a bijection between the set of pairs (a, b) satisfying the above conditions and the set of primitive integer solutions of the Pythagorean equation (1.1).

Proof. It is easy to verify that any triple of the form (1.2) solves the Pythagorean equation (1.1):

$$x^2 + y^2 = \left(a^2 - b^2\right)^2 + (2ab)^2 = a^4 + 2a^2b^2 + b^4 = \left(a^2 + b^2\right)^2 = z^2.$$

Clearly, x, y, z are positive integers. If $d = \gcd(x, y, z)$, then d divides $x + z = 2a^2$ and $z - x = 2b^2$. Since a and b are coprime, it follows that either $d = 1$ or $d = 2$. Since a and b have opposite parity, x is odd and thus the case $d = 2$ cannot occur. This shows that (x, y, z) is a primitive Pythagorean triple.

For the converse implication assume that (x, y, z) is a Pythagorean triple. Since x and y are coprime, and y is even, it follows that x and z are odd and coprime. Hence, $\frac{1}{2}(z + x)$ and $\frac{1}{2}(z - x)$ are coprime integers and, by (1.1),

$$\left(\frac{1}{2}y\right)^2 = \left(\frac{1}{2}(z + x)\right) \cdot \left(\frac{1}{2}(z - x)\right).$$

Since the factors on the right have no common divisor, both have to be squares; i.e., there are coprime positive integers a and b such that

$$a^2 = \frac{1}{2}(z + x) \quad \text{and} \quad b^2 = \frac{1}{2}(z - x)$$

(here we used the fundamental theorem of arithmetic). Further,

$$a + b \equiv a^2 + b^2 = z \equiv 1 \bmod 2,$$

so a and b have opposite parity. Now it is easy to deduce the parametrization (1.2).

It remains to show the one-to-one correspondence between pairs (a, b) and triples (x, y, z). If x and z are given, a^2 and b^2, and consequently a and b, are uniquely determined. Thus, different triples (x, y, z) correspond to different pairs (a, b). The theorem is proved. •

One important invention of mathematics is *considering numbers modulo m, so putting the infinitude of integers into a finite set of residues.* We used this idea of modular arithmetic only a tiny bit in the proof just given (when we investigated the parity) but later on we shall meet it several times.

1.3. Fermat's last theorem

Pierre de Fermat (1607(?)–1665) was a lawyer and government official in Toulouse, and, last but not least, a hobby mathematician. Often his year of birth is dated to be 1601; however, recent investigations make the above given date more reasonable; see Barner [**14**]. When Fermat died, he was one of the most famous mathematicians in Europe although he never had published any mathematical work; his reputation simply grew out of his extensive correspondence with other scientists. Fermat made important contributions to the very beginnings of analytic geometry, probability theory, and number theory. The reading of Diophantus' *Arithmetica*, in particular, the part on the Pythagorean equation, inspired Fermat to write in his copy of Diophantus' monograph:

> *It is impossible for a cube to be written as a sum of two cubes or a fourth power to be written as the sum of two fourth powers or, in general, for any number which is a power greater than the second to be written as a sum of two like powers. I have a truly marvelous demonstration of this proposition which this margin is too narrow to contain.* (cf. Singh [**153**])

In the modern language of algebra, he claimed to have a proof of

Fermat's last theorem. *All solutions of the equation*

$$(1.3) \qquad X^n + Y^n = Z^n$$

in integers x, y, z are **trivial**, *i.e., $xyz = 0$, whenever $n \geq 3$.*

Fermat never published a proof and, by the unsuccessful quest for a solution of Fermat's last theorem, mathematicians started to believe that Fermat actually had no proof. However, no counterexample was found. In fact, the above statement is the only result stated by Fermat which could not be proved for quite a long time, and so it became Fermat's *last* theorem. Only recently Wiles [**174**], supported by Taylor and the earlier works of many others, found a proof for Fermat's last theorem. The proof relies on a link to the theory of modular forms. We refer to Edwards [**52**] for the prehistory of attempts to solve Fermat's last theorem, Singh [**153**] for the amazing story of this problem and its final solution, and Washington [**169**] for a brief mathematical discussion of Wiles' breakthrough.

One may ask why the ancient Greeks considered only the quadratic case of the Fermat equation but not the general one. Greek mathematics was inspired by at most three-dimensional geometry and only in the late works of Greek mathematics higher powers occur. They also had an advanced knowledge on divisibility and prime numbers but it seems that they had no idea about the unique prime factorization of the integers.

The exponent in Fermat's equation is crucial. By Theorem 1.1, there are infinitely many solutions when $n = 2$, but by Wiles' proof there are only trivial solutions when $n \geq 3$. This observation due to Fermat is essential for the importance of Fermat's last theorem for diophantine analysis. It is the exponent n which defines the geometric character of the Fermat curve (1.3) and indeed, the corresponding geometric quantity called genus rules the solvability.

1.4. The method of infinite descent

The classification of the Pythagorean triples, Theorem 1.1, can be used to prove that the biquadratic case of Fermat's last theorem has only trivial solutions. However, we start with a slightly more general equation.

Theorem 1.2. *There are no positive integer solutions of the equation*
$$X^4 + Y^4 = Z^2.$$

We give Fermat's original and marvelous

Proof. Suppose that z is the least positive integer for which the equation
$$X^4 + Y^4 = z^2$$
has a solution in positive integers x, y. It follows that x and y are coprime since otherwise we can divide through $\gcd(x,y)^4$, contradicting the minimality of z. Thus at least one of x and y is odd. Since the squares modulo 4 are 0 and 1, it follows that
$$z^2 = x^4 + y^4 \equiv 1 \text{ or } 2 \mod 4.$$

A square cannot be congruent 2 mod 4, so z is odd, and only one of x and y is odd; the other one is even. Without loss of generality we may assume that y is even. Then, by (1.2) of Theorem 1.1,
$$x^2 = a^2 - b^2, \quad y^2 = 2ab, \quad z = a^2 + b^2,$$
where a and b are coprime positive integers of opposite parity. If a is even and b is odd, then $x^2 \equiv -1 \mod 4$, which is impossible. Thus, a is odd and b is even, say $b = 2c$ for some integer c. We observe that
$$\left(\frac{1}{2}y \right)^2 = ac,$$
where a and c are coprime. It follows that $a = u^2$ and $c = v^2$ with some positive coprime integers u, v, where u is odd (since a is odd). This leads to
$$\left(2v^2\right)^2 + x^2 = \left(u^2\right)^2,$$
where no two of the numbers $2v^2, x, u^2$ have a common factor. Applying once more Theorem 1.1 we obtain
$$2v^2 = 2AB \quad \text{and} \quad u^2 = A^2 + B^2,$$
where A and B are coprime positive integers. Dividing the v-equation by 2, we get (by the coprimality of A and B) the existence of some positive coprime integers $\mathcal{X}$ and $\mathcal{Y}$ such that $A = \mathcal{X}^2$ and $B = \mathcal{Y}^2$. Substituting this in the u-equation gives
$$\mathcal{X}^4 + \mathcal{Y}^4 = u^2,$$
which is another non-trivial solution of the diophantine equation under consideration. However,
$$u \leq u^2 = a \leq a^2 < a^2 + b^2 = z,$$
which contradicts the assumption that z was the least solution. This proves the assertion of the theorem. •

The method of proof is called **method of infinite descent**. The simple but ingenious idea of *constructing a smaller solution out of a given one can often be used for proving that certain diophantine equations have no integer solutions.*

It is obvious how Theorem 1.2 solves Fermat's last theorem in the case $n = 4$. It might be possible that Fermat had this argument in mind when he made his statement of having a proof for the general case. However, the case $n = 4$ in Fermat's last theorem is the only *easy* one. Odd exponents cannot be treated as above.

1.5. Cantor's paradise

Usually, the set of positive integers $\mathbb{N}$ is introduced by the Peano axioms. Adding the neutral element zero and the inverse elements with respect to addition, we obtain the set of integers $\mathbb{Z}$. Further, incorporating the inverse elements with respect to multiplication we get the field of rational numbers $\mathbb{Q}$. Hence, a number is said to be **rational** if it can be represented as a quotient of two integers, the denominator being non-zero; all other numbers (in the set $\mathbb{R}$ of real and the set $\mathbb{C}$ of complex numbers, respectively) are called **irrational**.

It is believed that we are living in a finite universe: there are about 10^{80} atoms in our universe. So our world can be described using only rational numbers. However, we cannot *understand* our world without a larger set of numbers. The Pythagorean equation (1.1) led to one of the great breakthroughs in ancient Greek mathematics. Hippasus, a pupil of Pythagoras, discovered that the set of rational numbers is too small for the simple geometry of triangles and squares. In fact, he proved that the length of the diagonal of a unit square is irrational:

$$\sqrt{2} = \sqrt{1^2 + 1^2} \notin \mathbb{Q}.$$

Nowadays this is taught at school and so we may omit a proof (a rather new and simple proof of this fact was given by Estermann; see Exercise 1.7). But for the Pythagoras school it was the death of its philosophy that all natural phenomena could be explained in integers. It is said that Pythagoras sentenced Hippasus to death by drowning (cf. Singh [**153**]). This unkind act could not stop the mathematical progress. In order to solve polynomial equations and to determine the merits in analysis, mathematicians invented various types of *new* numbers:

$$\mathbb{N} \subset \mathbb{Z} \subset \mathbb{Q} \subset \mathbb{R} \subset \mathbb{C}.$$

In fact, this is only the top of an iceberg; we shall learn about another type of numbers in Chapter 12. We refer the interested reader to the collection [**51**] of excellent surveys on numbers of all kinds. However, to begin with we shall only consider the set $\mathbb{R}$ of real numbers.

An infinite set is called **countable** if there exists a bijection onto $\mathbb{N}$; otherwise the set is said to be **uncountable**. It is easy to see that any union of countably many countable sets is again countable. $\mathbb{Q}$ is countable as shown by the following one-to-one mapping from $\mathbb{N}$ to the set of positive rationals:

$$\mathbb{Q}^+ \ni \frac{m}{n} \quad \longleftrightarrow \quad n + \frac{1}{2}(m + n - 1)(m + n - 2) \quad \text{for } m, n \in \mathbb{N}.$$

The real numbers represent a quite different type of infinite set than $\mathbb{Q}$.

Theorem 1.3. *The set $\mathbb{R}$ of real numbers is uncountable.*

This is a famous result of Cantor and it simply shows that almost all real numbers are irrational. Its proof relies on his marvelous **diagonalization argument**.

Proof. It suffices to prove the assertion for the subset of real numbers lying in the interval $(0, 1]$.

Suppose the contrary, that is, $\mathbb{R}$ is countable. Let $\{r_1, r_2, \ldots\}$ be a listing of all real numbers in $(0, 1]$. Using the decimal fraction expansion, every r_n can be written as

$$r_n = 0.a_{n1}a_{n2}a_{n3} \cdots,$$

where the digits a_{nk} are integers satisfying $0 \le a_{nk} \le 9$. If we assume additionally that we do not allow any infinite sequence of zeros at the end, this decimal expansion is unique (e.g., $0.1 = 0.09999\ldots$). Now define some $r = 0.b_1b_2b_3 \ldots$ by choosing $b_n \in \{1, \ldots, 8\}$ different from a_{nn} (the diagonal entry in our list) for each n. Then, r is a real number in $(0, 1)$ which does not appear in the list of the r_n (since r differs in the nth entry: $b_n \ne a_{nn}$). This is the desired contradiction. •

Theorem 1.3 marks the beginning of modern set theory. Hilbert once said in tribute to Cantor that *no one will drive us from the paradise that Cantor has created.*

1.6. Irrationality of e

One of the most fundamental functions in analysis (and natural sciences) is the exponential function given by the infinite series

$$\exp(x) = \sum_{k=0}^{\infty} \frac{x^k}{k!};$$

as usual, we will sometimes also write e^x for $\exp(x)$. A special role plays **Euler's number**

$$e := \exp(1) = \sum_{n=0}^{\infty} \frac{1}{n!} = 2.71828\,18284\,\ldots.$$

The exponential series converges very fast. For instance, taking into account the first 15 terms gives the approximation above.

Theorem 1.4. e *is irrational.*

This result dates back to Euler in 1737, resp. Lambert in 1760; however, their approach via continued fractions is rather difficult (we will return to this subject later). The following simple proof would easily have been possible in Euler's time.

Proof. Suppose the contrary; then there exist positive integers a, b such that $e = \frac{a}{b}$. Let m be an integer $\geq b$. Then b divides $m!$ and the number

$$\alpha := m! \left(e - \sum_{n=0}^{m} \frac{1}{n!} \right) = a\frac{m!}{b} - \sum_{n=0}^{m} \frac{m!}{n!}$$

is an integer (term by term). We have

$$\alpha = \sum_{n=m+1}^{\infty} \frac{m!}{n!} < \frac{1}{m+1} \sum_{k=0}^{\infty} \left(\frac{1}{m+1} \right)^{k}.$$

By the formula for the infinite geometric series, we can easily bound the right-hand side and find

$$0 < \alpha < \frac{1}{m+1} \cdot \frac{1}{1 - \frac{1}{m+1}} = \frac{1}{m} \leq 1.$$

Since the interval $(0, 1)$ is free of integers, this contradicts the fact that α is integral. The theorem is proved. •

This proof reveals an important principle in the diophantine toolbox: *the series converges so fast that the limit cannot be of a restricted arithmetic nature!*

The question whether a given real number is irrational might seem to be simple at first glance. Actually, this is a rather difficult problem. For instance, it is unknown whether the **Euler–Mascheroni constant** is irrational:

$$\gamma := \lim_{N \to \infty} \left(\sum_{n=1}^{N} \frac{1}{n} - \log N \right) = 0.57721 \ldots \notin \mathbb{Q} \ ?$$

1.7. Irrationality of π

Another important constant is the ratio π of the circumference to the diameter of a circle. We define

$$\pi = 3.14159\,26535\,89793 \ldots$$

to be the least positive root of the sine function. Our next aim is

Theorem 1.5. π and π^2 are irrational.

The first proof of the irrationality of π was given by Lambert in 1761, also by using continued fractions. The proof which we shall give now, as all other known proofs, is slightly more difficult than the one just given for e. This holds true for other questions concerning these two fundamental numbers. It seems that e has somehow more *structure* than π. Our short but tricky proof is due to Niven [**124**].

Proof. We start with some preliminaries. For $n \in \mathbb{N}$ define the function

(1.4) $$f_n(x) = \frac{1}{n!} x^n (1-x)^n.$$

It is obvious that

(1.5)
$$0 < f_n(x) < \frac{1}{n!} \qquad \text{for} \quad 0 < x < 1.$$

By the binomial theorem,

$$(1-x)^n = \sum_{j=0}^{n} \binom{n}{j} (-x)^j.$$

Since the binomial coefficients are integers (this follows immediately from their combinatorial meaning), we have

$$f_n(x) = \frac{1}{n!} \sum_{j=n}^{2n} c_j x^j,$$

where the c_j are integers (actually, they are equal to $\pm\binom{n}{j}$ but we do not need this information). The functions $f_n(x)$ share a symmetry of type $f(x) = f(1-x)$. Differentiation of this functional equation leads to

$$f_n^{(k)}(x) = (-1)^k f_n^{(k)}(1-x),$$

where $f^{(k)}$ denotes the kth derivative of f. Taking into account the Taylor series expansion (or dumb computation) we deduce

(1.6)
$$(-1)^k f_n^{(k)}(1) = f_n^{(k)}(0) = \begin{cases} 0 & \text{if } 0 \le k < n, \\ \frac{k!}{n!} c_k & \text{if } n \le k \le 2n. \end{cases}$$

Note that the values in (1.6) are all integers.

Now we are in the position to prove the theorem. Obviously, it suffices to show that π^2 is irrational. Assume that $\pi^2 = \frac{a}{b}$ with positive integers a and b. We consider the polynomial

$$F_n(x) := b^n \left(\pi^{2n} f_n(x) - \pi^{2n-2} f_n^{(2)}(x) \pm \ldots + (-1)^n f_n^{(2n)}(x) \right).$$

Since $b^n \pi^{2k} = b^{n-k} a^k \in \mathbb{Z}$ for $0 \le k \le n$, it follows from (1.6) that $F_n(0), F_n(1) \in \mathbb{Z}$. A short calculation shows

$$(F_n'(x) \sin(\pi x) - \pi F_n(x) \cos(\pi x))' = \pi^2 a^n f_n(x) \sin(\pi x).$$

Taking into account $\sin \pi = \sin 0 = 0$ this yields

$$\mathcal{I}_n := \pi a^n \int_0^1 f_n(r) \sin(\pi x)\, dx - F_n(0) + F_n(1).$$

In view of our previous observation it follows that $\mathcal{I}_n$ is an integer. On the other side with regard to (1.5) we get

$$0 < \mathcal{I}_n < \pi \frac{a^n}{n!}.$$

Since the exponential series for $\exp(a)$ converges, $n!$ grows faster than a^n as $n \to \infty$, and thus the right-hand side is <1 for sufficiently large n. This contradicts $\mathcal{I}_n \in \mathbb{Z}$ and the theorem is proved. $\bullet$

Again this proof is very interesting: *a problem concerning the arithmetic nature of a given real number is solved by the construction of an appropriate sequence of polynomials with respect to its analytic behavior!*

This method of proof can also be applied to prove the irrationality of the exponential function at any non-zero rational value (see Exercise 1.11).

1.8. Approximating with rationals

In 1682, the astronomer and mathematician Huygens (1629–1695) built an automatic planetarium. In one year Earth covers $359°45'40''30'''$ and Saturn covers $12°13'34''18'''$, which gives the ratio

$$\frac{77\,708\,431}{2\,640\,858} = 29.42544\ldots .$$

Huygens had to construct a gear mechanism which materializes this ratio well. It makes sense to search for approximations which allow a *good* approximation with only a *few* teeth for the gears. So Huygens was looking for *small* integers whose ratio is sufficiently close to the preceding one. The first idea for such an approximation might be $\frac{294}{10} = \frac{147}{5}$ coming from the decimal fraction expansion. However, this does not approximate sufficiently good; the error is $0.02544\ldots$, so more than two percent. Can we do better? Huygens could; he found the rational approximation $\frac{206}{7}$. The error of this approximation is

$$\frac{206}{7} - \frac{77\,708\,431}{2\,640\,858} = 0.00312\ldots ,$$

which is less than $40'$ in a century! How did Huygens find this excellent approximation?

Almost all real numbers are irrational. If we have to deal with such numbers, the situation is even worse than in Huygens' case. Computers cannot work with irrationals! In fact, a computer even has problems working with rational numbers; however, there are several strategies to overcome this problem (we will meet this theme in Section 12.8). Fortunately, for most problems it is sufficient to have an *approximate* solution. Since $\mathbb{Q}$ is dense in $\mathbb{R}$, it is natural to search for rational approximations. However, $\mathbb{Q}$ has the disadvantage of being very *thin* in $\mathbb{R}$.

We return to the famous constant π. It is interesting to see which rational approximations to π were used in ancient times:

- The Rhind Papyrus (≈ 1650 B.C.): $\pi \approx 4\left(\frac{8}{9}\right)^2 = 3.16\ldots$;
- Old Testament (≈ 1000 B.C.): $\pi \approx 3$;
- Archimedes (287–212 B.C.): $\pi \approx \frac{22}{7} = 3.142\ldots$;
- Tsu Chung Chi (≈ 500 A.D.): $\pi \approx \frac{355}{113} = 3.141\,59\,29\ldots$.

How could they find these approximations in those Dark Ages without computers? Their approaches used the underlying geometry.

We shall briefly sketch how Archimedes came to his approximation. He considered a circle of radius one, in which he inscribed a regular polygon of 96 sides, and circumscribed a regular polygon with the same number of sides, such that the first polygon has its vertices on the circle; the second

FIGURE 1.1. The pentagon inscribed in the unit circle has area $\frac{5}{2}\sin\frac{2\pi}{5} = 2.37764\ldots$, which gives a *poor* lower bound for π.

one, the midpoints of its edges. Comparing the perimeters led him to the inequality

(1.7)
$$\frac{223}{71} < \pi < \frac{22}{7}.$$

Alternatively, one can also consider the areas. This method of exhaustion can be used to find as good rational approximations to π as we please; however, this algorithm is not very efficient. Recently, Kanada & Takahashi computed π up to more than 206 billion digits, based on fast converging series, so calculus replaces geometry. Such a precision is beyond any use in applications (the Planck constant 10^{-33} is the smallest unit in quantum mechanics) but interesting from a mathematical point of view. To remember the first decimals, we recommend the rhyme

> *Now I want a drink, alcoholic of course, after the heavy lectures involving quantum mechanics!*

Since $\mathbb{Q}$ is dense in $\mathbb{R}$, for any real number there exist infinitely many rational approximations and we can approximate with any assigned degree of accuracy. But what are *good* and what are *bad* approximations among them? Thinking back to Huygens' gears, we find that a natural measure for a rational approximation is its denominator.

Let α be any real number; then we say that $\frac{p}{q} \in \mathbb{Q}$ with $q \geq 1$ is a **best approximation** to α if

(1.8)
$$|q\alpha - p| < |Q\alpha - P| \qquad \text{for all} \quad Q < q,$$

where $P, Q \in \mathbb{Z}$. Necessarily a best approximation $\frac{p}{q}$ is a reduced fraction (that is, p and q are coprime). Dividing inequality (1.8) by q shows

$$\left| \alpha - \frac{p}{q} \right| < \frac{Q}{q} \left| \alpha - \frac{P}{Q} \right| < \left| \alpha - \frac{P}{Q} \right|$$

(since $Q < q$). Consequently, a best approximation $\frac{p}{q}$ to α is the nearest rational number with denominator $\leq q$. However, the converse does not hold as we shall see in the following section.

The first best approximations to π are

(1.9) $$\frac{3}{1}, \frac{22}{7}, \frac{333}{106}, \frac{355}{113}, \frac{1\,03993}{33102}, \cdots \quad \rightarrow \quad \pi.$$

It is remarkable that the fractions given by Archimedes and Tsu Chung Chi are best approximations. This means that they could not do better than they did. Given a real number we want to approximate, there is usually no geometric information which we can use to find an appropriate rational approximation (as we did in the case of π). So we are faced with the problem of finding an *efficient* and *universal* algorithm which provides the best approximations to any given real number.

1.9. Linear diophantine equations

In honor of Diophantus we speak about

- **Diophantine approximations** when we search for rational approximations to rational or irrational numbers;
- **Diophantine equations** when we investigate polynomial equations for solutions in integers (or rationals).

As we shall show now these areas are linked in both directions.

We consider linear diophantine equations in two variables. For instance, we may ask for solutions of the equation

(1.10) $$106X - 333Y = 1$$

in integers; here and in the sequel we denote variables by capitals and corresponding solutions by small letters. One approach to answering this question offers Euclid's algorithm.

We recall some facts from elementary number theory. By **division with remainder**, for any positive integers a, b with $b \leq a$ there exist integers q, r such that

$$a = bq + r \qquad \text{with} \quad 0 \leq r < b.$$

Now define $r_{-1} := a, r_0 := b$. Then, successive application of division with remainder yields the **Euclidean algorithm**:

$$\text{For} \quad n = 0, 1, \ldots \quad \text{do}$$
(1.11) $$r_{n-1} = q_{n+1}r_n + r_{n+1} \qquad \text{with} \quad 0 \leq r_{n+1} < r_n.$$

Since the sequence of remainders is strictly decreasing, the algorithm terminates, and by simplest divisibility properties the last non-vanishing remainder r_m is equal to the greatest common divisor of a and b,

$$r_m = \gcd(a, b).$$

It should be noted that the Euclidean algorithm is *very fast*; more precisely, the number of steps m is bounded by a polynomial in the input length.

Reading the Euclidean algorithm backwards, we can substitute any r_{n+1} in terms of r_{n-1} and r_n one after the other down to $n = 0$. This yields a representation of r_m as a linear combination of $r_{-1} = a$ and $r_0 = b$. Thus we get an explicit integral solution of the linear equation

$$(1.12) \qquad\qquad bX - aY = \gcd(a, b),$$

say x_0, y_0. It is easily seen that then all integral solutions to the latter diophantine equation are given by

$$(1.13) \qquad \binom{x}{y} = \binom{x_0}{y_0} + k\frac{1}{\gcd(a,b)}\binom{a}{b} \qquad \text{for} \quad k \in \mathbb{Z}.$$

From here it is only a small step to Bezout's theorem:

Theorem 1.6. *The linear diophantine equation*

$$(1.14) \qquad\qquad bX - aY = c$$

with integers a, b, c is solvable if and only if $\gcd(a, b)$ divides c; in the case of solvability, the set of solutions is given by (1.13).

Proof. Given any integer solution x, y of (1.12), if $\gcd(a, b)$ divides c, then the numbers

$$\mathcal{X} = \frac{c}{\gcd(a,b)}\, x \qquad \text{and} \qquad \mathcal{Y} = \frac{c}{\gcd(a,b)}\, y$$

solve (1.14). For the converse implication simply note that, for any integers x, y, the number $ax + by$ is divisible by $\gcd(a, b)$. The theorem is proved. •

We return to our example (1.10). Applying the Euclidean algorithm shows

$$22 \cdot 106 - 7 \cdot 333 = 1.$$

With regard to our observations above we find that the set of integer solutions to (1.10) is given by

$$\binom{x}{y} = \binom{22}{7} + k\binom{333}{106} \qquad \text{for} \quad k \in \mathbb{Z}.$$

It might be a little surprising to see that the solutions x, y of (1.10) yield *good* approximations $\frac{x}{y}$ to $\frac{333}{106}$. We may rewrite each solution as

$$(1.15) \qquad\qquad \frac{x}{y} = \frac{333}{106} + \frac{1}{106y},$$

and since the second term is rather small, tending to zero as $|y| \to \infty$, the solutions x, y to (1.10) yield better and better approximations to $\frac{333}{106}$; of course, having Huygens' approximation problem in mind, we are only interested in approximations with a denominator less than 106. Any fraction $\frac{P}{Q}$ with $1 \le Q < 106$ satisfies

$$\left| Q\frac{333}{106} - P \right| = Q\left| \frac{333}{106} - \frac{P}{Q} \right| = Q\frac{|106P - 333Q|}{106Q} \ge \frac{1}{106},$$

equality holding if and only if P, Q is a solution of (1.10). Thus, we cannot approximate $\frac{333}{106}$ better than with a fraction coming from a solution of (1.10). Moreover, the solution x, y with minimal $|y|$ yields a best approximation $\frac{x}{y}$ to $\frac{333}{106}$, here $\frac{22}{7}$ (notice that both rationals are best approximations to π). Of course, this holds more generally for linear diophantine equations of the form (1.12).

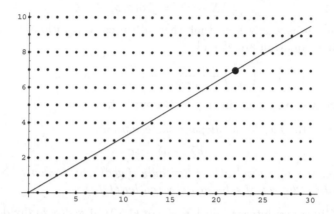

FIGURE 1.2. Integer lattice points (x, y) on the straight line $106X - 333Y = 1$ provide good rational approximations $\frac{x}{y}$ to $\frac{333}{106}$.

In some sense, we have replaced the diophantine equation by an appropriate *diophantine inequality*. So in place of using the Euclidean algorithm backwards, we can also solve the linear diophantine equation (1.12) by searching for best approximations to $\frac{a}{b}$. This was first discovered by Indian mathematicians, namely, Aryabhata around 550 A.D. and Bhaskara around 1150. However, this example marks only the very beginning of the interplay between diophantine approximations and diophantine equations: *certain diophantine equations can be investigated by studying a related problem in the theory of diophantine approximations!*

Exercises

1.1. *How old was Diophantus when he died?*

1.2. *For a Pythagorean triple (x, y, z), show that xyz is divisible by* 60.

1.3.* Consider the recursion defined by $a_1 = 3, c_1 = 5$,

$$a_{n+1} = 3a_n + 2c_n + 1 \qquad \text{and} \qquad c_{n+1} = 4a_n + 3c_n + 2.$$

Prove that $(a_n, a_n + 1, c_n)$ is a Pythagorean triple. Show that this recursion yields all Pythagorean triples of the form $(a, a + 1, c)$.

The first assertion is due to Ryden and the second one was found by Hering; for a more general recursion which constructs *all* primitive Pythagorean triples out of $(3, 4, 5)$, see Gollnick et al. [**67**].

1.4. *Show that it suffices to prove Fermat's last theorem for exponents n being prime and $n = 4$.*

1.5.* *i) Show that the equation*

$$X^4 - 4Y^4 = \pm Z^2$$

has no solutions in positive integers.

Hint: Use Fermat's method of infinite descent.

ii) Deduce from i) that the area of a Pythagorean triangle is not the square of an integer.

1.6.* *Show that $x = y = z = 0$ is the only integral solution of the equation*

$$X^3 + 2Y^3 + 4Z^3 - 34XYZ = 0.$$

1.7. *Let S be the set of those positive integers n for which $n\sqrt{2}$ is an integer. If $\sqrt{2}$ were rational, S would be not empty. For the least element of S, say k, consider the number $(\sqrt{2} - 1)k$. Deduce from*

$$(\sqrt{2} - 1)k\sqrt{2} = 2k - k\sqrt{2}$$

that the set S is empty and that $\sqrt{2}$ is irrational. Generalize this argument! This idea for proving the irrationality of $\sqrt{2}$ is due to Estermann.

1.8. *Prove that $\mathbb{Q}$ is dense in $\mathbb{R}$.*

1.9.* *Using Archimedes' exhaustion method, improve inequality (1.7). Prove that the area of the regular n-gon with vertices on the unit circle is equal to $\frac{n}{2} \sin \frac{2\pi}{n}$ and show that this tends to π as $n \to \infty$.*

1.10.* For computation of the first digits of π it is convenient to make use of analysis. **Machin's formula** states

$$\frac{\pi}{4} = 4 \arctan\left(\frac{1}{5}\right) - \arctan\left(\frac{1}{239}\right),$$

where arcus tangent is given by the power series

$$\arctan(x) = x - \frac{x^3}{3} + \frac{x^5}{5} \mp \ldots = \sum_{n=0}^{\infty} (-1)^n \frac{x^{2n+1}}{2n+1}.$$

Deduce Machin's formula from the addition formula of the tangent,

$$\tan(x + y) = \frac{\tan(x) + \tan(y)}{1 - \tan(x)\tan(y)}$$

and compute the first ten digits of the decimal fraction of π.

1.11.* With the same notation as in the proof of Theorem 1.5 let q be a positive integer and define

$$G_n(x) = q^{2n} f_n(x) - q^{2n-1} f_n'(x) \pm \ldots + f_n^{(2n)}(x).$$

Prove that $(\exp(qx)G_n(x))' = q^{2n+1}\exp(qx)f_n(x)$, and deduce

$$\int_0^1 q^{2n+1} f_n(x)\exp(qx)\,dx = \exp(q)G_n(1) - G_n(0).$$

Following the proof of Theorem 1.5, use the above identities to prove that $\exp(q)$ is irrational for any $0 \neq q \in \mathbb{Q}$.

1.12. For non-negative integers k, the **Fermat numbers** are defined by $F_k = 2^{2^k} + 1$.

Show that $\gcd(F_k, F_\ell) = 1$ *for* $k \neq \ell$. *Deduce from this coprimality that there exist infinitely many prime numbers.*

Fermat numbers are important with respect to the old question which regular n-gon can be constructed with ruler and compass (more details can be found in Section 9.6).

1.13.* From an algorithmic point of view it is sometimes an advantage to use a modified version of the Euclidean algorithm which works with the nearest integer.

Rewrite the Euclidean algorithm by replacing (1.11) by

$$r_{n-1} = q_{n+1} r_n + r_{n+1} \quad \text{with} \quad r_{n+1} \in \mathbb{Z}, \ |r_{n+1}| \leq \frac{1}{2}|r_n|.$$

Show that this algorithm terminates, is at least as fast as the Euclidean algorithm, and also produces the greatest common divisor.

1.14. *Prove that there are only finitely many best approximations $\frac{x}{y}$ to a given $\frac{a}{b} \in \mathbb{Q}$. Show that the best approximation with maximal y corresponds to the solution $x, y \in \mathbb{Z}$ of (1.12) with minimal $|y|$.*

1.15. *Using the Euclidean algorithm solve the linear diophantine equation*

$$77\,708\,431X - 2\,640\,858Y = 1.$$

This provides a best approximation $\frac{p_1}{q_1}$ to $\frac{77\,708\,431}{2\,640\,858}$. Now continue these investigations for the linear equations $q_n X - p_n Y = 1$ for $n = 1, 2, \ldots$, where $\frac{p_n}{q_n}$ is the best approximation obtained by the preceding linear equation. Recover Huygens' approximation $\frac{206}{7}$ to $\frac{77\,708\,431}{2\,640\,858}$ from Section 1.8.

This provides an algorithm for finding the best approximations to a given rational number.

1.16. *Write a computer program which computes the best approximations to a given real number below a given magnitude for the denominators. Use this to find the best approximations $\frac{p}{q}$ to π with $q < 10^7$.*

1.17. From chemistry we know the fermentation reaction (important for producing alcohol) in the variables X, Y, Z:

$$X \cdot C_6H_{12}O_6 \quad \longrightarrow \quad Y \cdot C_2H_5OH + Z \cdot CO_2.$$

Which quantities x, y, z for the ingredients can be chosen for this reaction?

1.18.* *Find integral solutions x, y, z to the linear equation*

$$17X - 11Y + 12Z = 1.$$

Classify all solutions! Show how to proceed with a linear diophantine equation in n variables.

Hint: Introducing a further variable W, investigate the system of linear equations $W = 17X - 11Y$ and $W + 12Z = 1$.

CHAPTER 2

Classical approximation theorems

So far we have proved only isolated results. Now it is time for theory! We start with the theory of diophantine approximations. In this chapter we shall prove classical theorems due to Dirichlet, Kronecker and Hurwitz and give some of their amazing applications (criteria for irrationality and uniform distribution). Further, with the pigeonhole principle and the Farey sequence we will learn two more simple but important tools in our diophantine toolbox.

2.1. Dirichlet's approximation theorem

It is much more interesting to approximate irrational numbers than rationals. As we have seen in Section 1.9, the set of best approximations to a given rational is finite. As we shall show now, the situation is rather different for irrational numbers.

In 1842, Dirichlet proved

Theorem 2.1. *If α is irrational, then there exist infinitely many rational numbers $\frac{p}{q}$ such that*

$$(2.1) \qquad \left| \alpha - \frac{p}{q} \right| < \frac{1}{q^2}.$$

This property characterizes irrational numbers; i.e., a rational α allows only finitely many approximations $\frac{p}{q}$ with (2.1).

The qualitative question whether a given real number α is irrational or not, depends on a quantitative property of α, namely, on the quality of rational approximations to α. Multiplying inequality (2.1) with q shows that, indeed, there are infinitely many best approximations to a given irrational α.

The following original proof of Dirichlet's approximation theorem relies on the so-called **pigeonhole principle** (or **box principle**) which states that if $n+1$ objects are distributed into n boxes, then at least one box contains at least two objects; this is easily seen by a contradiction argument or by induction.

Proof. We denote the **integral part** and the **fractional part** of a real number x by

$$[x] = \max\{z \in \mathbb{Z} : z \le x\} \qquad \text{and} \qquad \{x\} = x - [x],$$

respectively; in some literature $[.]$ is also called **Gauss bracket**. Let Q be a positive integer. The numbers

$$0, \{\alpha\}, \{2\alpha\}, \ldots, \{Q\alpha\}$$

define $Q+1$ points distributed among the Q disjoint intervals

$$\left[\frac{j-1}{Q}, \frac{j}{Q}\right) \qquad \text{for} \quad j = 1, \ldots Q.$$

By the pigeonhole principle there has to be at least one interval which contains at least two numbers $\{k\alpha\} \geq \{\ell\alpha\}$, say, with $0 \leq k, \ell \leq Q$ and $k \neq \ell$. It follows that

$$\begin{aligned}
(2.2) \qquad \{k\alpha\} - \{\ell\alpha\} &= k\alpha - [k\alpha] - \ell\alpha + [\ell\alpha] \\
&= \{(k-\ell)\alpha\} + \underbrace{[(k-\ell)\alpha] + [\ell\alpha] - [k\alpha]}_{\in \mathbb{Z}}.
\end{aligned}$$

Since $\{k\alpha\} - \{\ell\alpha\}$ lies in the interval $[0, \frac{1}{Q})$, the integral parts in (2.2) add up to zero. Setting $q = k - \ell$ we obtain

$$\{q\alpha\} = \{k\alpha\} - \{\ell\alpha\} < \frac{1}{Q}.$$

With $p := [q\alpha]$ it follows that

$$(2.3) \qquad \left| \alpha - \frac{p}{q} \right| = \frac{|q\alpha - p|}{q} = \frac{\{q\alpha\}}{q} < \frac{1}{qQ},$$

which implies the estimate (2.1) (since $q < Q$).

Now suppose that α is irrational and that there exist only finitely many solutions $\frac{p_1}{q_1}, \ldots, \frac{p_n}{q_n}$ to (2.1). Since $\alpha \notin \mathbb{Q}$, we can find a Q such that

$$\left| \alpha - \frac{p_j}{q_j} \right| > \frac{1}{Q} \qquad \text{for} \quad j = 1, \ldots, n,$$

contradicting (2.3).

Finally, assume that α is rational, say $\alpha = \frac{a}{b}$ with $a \in \mathbb{Z}$ and $b \in \mathbb{N}$. If $\alpha = \frac{a}{b} \neq \frac{p}{q}$, then

$$(2.4) \qquad \left| \alpha - \frac{p}{q} \right| = \frac{|aq - bp|}{bq} \geq \frac{1}{bq},$$

and (2.1) involves $q < b$. This proves that there are only finitely many $\frac{p}{q}$ with (2.1). The theorem is proved. •

The proof is inefficient. Yet we cannot compute best approximations without big computational effort.

2.2. A first irrationality criterion

Dirichlet's approximation theorem leads to a first irrationality criterion.

Theorem 2.2. *If there are infinitely many coprime solutions p, q of*

$$(2.5) \qquad \left| \alpha - \frac{p}{q} \right| < \frac{1}{q^{1+\delta}},$$

with a fixed $\delta > 0$, then α is irrational.

Proof. We assume that α is rational, say $\alpha = \frac{a}{b}$ with $a \in \mathbb{Z}$ and $b \in \mathbb{N}$. Then we shall show that both the set of p's and the set of q's satisfying (2.5) are finite. By contraposition, this gives the assertion of the theorem.

Combining (2.4) from the proof of Theorem 2.1 with (2.5) we get the inequality

$$\frac{1}{bq} < \frac{1}{q^{1+\delta}},$$

which implies that $q < b^{\frac{1}{\delta}}$. So there are only finitely many possible q's. It remains to be shown that the same holds true for the p's. For any given p, q with (2.5) we consider fractions of the form $\frac{p(t)}{q}$, where $p(t) = p + t$ with some integer variable t. If any of these fractions satisfies (2.5), it follows from the triangle inequality that

$$\frac{|t|}{q} = \left| \frac{t}{q} + \underbrace{\frac{p}{q} - \frac{a}{b} - \left(\frac{p}{q} - \frac{a}{b} \right)}_{=0} \right| \leq \left| \frac{p(t)}{q} - \frac{a}{b} \right| + \left| \frac{p}{q} - \frac{a}{b} \right| < \frac{2}{q^{1+\delta}}.$$

Thus, $|t| < 2q^{-\delta} \leq 2$ (since $q \geq 1$). Hence, for any one of the finitely many q's there are at most three different values of t for which $p(t)$ can satisfy (2.5). So we are done. •

This simple criterion is surprisingly strong. For example,

$$(2.6) \qquad \sum_{n=1}^{\infty} 10^{-n!} = 0.11000\,10000\ldots$$

is irrational. In order to prove this one only has to cut the series at one of the 1's which leads immediately to a candidate $\frac{p}{q}$ that satisfies (2.5) for some positive δ.

2.3. The order of approximation

We say that α is **approximable by rationals of order** κ if there exists a positive constant $c(\alpha)$, depending only on α, such that

$$\left| \alpha - \frac{p}{q} \right| < \frac{c(\alpha)}{q^{\kappa}}$$

has an infinity of rational solutions $\frac{p}{q}$. In a sense, κ indicates the speed of convergence of the "sequence" $\frac{p}{q}$ to α. Taking into account our previous observations we see that

- any rational α is approximable of order 1, and to no higher order (by Exercises 1.14 and Theorem 2.1);
- any irrational α is at least approximable of order 2 (by Theorem 2.1).

It is a natural question to ask for improvements of Dirichlet's theorem. Khintchine [90] proved in the 1930s a remarkable result which states that the set of numbers which allow a stronger approximation has **measure zero**; i.e., given any positive ε, the set can be covered by a countable number of intervals of total length less than ε.

Theorem 2.3. *Suppose that ψ is a positive function such that*

$$\sum_{q=1}^{\infty} \psi(q)$$

converges. Then for almost all α (i.e., the set of exceptional α has measure zero), there is only a finite number of solutions $p, q \in \mathbb{Z}$ to the inequality

$$(2.7) \qquad |q\alpha - p| < \psi(q).$$

We are only interested in the fractional part of $q\alpha$, that is, the residue class of $q\alpha$ modulo $\mathbb{Z}$. For this purpose it is useful to represent numbers on a circle instead of on a straight line. In order to eliminate integral parts we may take the complex unit circle and represent a real number x by the point whose distance from 1 measured round the circumference in counter-clockwise direction is $2\pi x$. Then numbers which differ by an integer, as x and $\{x\}$, are represented by the same point of the circumference. This so-called **circular representation** automatically neglects the integer part $[x]$ of x. In other words, the mapping

$$x \mapsto \exp(2\pi i x)$$

is an isomorphism between the **circle group** $\mathbb{R}/\mathbb{Z}$ and the group of complex numbers of absolute value one.

Proof of Theorem 2.3. Given $\varepsilon > 0$, we can find an integer Q such that

$$\sum_{q \geq Q} \psi(q) < \frac{\varepsilon}{2}.$$

Without loss of generality we may restrict ourselves to numbers α lying in the interval $[0, 1)$. Consider those α for which the inequality (2.7) has infinitely many solutions. For each $q \geq Q$ consider the intervals of length $2\frac{\psi(q)}{q}$ centered around the rational numbers $\frac{0}{q}, \frac{1}{q}, \ldots, \frac{q-1}{q}$ with respect to $\mathbb{R}/\mathbb{Z}$. Consequently, each α will lie in one of these intervals since

$$\left| \alpha - \frac{p}{q} \right| = \frac{|q\alpha - p|}{q} < \frac{\psi(q)}{q}.$$

The measure of these intervals is

$$\sum_{q \geq Q} q \cdot \frac{2\psi(q)}{q} < \varepsilon,$$

which proves the theorem. •

For example, we may take $\psi(q) = q^{-1}(\log q)^{-1-\varepsilon}$. Thus almost all numbers cannot be approximated by an order $2 + \varepsilon$. On the contrary, for a positive function ψ such that

$$\sum_{q=1}^{\infty} \psi(q)$$

diverges, Khintchine proved that for almost all α, there exist infinitely many solutions $p, q \in \mathbb{Z}$ to the inequality (2.7). A more subtle characterization of real numbers with respect to the order of approximation seems to be a rather difficult task.

2.4. Kronecker's approximation theorem

As we have seen above, for any given real α we can find integers q for which $q\alpha$ differs from an integer by as little as we please. Kronecker's celebrated approximation theorem from 1891 considers the inhomogeneous case.

Theorem 2.4. *If α is irrational, $\eta \in \mathbb{R}$ is arbitrary; then for any $N \in \mathbb{N}$ there exist $Q \in \mathbb{N}$ with $Q > N$ and $P \in \mathbb{Z}$ such that*

$$|Q\alpha - P - \eta| < \frac{3}{Q}.$$

Proof. By Theorem 2.1 there are integers $q > 2N$ and p such that

$$|q\alpha - p| < \frac{1}{q}.$$

Suppose that m is the integer, or one of the two integers, for which

$$|q\eta - m| \leq \frac{1}{2}.$$

In view of Theorem 1.6 we can write $m = px - qy$, where x and y are integers and $|x| \leq \frac{1}{2}q$. Since

$$q(x\alpha - y - \eta) = x(q\alpha - p) - (q\eta - m),$$

we find

$$|q(x\alpha - y - \eta)| < \frac{1}{2}q \cdot \frac{1}{q} + \frac{1}{2} = 1.$$

Setting $Q = q + x$ and $P = p + y$ yields

$$N < \frac{1}{2}q \leq Q \leq \frac{3}{2}q,$$

and thus

$$|Q\alpha - P - \eta| \leq |x\alpha - y - \eta| + |q\alpha - p| < \frac{1}{q} + \frac{1}{q} = \frac{2}{q} \leq \frac{3}{Q}.$$

This is the inequality of the theorem. •

Kronecker's approximation theorem provides information on topological properties of sequences with respect to the circular representation. We start with dense sequences. A sequence of real numbers α_n is said to lie **dense** in an interval $[a, b]$ if for *any* open neighborhood $\mathcal{U}$ of *any* point of $[a, b]$, there is an element $\alpha_n \in \mathcal{U}$. Recall that $\{\alpha_n\}$ denotes the fractional part of α_n, i.e., $\{\alpha_n\} = \alpha_n - [\alpha_n]$. By virtue of Kronecker's approximation theorem the sequence defined by $\alpha_n = \{n\alpha\}$ lies dense in $[0, 1)$ if and only if α is irrational. This gives another characterization of irrational numbers.

2.5. Billiard

Kronecker's approximation theorem has an interesting and entertaining consequence for billiard. This application is due to König & Szücs [**93**] and is known as the problem of the reflected light ray in plane geometry. The sides of a square are reflecting mirrors. A ray of light leaves a point inside the square and is reflected by the mirrors. What can be said about the path the ray of light will take?

Let γ be the angle between a side of the square and the initial direction. Without loss of generality we may assume that the square has edges of length 1, so we may think of the square as $[0, 1)^2 \subset \mathbb{R}^2$. The path of the ray of light is periodic if and only if the straight line is modulo $\mathbb{Z}^2$ the union of finitely many straight lines. On the contrary, the path is dense if the straight line lies dense in $(\mathbb{R}/\mathbb{Z})^2$. The straight line is given by

$$y = \alpha x + \beta,$$

where $\alpha = \tan \gamma$ and β is some real number.

First, assume that α is rational, say equal to $\frac{p}{q}$, where p and q are coprime integers. Consequently, this straight line is invariant under transformations

$$\begin{pmatrix} x \\ y \end{pmatrix} \mapsto \begin{pmatrix} x \\ y \end{pmatrix} + k \begin{pmatrix} q \\ p \end{pmatrix} \quad \text{for } k \in \mathbb{Z}.$$

Now assume that α is irrational. Given any point $(x_1, y_1) \in \mathbb{R}^2$ and any positive ε, Kronecker's approximation theorem, applied with $\eta = -y_1 + \beta + \alpha x_1$, yields integers P, Q such that

$$|y_1 + P - (\alpha(x_1 + Q) + \beta)| = |\underbrace{y_1 - \beta - \alpha x_1}_{=-\eta} + P - Q\alpha| < \varepsilon.$$

Hence, (x_1, y_1) and the point $(x_1, \alpha(x_1 + Q) + \beta)$ lie modulo $\mathbb{Z}^2$ as close as we please.

FIGURE 2.1. The way of reflected rays of light, one with an irrational tangent, the other one with a rational tangent.

Thus, the path of the ray of light is closed and periodic if and only if the angle between a side of the square and the initial direction of the ray has a rational tangent, that is, $\alpha = \tan \gamma \in \mathbb{Q}$; otherwise the ray of light passes arbitrarily near to every point of the square.

2.6. Uniform distribution

We say that a sequence $(\alpha_n)_n$ is **uniformly distributed modulo** 1 if the proportion of the fractional parts $\{\alpha_n\}$ which lie in an arbitrary subinterval $[a, b)$ of $[0, 1)$ coincides with the length of $[a, b)$; more precisely, for any $0 \le a < b < 1$,

$$\lim_{N \to \infty} \frac{1}{N} \#\{n \le N \; : \; \{\alpha_n\} \in [a, b)\} = b - a.$$

The notion of uniform distribution has its origin in probability theory.

It is easy to see that a uniformly distributed sequence modulo 1 lies dense in $[0, 1)$ a fortiori. However, the converse implication does not hold: uniform distribution is a stronger concept than denseness. For instance, consider the sequence of the fractional parts of $\log n$, where here and in the sequel $\log n$ is the logarithm to base e. This sequence lies dense in $[0, 1)$ but it is easily seen that it is not uniformly distributed modulo 1. It is yet unproved whether the sequence of numbers $\exp(n)$ for $n \in \mathbb{N}$ is uniformly distributed modulo 1.

Theorem 2.5. *For irrational* α, *the sequence* $(n\alpha)_n$ *is uniformly distributed modulo* 1.

Proof. Let ε be a small positive quantity less than $\frac{1}{10}$. In view of Kroneckers's theorem 2.4 there exists an integer m such that $0 < \{m\alpha\} =: \delta < \varepsilon$. Put $\Delta := [\frac{1}{\delta}]$ and, for $0 \le d \le \Delta$, define the intervals $\mathcal{I}_d$ by the inequalities

$$\{dm\alpha\} < x \le \{(d+1)m\alpha\}.$$

The interval $\mathcal{I}_\Delta$ extends beyond the point 1, but we may use the circular representation. For $N \in \mathbb{N}$, denote by $\lambda_d(N)$ the number of $\{\alpha\}, \{2\alpha\}, \dots, \{N\alpha\}$, which lie in the interval $\mathcal{I}_d$. If $\{n\alpha\} \in \mathcal{I}_0$ for some positive integer n, then $\{(n + dm)\alpha\} \in \mathcal{I}_d$ and vice versa. Hence, for $N > dm$,

$$\lambda_d(N) - \lambda_d(dm) = \lambda_0(N - dm).$$

Obviously, $\lambda_d(dm) \le dm$ and $\lambda_0(N - dm) \ge \lambda_0(N) - dm$. Therefore

$$\lambda_0(N) - dm \le \lambda_d(N) \le \lambda_0(N) + dm,$$

which implies

(2.8) $$\lim_{N \to \infty} \frac{\lambda_d(N)}{\lambda_0(N)} = 1 \qquad \text{for} \quad 0 \le d \le \Delta.$$

Now

$$\sum_{d=0}^{\Delta-1} \lambda_d(N) \le N \le \sum_{d=0}^{\Delta} \lambda_d(N).$$

In view of (2.8) we get

$$(\Delta + o(1))\lambda_0(N) \le N \le (\Delta + 1 + o(1))\lambda_0(N),$$

where as usual $f(x) = o(g(x))$ with a positive function $g(x)$ denotes that

$$\lim_{x \to \infty} \frac{|f(x)|}{g(x)} = 0.$$

It follows that

$$(2.9) \qquad \frac{1}{\Delta + 1} \leq \liminf_{N \to \infty} \frac{\lambda_0(N)}{N} \leq \limsup_{N \to \infty} \frac{\lambda_0(N)}{N} \leq \frac{1}{\Delta}.$$

If $\mathcal{I} = [a, b)$ with $b - a \geq \varepsilon$, there exist positive integers u and v such that

$$0 \leq \{um\alpha\} \leq a \leq \{(u+1)m\alpha\} \leq \{(u+v)m\alpha\} \leq b < \{(u+v+1)m\alpha\},$$

and

$$\sum_{d=u+1}^{u+v-1} \lambda_d(N) \leq N_{\mathcal{I}} \leq \sum_{d=u}^{u+v} \lambda_d(N),$$

where $N_{\mathcal{I}}$ is the number of $\{n\alpha\}$ with $n \leq N$ which fall into the interval $\mathcal{I}$, i.e.,

$$N_{\mathcal{I}} = \#\{n \leq N : \{\alpha_n\} \in [a, b)\}.$$

By (2.8)

$$v - 1 \leq \liminf_{N \to \infty} \frac{N_{\mathcal{I}}}{\lambda_0(N)} \leq \limsup_{N \to \infty} \frac{N_{\mathcal{I}}}{\lambda_0(N)} \leq v + 1.$$

This gives in (2.9)

$$\frac{v - 1}{\Delta + 1} \leq \liminf_{N \to \infty} \frac{N_{\mathcal{I}}}{N} \leq \limsup_{N \to \infty} \frac{N_{\mathcal{I}}}{N} \leq \frac{v + 1}{\Delta}.$$

Since

$$\delta \Delta \leq 1 \leq \delta(\Delta + 1) \qquad \text{and} \qquad \delta(v - 1) < b - a < \delta(v + 1),$$

we deduce

$$\frac{b - a - 2\delta}{1 + \delta} \leq \liminf_{N \to \infty} \frac{N_{\mathcal{I}}}{N} \leq \limsup_{N \to \infty} \frac{N_{\mathcal{I}}}{N} \leq \frac{b - a + 2\delta}{1 - \delta}.$$

The quantities ε and δ can be chosen as small as we please; thus the theorem is proved. •

It is easy to see that $\sqrt{m}$ is rational if and only if m is a perfect square. Thus, Theorem 2.5 implies that the sequence $(n\sqrt{m})_n$ is uniformly distributed modulo 1 whenever m is not a perfect square.

Uniform distribution is an interesting subject with many important applications. For example, the Monte Carlo method for numerical integration relies on randomly chosen points in the range of integration in which the integrand in question is evaluated. If these points are uniformly distributed, the sum of these values is an approximate substitute for the integral. For this and more we refer the interested reader to Hlawka [83] or Kuipers & Niederreiter [96]. The city of Monte Carlo is famous for its elegant gambling casinos. The name Monte Carlo method is a tribute to the celebration of the *laws of chance* played there.

2.7. The Farey sequence

The Farey sequence was introduced by Haros in 1802 and (independently) by Farey in 1816. However, Cauchy was the first who studied it systematically.

For any positive integer n the **Farey sequence** $\mathcal{F}_n$ **of order** n is the ordered list of all reduced fractions in the unit interval having denominators $\leq n$:

$$\mathcal{F}_n = \left\{ \frac{a}{b} \in \mathbb{Q} : 0 \leq a \leq b \leq n \quad \text{with} \quad (a,b) = 1 \right\}.$$

For example,

$$(2.10) \quad \mathcal{F}_1 = \left\{ \frac{0}{1}, \frac{1}{1} \right\} \subset \mathcal{F}_2 = \left\{ \frac{0}{1}, \frac{1}{2}, \frac{1}{1} \right\} \subset \mathcal{F}_3 = \left\{ \frac{0}{1}, \frac{1}{3}, \frac{1}{2}, \frac{2}{3}, \frac{1}{1} \right\} \subset \cdots .$$

Clearly, $\mathcal{F}_n \subset \mathcal{F}_{n+1}$ (here we wrote the new members in bold face). Each rational in the unit interval will eventually occur in the Farey sequence, and hence the Farey sequence is building $\mathbb{Q}$ modulo $\mathbb{Z}$. In particular, this construction proves once more that $\mathbb{Q}$ is countable.

How large is $\mathcal{F}_n$? The number of Farey fractions in $\mathcal{F}_n$ is related to **Euler's totient** $\varphi(b)$ which counts the number of positive coprime integers $a \leq b$, i.e.,

$$\varphi(b) = \sharp\{1 \leq a \leq b : \gcd(a,b) = 1\}.$$

Before we answer the question on the size of $\mathcal{F}_n$ we introduce the useful **Landau–Vinogradov symbols**. We write

$$f(x) = O(g(x)) \qquad \text{and} \qquad f(x) \ll g(x),$$

respectively, where $g(x)$ is a positive function, if

$$\limsup_{x \to \infty} \frac{|f(x)|}{g(x)} \quad \text{is bounded,}$$

or equivalently, if there exists a positive constant C such that $|f(x)| \leq Cg(x)$ for all sufficiently large x. Using this notation and with some knowledge on arithmetic functions one can prove the asymptotic formula

$$(2.11) \qquad \sharp\mathcal{F}_n = 1 + \sum_{b \leq n} \varphi(b) = \frac{3}{\pi^2} n^2 + O(n \log n).$$

Thus, $\mathcal{F}_n$ is growing quadratically in n.

Two consecutive elements in $\mathcal{F}_n$ are called **neighbors**. Of course, this relationship breaks down as $n \to \infty$. Nevertheless, neighborhood is a fundamental concept in the theory of Farey fractions.

Theorem 2.6. *For any neighbors $\frac{a}{b} < \frac{c}{d}$ in $\mathcal{F}_n$,*

$$bc - ad = 1.$$

In particular, under the assumptions of the theorem

$$\frac{c}{d} - \frac{a}{b} = \frac{bc - ad}{bd} = \frac{1}{bd}.$$

This shows that $\mathcal{F}_n$ is not equidistantly distributed, but in a diophantine sense, *optimally* distributed, namely, with regard to the size of the denominators. Remarkably, the Farey sequence is uniformly distributed in $[0,1]$ (see [**96**]). For an interesting and somewhat surprising correspondence between the distribution of the Farey fractions in $\mathcal{F}_n$ and the distribution of prime numbers see Appendix A.2.

Proof. Consider the diophantine equation

$$bX - aY = 1.$$

Since a and b are coprime, by Theorem 1.6, there exists a solution $x, y \in \mathbb{Z}$ with $n - b < y \leq n$ (that we can restrict y lying in the prescribed interval follows immediately from (1.13)). Consequently, also x and y are coprime, and therefore $\frac{x}{y} \in \mathcal{F}_n$. It is clear that

$$\text{(2.12)} \qquad \frac{x}{y} - \frac{a}{b} = \frac{1}{by}.$$

Now suppose that $\frac{c}{d} < \frac{x}{y}$; then

$$\frac{x}{y} - \frac{c}{d} = \frac{dx - cy}{dy} \geq \frac{1}{dy}.$$

Further,

$$\frac{c}{d} - \frac{a}{b} = \frac{bc - ad}{bd} \geq \frac{1}{bd}.$$

All these estimates imply

$$\frac{x}{y} - \frac{a}{b} = \frac{x}{y} - \underbrace{\frac{c}{d} + \frac{c}{d}}_{=0} - \frac{a}{b} \geq \frac{1}{dy} + \frac{1}{bd} = \frac{y+b}{bdy} > \frac{n}{bdy}.$$

In view of (2.12) it follows that $n < d$, contradicting $\frac{c}{d} \in \mathcal{F}_n$. Thus we have $\frac{c}{d} = \frac{x}{y}$ which proves the theorem. •

2.8. Mediants and Ford circles

The proof of the previous theorem gives a rule for the computation of the successor of a Farey fraction $\frac{a}{b}$ in $\mathcal{F}_n$. This successor is also related to the former *right* neighbor of $\frac{a}{b}$. We define the **mediant** of $\frac{a}{b}, \frac{c}{d} \in \mathcal{F}_n$ by $\frac{a+c}{b+d}$ (like adding two fractions the wrong way). If $\frac{a}{b} < \frac{c}{d}$, then it is easily seen that

$$\frac{a}{b} < \frac{a+c}{b+d} < \frac{c}{d}.$$

The mediant is a mean value, but different to other more common mean values like the arithmetic or the geometric mean value: the mediant is not monotonic, e.g.,

$$\frac{1}{6} < \frac{1}{3} \quad \text{and} \quad \frac{23}{36} < \frac{2}{3}, \quad \text{but} \quad \frac{1+23}{6+36} = \frac{4}{7} > \frac{1}{2} = \frac{1+2}{3+3}.$$

This phenomenon is well-known in statistics where it is called **Simpson's paradox** (though it is not paradox). At first glance it might be surprising; however, Simpson's paradox appears quite often.

The importance of the notion of mediants becomes clear by investigating (2.10). The process of taking mediants builds the Farey sequence out of $\frac{0}{1}$ and $\frac{1}{1}$.

Theorem 2.7. *The fractions which belong to $\mathcal{F}_n$ but not to $\mathcal{F}_{n-1}$ are mediants of elements of $\mathcal{F}_{n-1}$.*

In particular, the property for two consecutive Farey fractions $\frac{a}{b}, \frac{c}{d}$ of being neighbors gets lost in $\mathcal{F}_n$ if $n = b + d$.

Proof. With regard to Theorem 2.6 for consecutive Farey fractions we have

$$\frac{a}{b} < \frac{x}{y} < \frac{c}{d}$$

in $\mathcal{F}_n$ the identities

$$bx - ay = 1 \qquad \text{and} \qquad cy - dx = 1.$$

Multiplying this with c and a, resp. d and b, yields

$$x(bc - ad) = a + c \qquad \text{and} \qquad y(bc - ad) = b + d.$$

This gives

$$\frac{x}{y} = \frac{x(bc - ad)}{y(bc - ad)} = \frac{a + c}{b + d},$$

which we had to prove. •

The Farey sequence is related to an amazing geometry, discovered by Ford [**62**] in the 1930s. We may embed the unit interval $[0, 1]$ into the complex plane $\mathbb{C}$ and define for each $\frac{a}{b} \in \mathcal{F}_n$ the so-called **Ford circle**

$$\mathcal{C}\left(\frac{a}{b}\right) = \left\{ z \in \mathbb{C} : \left| z - \left(\frac{a}{b} + \frac{i}{2b^2} \right) \right| = \frac{1}{2b^2} \right\},$$

where, as usual, $i := \sqrt{-1}$ stands for the imaginary unit in $\mathbb{C}$.

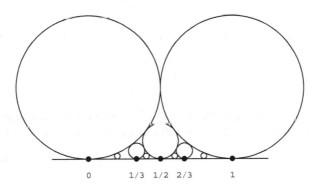

$$\begin{array}{ccccc} 0 & & 1/3 \; 1/2 \; 2/3 & & 1 \end{array}$$

FIGURE 2.2. Ford circles provide an interesting sphere packing. Moreover, they can be used to find visually good rational approximations to a given real number (see the front page).

We return to our problem of finding *explicitly* good approximations to a given α in the unit interval. If we want to get a rational approximation with denominator $\leq n$, then we have only to look for the Farey fraction $\frac{a}{b} \in \mathcal{F}_n$ with minimal distance to α. Further approximations can be found by taking appropriate mediants. It is obvious how one can use Ford circles to localize visually these Farey fractions.

The Farey dissection of the continuum is a very useful tool in the theory of diophantine approximations. It provides not only an approach to explicit approximations but also an improvement of Dirichlet's approximation theorem, namely, Hurwitz' approximation theorem.

2.9. Hurwitz' theorem

The **golden section** (resp. **golden ratio**) is given by

$$G = \frac{1}{2}(\sqrt{5} + 1) = 1.61803\,39890\,\ldots\,.$$

This famous number appears in arts and architecture.* The first best approximations to G are given by

$$\frac{2}{1}, \frac{3}{2}, \frac{5}{3}, \frac{8}{5}, \frac{13}{8}, \frac{21}{13}, \frac{34}{21}, \cdots \quad \rightarrow \quad G.$$

There is a hidden law inside this sequence of best approximations $\frac{p}{q}$; if the reader is not aware of it, we will solve this riddle later. A short computation shows that the quantities

$$q|qG - p|$$

seem to converge to some limit $0.447\ldots$. If so, we could slightly improve on what an application of Dirichlet's approximation theorem 2.1 gives for $\alpha = G$.

In 1891, Hurwitz sharpened Dirichlet's approximation theorem by showing

Theorem 2.8. *If α is irrational, then there exist infinitely many rational numbers $\frac{p}{q}$ such that*

$$\left| \alpha - \frac{p}{q} \right| < \frac{1}{\sqrt{5}q^2};$$

the constant $\sqrt{5}$ is best possible, i.e., if $\sqrt{5}$ is replaced by any larger constant, then there are only finitely many solutions p, q in the case $\alpha = G$.

Proof. Suppose that $\frac{a}{b} < \frac{c}{d}$ are those neighbors in $\mathcal{F}_n$ for which

(2.13) $$\frac{a}{b} < \alpha < \frac{c}{d}.$$

---

*For example, the orthodox church at the red square in Moscow contains a lot of *golden* proportions; the interested reader can have a look at http://nauka.relis.ru/cgi/nauka.pl?52+0306+52306082+HTML.

Without loss of generality we assume that $\alpha > \frac{a+c}{b+d}$ (otherwise we may replace α by $1 - \alpha$). If now

$$\alpha - \frac{a}{b} \geq \frac{1}{\sqrt{5}b^2}, \quad \alpha - \frac{a+c}{b+d} \geq \frac{1}{\sqrt{5}(b+d)^2},$$

(2.14)
$$\text{and} \quad \frac{c}{d} - \alpha \geq \frac{1}{\sqrt{5}d^2},$$

then we obtain by adding these inequalities

$$\frac{c}{d} - \frac{a}{b} \geq \frac{1}{\sqrt{5}}\left(\frac{1}{b^2} + \frac{1}{d^2}\right) = \frac{1}{\sqrt{5}}\frac{b^2+d^2}{b^2d^2}$$

and

$$\frac{c}{d} - \frac{a+c}{b+d} \geq \frac{1}{\sqrt{5}}\left(\frac{1}{d^2} + \frac{1}{(b+d)^2}\right) = \frac{1}{\sqrt{5}}\frac{(b+d)^2+d^2}{d^2(b+d)^2}.$$

On the other side, Theorem 2.6 and 2.7 give

$$\frac{c}{d} - \frac{a}{b} = \frac{1}{bd} \quad \text{and} \quad \frac{c}{d} - \frac{a+c}{b+d} = \frac{1}{d(b+d)}.$$

Putting these estimates together, we obtain

$$\sqrt{5}bd \geq b^2 + d^2 \quad \text{and} \quad \sqrt{5}(b+d)d \geq (b+d)^2 + d^2.$$

Adding both inequalities gives

$$\sqrt{5}(d^2 + 2bd) \geq 3d^2 + 2bd + 2b^2,$$

resp.

$$0 \geq 4b^2 - 4(\sqrt{5}-1)bd + (5 - 2\sqrt{5} + 1)d^2 = (2b - (\sqrt{5}-1)d)^2,$$

which is impossible since $b, d \in \mathbb{Z}$ but $\sqrt{5} \notin \mathbb{Q}$. Thus, in view of (2.14), at least one of the inequalities

$$\left|\alpha - \frac{a}{b}\right| < \frac{1}{\sqrt{5}b^2}, \quad \left|\alpha - \frac{a+c}{b+d}\right| < \frac{1}{\sqrt{5}(b+d)^2}, \quad \text{and} \quad \left|\alpha - \frac{c}{d}\right| < \frac{1}{\sqrt{5}d^2}$$

holds. Recall that we started with inequality (2.13) relating α to its neighbors in $\mathcal{F}_n$. Since α is irrational, the argument above yields the existence of infinitely many such rational approximations by sending $n \to \infty$. The first assertion of the theorem is proved.

For the second one we consider the polynomial

$$P(X) := X^2 - X - 1 = (X - G)(X + g),$$

where $g := \frac{1}{2}(\sqrt{5} - 1)$. Assume that, for some positive constant C,

$$\left|G - \frac{p}{q}\right| < \frac{1}{Cq^2}$$

with $p \in \mathbb{Z}, q \in \mathbb{N}$. Then

$$
\left| P\left(\frac{p}{q}\right) \right| = \left| G - \frac{p}{q} \right| \cdot \left| g + \frac{p}{q} \right| = \left| G - \frac{p}{q} \right| \cdot \left| g \underbrace{+G - G}_{=0} + \frac{p}{q} \right|
$$

$$
\leq \left| G - \frac{p}{q} \right| \left(g + G + \left| G - \frac{p}{q} \right| \right) < \frac{\sqrt{5}}{Cq^2} + \frac{1}{C^2 q^4}.
$$

Since the polynomial P is irreducible, we get on the other side

$$
\left| P\left(\frac{p}{q}\right) \right| = \frac{|p^2 + pq - q^2|}{q^2} \geq \frac{1}{q^2}.
$$

Comparing both estimates for $P(\frac{p}{q})$, we obtain

$$
C < \sqrt{5} + \frac{1}{Cq^2}.
$$

This implies $C \leq \sqrt{5}$ by letting $q \to \infty$. This proves the second assertion of the theorem. •

Reviewing the proof we can argue the same way for g in place of G by substituting $X \mapsto -X$. Then we have to consider the polynomial $P(-X) = (X+G)(X-g)$, and we obtain the same bound $\sqrt{5}$ for g. Indeed, this bound depends only on the minimal polynomial of G, and this is the same for its conjugate root $-g$. Following the lines of the proof of Hurwitz' theorem this observation leads to

Theorem 2.9. *Suppose that α is the irrational zero of a quadratic polynomial $aX^2 + bX + c$, where $a, b, c \in \mathbb{Z}$. Then, for $C > \sqrt{b^2 - 4ac}$, the inequality*

$$
\left| \alpha - \frac{p}{q} \right| < \frac{1}{Cq^2}
$$

has only finitely many solutions $p, q \in \mathbb{Z}$.

2.10. Padé approximation

We conclude this chapter with a short presentation of Padé approximation, an important generalization of diophantine approximations which dates back at least to Lagrange. The first systematic study was done by Padé at the end of the nineteenth century.

Instead of searching for $\mathbb{Q}$-approximations of elements of $\mathbb{R}$ we ask for the analogue for functions; i.e., we want to approximate an element of $\mathbb{R}[[x]]$ by elements of $\mathbb{R}(x)$. More precisely, given a power series

$$
(2.15) \qquad\qquad \alpha(x) = \sum_{k=0}^{\infty} a_k x^k
$$

in (for the sake of simplicity) one variable x with real coefficients, we search for a rational function $r(x)$ with coefficients from $\mathbb{R}$ which approximates $\alpha(x)$ *well*. Any rational function $r \in \mathbb{R}(x)$ has a representation as quotient of two polynomials. Thus we may formulate our approximation problem as follows. Given a power series (2.15), find polynomials $p, q \in \mathbb{R}[x]$ of degree μ and ν, respectively, such that the power series representation of

$$q(x)\alpha(x) - p(x)$$

contains no powers of degree $\leq \mu + \nu$.

For negative k define $a_k = 0$. Given non-negative integers μ, ν, the ansatz

$$
\begin{aligned}
p(x) &= p_0 + p_1 x + \ldots + p_\mu x^\mu, \\
q(x) &= q_0 + q_1 x + \ldots + q_\nu x^\nu,
\end{aligned}
$$

gives

(2.16) $$p_n = \sum_{j=0}^{\nu} q_j a_{n-j} \quad \text{for} \quad 0 \leq n \leq \mu,$$

(2.17) $$0 = \sum_{j=0}^{\nu} q_j a_{n-j} \quad \text{for} \quad \mu + 1 \leq n \leq \mu + \nu.$$

Note that in the case $\nu = 0$ the condition in (2.17) is empty. Also (2.17) is a system of ν homogeneous linear equations in the $\nu + 1$ unknowns $q_0, \ldots, q_\nu$. Hence, there always exists a solution to (2.17), and this solution, $q_0, \ldots, q_\nu$ say, determines $p_0, \ldots, p_\mu$ in (2.16). It thus follows that the approximation problem in question is always solvable. We get

$$\alpha(x) - \frac{p(x)}{q(x)} = O(x^{\nu+1}),$$

which is small in a neighborhood of $x = 0$. The rational function

$$r(x) = \frac{p(x)}{q(x)}$$

is called **Padé approximation** to $\alpha(x)$ of order $\nu + 1$. Indicating the solution p, q to given degrees μ, ν by $p_{\mu,\nu}, q_{\mu,\nu}$, we may define the infinite matrix

$$\left(\frac{p_{\mu,\nu}(x)}{q_{\mu,\nu}(x)} \right)_{\mu,\nu=1,2,\ldots} .$$

This matrix is called the **Padé table** to $\alpha(x)$. Note that its entries are in general not uniquely determined.

We shall give an example. Consider the exponential function

$$\exp(x) = 1 + x + \frac{x^2}{2!} + \ldots .$$

After a short computation we get the approximation

(2.18) $$\exp(x) - \frac{1 + \frac{1}{2}x + \frac{1}{6}\frac{x^2}{2!}}{1 - \frac{1}{2}x + \frac{1}{6}\frac{x^2}{2!}} = O(x^3).$$

The pattern of the exponential series is still to be seen in the Padé approximation; indeed, in the case of the exponential function one can find an explicit formula for all Padé approximations. In view of the fast convergence of the exponential series the Padé approximation (2.18) yields even a good rational approximation for slightly larger x, e.g.,

$$\left(\exp(x) - \frac{1 + \frac{1}{2}x + \frac{1}{6}\frac{x^2}{2!}}{1 - \frac{1}{2}x + \frac{1}{6}\frac{x^2}{2!}}\right)\Bigg|_{x=1} = e - \frac{19}{7} = 0.00399\ldots .$$

Padé approximations are important in approximation theory but also in the theory of differential equations. For more details we refer to Perron [**127**]; historical details can be found in Brezinski [**30**].

Exercises

2.1. The least common multiple lcm is closely related to the greatest common divisor.

i) Given two integers a, b, show that $ab = \gcd(a, b) \cdot \text{lcm}[a, b]$.

ii) For arbitrary positive integers a, b prove that

$$a + b \le \gcd(a, b) + \text{lcm}[a, b].$$

When do we have equality?

2.2.* Let $\alpha \in (0, 1)$ be irrational. Then the associated **Beatty sequences** are defined by

$$a_n = \left[\frac{n}{\alpha}\right] \quad \text{and} \quad b_n = \left[\frac{n}{1 - \alpha}\right] \quad \text{for} \quad n \in \mathbb{N}.$$

Prove that both sequences together take every positive integer value exactly once.

This provides an interesting tiling of $\mathbb{N}$. Color the positive integer m red or blue according to being represented by some a_n or some b_n.

2.3.* The next topic provides a bijection between $\mathbb{N}$ and $\mathbb{Q}^+$ different from the usual one, and much more convenient, from Calkin & Wilf [**34**]. Starting with the initial value $\frac{1}{1}$ construct recursively a tree by

$$\frac{a}{b} \mapsto \frac{a}{a + b}, \frac{a + b}{b}.$$

The **Calkin–Wilf sequence** is then given by reading this tree line by line from the top,

$$\frac{1}{1}, \frac{1}{2}, \frac{2}{1}, \frac{1}{3}, \frac{3}{2}, \frac{2}{3}, \frac{3}{1}, \frac{1}{4}, \frac{4}{3}, \frac{3}{5}, \ldots .$$

Show that the successors of any reduced fraction in the Calkin–Wilf sequence are reduced too. Further, prove that the Calkin–Wilf sequence takes any positive rational value exactly once!

Hint: Suppose that not all rationals are represented. Then there exists $\frac{a}{b}$ with the least $a + b$ among the ones not present. Find a contradiction to the minimality of $a + b$.[†]

[†]A nice discussion of this amazing topic can be found on the Internet under "Friday, March 27, 2004," http ://web.comlab.ox.ac.uk/oucl/research/areas/ap/minutes/.

2.4. Let $\mathcal{M} \subset \{1, 2, \ldots, 2n\}$ with cardinality $\sharp\mathcal{M} = n + 1$.

i) Show that there are at least two elements in $\mathcal{M}$ which are coprime.

ii) Prove that there are at least two elements in $\mathcal{M}$ such that one divides the other.

Hint: Use the pigeonhole principle.

2.5.* *Given any positive function $\psi(q)$ defined on $\mathbb{N}$, prove the existence of an irrational number α for which the inequality*

$$\left| \alpha - \frac{p}{q} \right| < \psi(q)$$

has infinitely many solutions $\frac{p}{q} \in \mathbb{Q}$.

Hint: Consider numbers of the form $\alpha = \sum_{n=1}^{\infty} (-1)^n 2^{-m_n}$ and try to find a sequence of integers m_n depending on the values of ψ at the integers 2^{m_n}. More help can be found in [**149**].

2.6. This exercise deals with a multidimensional generalization of Dirichlet's approximation theorem. One can prove it along the lines of the proof of the one-dimensional case.

Prove that if $\alpha_1, \ldots, \alpha_n$ are arbitrary real numbers, then the system of inequalities

$$\left| \alpha_j - \frac{p_j}{q} \right| < q^{-1-\frac{1}{n}} \qquad for \quad j = 1, \ldots, n$$

has at least one solution $p_1, \ldots, p_n, q \in \mathbb{Z}$; if at least one α_j is irrational, then it has an infinitude of solutions.

2.7. A real number is rational if and only if its decimal fraction expansion is eventually periodic. This yields the irrationality of (2.6). However, Theorem 2.2 applies to more general numbers of this type (which cannot be proved that easily by investigating their decimal fraction).

Using Theorem 2.2, give a rigorous proof of the irrationality of (2.6). Show more generally that

$$\sum_{n=1}^{\infty} b^{-n!}$$

is irrational whenever $2 \le b \in \mathbb{N}$.

2.8. *i) Let p be a prime number. Show that $\log p$ is irrational.*

ii) Show that the logarithms of the prime numbers are linearly independent.

Hint: Use the unique prime factorization of the integers.

2.9.* This is a multidimensional version of Kronecker's approximation theorem.

Let $1, \alpha_1, \ldots, \alpha_n$ be linearly independent over $\mathbb{Q}$, $\eta_1, \ldots, \eta_n$ be arbitrary real numbers, and N and ε be positive. Then there exist integers $P_1, \ldots, P_n$ and $Q > N$ such that

$$|Q\alpha_j - P_j - \eta_j| < \varepsilon \qquad for \quad j = 1, \ldots, n.$$

Hint: Use induction on n. By Exercise 2.6 there exist integers $p_1, \ldots, p_n$ and $q > 0$ such that $|q\alpha_j - p_j| < \varepsilon$ for $1 \le j \le n$. Now apply the induction hypothesis with $n - 1$ for n to the system

$$\alpha_j' := \frac{q\alpha_j - p_j}{q\alpha_n - p_n} \qquad \text{and} \qquad \eta_j' = \eta_j - \eta_n \alpha_j'.$$

The above proof is by Estermann (see also [**76**]). The multidimensional version of Kronecker's approximation theorem may be regarded as a mathematical version of Murphy's law: *everything that can happen will happen.*

2.10. *For $n \in \mathbb{N}$, prove that the sequence of numbers $\{\log n\}$ is dense in $[0,1)$. Further, show that $\{\log n\}$ takes more often small values and deduce that the sequence of numbers $\log n$ is not uniformly distributed modulo 1.*

2.11. The **Möbius μ-function** $\mu(n)$ is equal to 1 for $n = 1$, equal to $(-1)^\nu$ if n is the product of ν distinct prime numbers, and equal to 0 otherwise, i.e., when n has a quadratic divisor $d > 1$. An arithmetic function $f : \mathbb{N} \to \mathbb{C}$ is said to be **multiplicative** if

$$f(m \cdot n) = f(m) \cdot f(n) \qquad \text{for} \quad m, n \in \mathbb{N}.$$

Prove that $\mu(n)$ is multiplicative. Deduce the formula

$$\sum_{d|n} \mu(d) = \begin{cases} 1 & \text{if } n = 1, \\ 0 & \text{otherwise.} \end{cases}$$

Hint: By the multiplicativity it suffices to prove the latter identity only when n is the power of a prime.

2.12.* This exercise is a continuation of the preceding one.

i) For Euler's totient, show that

$$\varphi(n) = n \sum_{d|n} \frac{\mu(d)}{d} = n \prod_{p|n} \left(1 - \frac{1}{p}\right).$$

Deduce that

$$\sum_{n \le N} \varphi(n) = \sum_{bd \le N} b\mu(d) = \frac{N^2}{2} \sum_{d=1}^{\infty} \frac{\mu(d)}{d^2} + O(N \log N).$$

Note that

(2.19) $$\zeta(2) := \sum_{n=1}^{\infty} \frac{1}{n^2} = \frac{\pi^2}{6};$$

the convergence of this infinite series is clear by Riemann's convergence criterion. Formula (2.19) will be proved in Exercise 4.4.

ii) Deduce from (2.19) that

$$\sum_{d=1}^{\infty} \frac{\mu(d)}{d^2} = \prod_p \left(1 - \frac{1}{p^2}\right) = \frac{6}{\pi^2}.$$

Use this and i) to prove the asymptotic formula (2.11).

2.13. *Prove that the probability that two randomly chosen positive integers are coprime is*

$$\lim_{N \to \infty} \frac{\#\{(d,n) \in \mathbb{N}^2 \, : \, d,n \leq N, \, \gcd(d,n) = 1\}}{\#\{(d,n) \in \mathbb{N}^2 \, : \, d,n \leq N\}} = \frac{6}{\pi^2} = 0.60792 \ldots .$$

2.14. *If $\frac{a}{b} < \frac{c}{d}$ with $bc - ad = 1$, then prove that any fraction in between $\frac{a}{b}$ and $\frac{c}{d}$ has a denominator $\geq b + d$. Deduce that $\frac{a}{b}$ and $\frac{c}{d}$ are neighbors in $\mathcal{F}_{b+d-1}$. Compute the neighbors of $\frac{13}{121}$ in $\mathcal{F}_{121}$ and in $\mathcal{F}_{212}$.*

2.15. *What are the minimal and the maximal distance of neighbors in $\mathcal{F}_n$?*

2.16.* *Show that for any given pair $\frac{A}{B} < \frac{C}{D}$ there exist infinitely many pairs $\frac{a}{b} < \frac{A}{B}, \frac{c}{d} < \frac{C}{D}$ such that*

$$\frac{a+c}{b+d} > \frac{A+B}{C+D}.$$

2.17. *Show that if $\frac{c}{d}$ is a neighbor of $\frac{a}{b}$, then all neighbors $\frac{p}{q}$ to $\frac{a}{b}$ are given by*

$$\frac{p}{q} = \frac{na + c}{nb + d} \qquad \text{for} \quad n \in \mathbb{Z}.$$

Note that the Ford circles corresponding to these fractions form a ring around the Ford circle $\mathcal{C}$ of $\frac{a}{b}$, all of them tangent, which completely surrounds $\mathcal{C}$.

2.18.* *Prove that two distinct Ford circles have an empty intersection of the interiors, and are tangent if and only if they are associated to neighbors in the Farey sequence.*

Further interesting results on Ford circles can be found in Rieger [133].

2.19.* *Prove Theorem 2.9. What does this imply for the quantities $q|q\alpha - p|$? Compare the result with the observation on best approximations to G from the beginning of Section 2.9.*

2.20.* Let $f(t)$ be a polynomial of degree m.
Prove that

$$(-1)^m x^{m+1} \int_0^1 f(t) \exp(tx) \, dt \;=\; \exp(x)(f^{(m)}(1) \mp \ldots + (-1)^m f(1)x^m)$$

$$- (f^{(m)}(0) \mp \ldots + (-1)^m f(0)x^m).$$

Evaluate this formula in the case $f(t) = t^\mu (1-t)^\nu$, $m = \mu + \nu$. Deduce that the Padé table to the exponential function is given by

$$\left(\frac{1 + \frac{\nu}{\mu+\nu}x + \ldots + \frac{(\nu(\nu-1)\cdot\ldots\cdot 2\cdot 1}{(\mu+\nu)\cdot\ldots\cdot(\mu+1)} \frac{x^\nu}{\nu!}}{1 - \frac{\nu}{\mu+\nu}x \pm \ldots + (-1)^\mu \frac{\mu(\mu-1)\cdot\ldots\cdot 2\cdot 1}{(\mu+\nu)\cdot\ldots\cdot(\nu+1)} \frac{x^\nu}{\nu!}} \right)_{\mu,\nu=1,2,\ldots} .$$

CHAPTER 3

Continued fractions

The powerful tool of continued fractions was systematically studied for the first time by Huygens in the seventeenth century. These fractions appear in a natural way by means of the Euclidean algorithm and may be used to construct the set of real numbers out of the set of rationals. With respect to approximation, continued fractions may be regarded as substitutes for Farey fractions. They provide best approximations in a rather quick and easy way.

3.1. The Euclidean algorithm revisited and calendars

Recall the intimate relation between the approximations to a given rational $\frac{a}{b}$ and the solutions of linear diophantine equations (1.12), obtained by the Euclidean algorithm. Let $r_{-1} = a$ and $r_0 = b$. We may rewrite the Euclidean algorithm (1.11) as follows:

For $n = 0, 1, \ldots, m$ do

$$(3.1) \qquad \frac{r_{n-1}}{r_n} = \left[\frac{r_{n-1}}{r_n}\right] + \frac{r_{n+1}}{r_n} \qquad \text{with} \quad 0 \le r_{n+1} < r_n.$$

Recall that the last non-vanishing r_m is the greatest common divisor of a and b. Setting $a_n = \left[\frac{r_{n-1}}{r_n}\right]$, we obtain

$$\frac{a}{b} = \frac{r_{-1}}{r_0} = a_0 + \left(\frac{r_0}{r_1}\right)^{-1} = a_0 + \cfrac{1}{a_1 + \left(\dfrac{r_1}{r_2}\right)^{-1}} = \ldots .$$

The first of these identities yields the integral part as an approximation to the rational $\frac{a}{b}$. Using later identities, we obtain better and better approximations, and among them we can even find *all* best approximations. We will prove this later. To begin with we give an example.

A solar year has approximately

$$365 \text{ days } 5 \text{ hours } 48 \text{ minutes and } 45.8 \text{ seconds} \quad \approx \quad 365 + \frac{419}{1730} \text{ days}.$$

Unfortunately, this is not an integer, so how can we create a *good* calendar? Applying the Euclidean algorithm to 1730 and 419 yields

$$1730 = 4 \cdot 419 + 54,$$
$$419 = 7 \cdot 54 + 41,$$
$$54 = 1 \cdot 41 + 13,$$
$$\ldots$$

With regard to (3.1) we find

$$\frac{1730}{419} = 4 + \frac{54}{419},$$

resp.

$$365 + \frac{419}{1730} = 365 + \left(\frac{1730}{419}\right)^{-1} \approx 365 + \frac{1}{4},$$

which is nothing other than Caesar's calendar, i.e., a leap year each fourth year. By the full Euclidean algorithm we get

$$365 + \frac{419}{1730} = 365 + \cfrac{1}{4 + \cfrac{1}{7 + \cfrac{1}{1 + \cfrac{1}{3 + \cfrac{1}{6 + \cfrac{1}{2}}}}}}.$$

Using this without the last fraction $\frac{1}{2}$ gives the better approximation

$$365 + \frac{194}{801} \approx 365 + \frac{419}{1730}.$$

This is our present calendar, the Gregorian calender, introduced by Pope Gregor XIII in 1582. Roughly speaking, within periods of 800 years 6 (= 200 − 194) of the leap years of the Julian calender are switched back to normal.

3.2. Finite continued fractions

We call

$$a_0 + \cfrac{1}{a_1 + \cfrac{1}{a_2 + \cfrac{1}{\ddots + \cfrac{1}{a_{m-1} + \cfrac{1}{a_m}}}}}$$

a **finite continued fraction** (in German *Kettenbruch*). The a_n are called **partial quotients**. To overcome the typographical nightmare of writing continued fractions, we denote for brevity the continued fraction above by

$$[a_0, a_1, a_2, \ldots, a_m].$$

First, we shall consider $[a_0, \ldots, a_m]$ as a function in independent variables $a_0, \ldots, a_m$. We find

$$[a_0] = a_0 \, , \quad [a_0, a_1] = \frac{a_1 a_0 + 1}{a_1} \, ,$$

and

$$[a_0, a_1, a_2] = \frac{a_2 a_1 a_0 + a_2 + a_0}{a_2 a_1 + 1}.$$

By induction on n, we get

$$(3.2) \qquad [a_0, a_1, \ldots, a_n] = \left[a_0, a_1, \ldots, a_{n-1} + \frac{1}{a_n} \right]$$

and

$$[a_0, a_1, \ldots, a_n] = a_0 + \frac{1}{[a_1, \ldots, a_n]} = [a_0, [a_1, \ldots, a_n]].$$

For $n \leq m$ we call $[a_0, a_1, \ldots, a_n]$ the n**th convergent** to $[a_0, a_1, \ldots, a_m]$. Further put

$$(3.3) \quad \begin{cases} p_{-1} = 1, \ p_0 = a_0, \quad \text{and} \quad p_n = a_n p_{n-1} + p_{n-2}, \\[2mm] q_{-1} = 0, \ q_0 = 1, \quad \text{and} \quad q_n = a_n q_{n-1} + q_{n-2}. \end{cases}$$

The computation of the convergents is easily ruled by means of

Theorem 3.1. *For* $0 \leq n \leq m$*, the functions* p_n, q_n *satisfy*

$$\frac{p_n}{q_n} = [a_0, a_1, \ldots, a_n].$$

Proof by induction on n. The cases $n = 0$ is obvious. The case $n = 1$ is easily computed by

$$[a_0, a_1] = \frac{a_1 a_0 + 1}{a_1} = \frac{p_1}{q_1}.$$

Now assume that the formula in question holds for n. In view of (3.2)

$$[a_0, a_1, \ldots, a_n, a_{n+1}] = \left[a_0, a_1, \ldots, a_n + \frac{1}{a_{n+1}} \right].$$

By the recursion formulae for the p_n, q_n, the latter expression is equal to

$$\frac{\left(a_n + \frac{1}{a_{n+1}} \right) p_{n-1} + p_{n-2}}{\left(a_n + \frac{1}{a_{n+1}} \right) q_{n-1} + q_{n-2}} = \frac{(a_{n+1} a_n + 1) p_{n-1} + a_{n+1} p_{n-2}}{(a_{n+1} a_n + 1) q_{n-1} + a_{n+1} q_{n-2}}$$

$$= \frac{a_{n+1} p_n + p_{n-1}}{a_{n+1} q_n + q_{n-1}} = \frac{p_{n+1}}{q_{n+1}},$$

which proves the theorem. •

A simple application of this is

Theorem 3.2. *For* $1 \leq n \leq m$*,*

$$p_n q_{n-1} - p_{n-1} q_n = (-1)^{n-1},$$

and

$$p_n q_{n-2} - p_{n-2} q_n = (-1)^n a_n.$$

Proof. In view of Theorem 3.1 we have

$$
\begin{aligned}
p_n q_{n-1} - p_{n-1} q_n &= (a_n p_{n-1} + p_{n-2}) q_{n-1} - p_{n-1}(a_n q_{n-1} + q_{n-2}) \\
&= -(p_{n-1} q_{n-2} - p_{n-2} q_{n-1}).
\end{aligned}
$$

Repeating this argument with $n-1, n-2, \ldots, 2, 1$ proves the first assertion. Similarly,

$$
\begin{aligned}
p_n q_{n-2} - p_{n-2} q_n &= (a_n p_{n-1} + p_{n-2}) q_{n-2} - p_{n-2}(a_n q_{n-1} + q_{n-2}) \\
&= a_n(p_{n-1} q_{n-2} - p_{n-2} q_{n-1}),
\end{aligned}
$$

and so the second assertion follows from the first one. •

We now assign numerical values to the partial quotions a_n, and so to the associated continued fractions $[a_0, \ldots, a_n]$. We suppose that $a_0 \in \mathbb{Z}$, that the a_n are positive integers for $1 \le n < m$, and that $a_m \ge 1$. Then it follows from Theorem 3.1 that p_n and q_n are integers for $n < m$, and the first assertion of Theorem 3.2 implies that they are coprime.

Now let α be any rational number. Then there exist coprime integers a and b with $b > 0$ such that $\alpha = \frac{a}{b}$. It then follows from the variation of the Euclidean algorithm (3.1), applied to $r_{-1} = a$ and $r_0 = b$, that α has a representation as a finite continued fraction, namely,

$$
\frac{a}{b} = [a_0, a_1, a_2, \ldots, a_m], \qquad \text{where} \quad a_n = \left[\frac{r_{n-1}}{r_n} \right]
$$

in the notation of (3.1); note that $r_n < r_{n-1}$ by division with reminder, and so the numbers a_n are positive integers for $1 \le n \le m$. The length of the continued fraction expansion is the number of steps of the Euclidean algorithm applied to the numerator and denominator of the reduced fraction α. In fact, the problem of analyzing the average running time for the Euclidean algorithm is made by a statistical theory of continued fraction expansions.

The continuous fraction expansion is not unique since

$$
[a_0, a_1, a_2, \ldots, a_m] = [a_0, a_1, a_2, \ldots, a_m - 1, 1];
$$

however, we can make it unique by the request $a_m \ge 2$. Alternatively, we could make it unique by allowing only continued fractions of even (resp. odd) length.

We collect our observations in

Theorem 3.3. *Any rational number has a representation as a finite continued fraction; this representation is unique if we request the last partial quotient to be greater than one.*

3.3. Interlude: Egyptian fractions

A fraction of the form $\frac{1}{x}$, where x is a positive integer, is called **Egyptian fraction** (or **unit fraction**). Egyptian fractions were an important tool in ancient Egypt mathematics for computations in practical life. The open

Erdös–Strauss conjecture asks whether for any positive integer $n \geq 2$ there exist positive integers x, y, z such that

$$\frac{4}{n} = \frac{1}{x} + \frac{1}{y} + \frac{1}{z}.$$

The Erdös–Strauss conjecture can be checked numerically. There exist several fast algorithms which compute for a given rational $\frac{a}{b}$ a representation as a sum of Egyptian fractions; one of them we shall now present.

Given $\frac{a}{b}$, let x be the least positive integer such that $\frac{1}{x} \leq \frac{a}{b}$. If the latter inequality is not an equality, then let y be the least positive integer such that

$$\frac{1}{y} \leq \frac{a}{b} - \frac{1}{x},$$

and so on. It is not too difficult to see that this procedure produces a sequence of fractions whose numerators are strictly decreasing, and so this algorithm terminates, i.e., we end up with a positive integer z such that

$$\frac{1}{z} = \frac{a}{b} - \frac{1}{x} - \dots.$$

It is clear how we can transform this into the desired representation of $\frac{a}{b}$ as a sum of Egyptian fractions. This is the so-called **greedy algorithm**, and the name has its origin from the fact that in each step we subtract as much as possible. For instance, the greedy algorithm computes

$$\frac{4}{23} = \frac{1}{6} + \frac{1}{138}.$$

Rewriting $\frac{1}{6}$ as the sum of two $\frac{1}{12}$ we see that the Erdös–Strauss conjecture is true for $n = 23$. The greedy algorithm is quite often not an appropriate algorithm to produce a *short* sum of Egyptian fractions, for example, one may try $\frac{31}{311}$; however, it is an appropriate algorithm for fractions of the form $\frac{4}{n}$.

Following Elsner, Sander & Steuding [**55**], we shall investigate which continued fractions allow a representation as a sum of two Egyptian fractions. We shall prove that, given an arbitrary sequence $a_2, \dots, a_m$ of positive integers, we can find a continued fraction of length m which contains the given sequence of the a_n as partial quotients and is the sum of two Egyptian fractions. But first we introduce some convenient notation.

For computations with continued fractions a certain abbreviation is of some practical use. Given a sequence $a_1, \dots, a_m$ of positive integers, the **Muir symbol** A_n is

$$A_n = \langle a_n, a_{n+1}, \dots, a_m \rangle$$

for $n = 1, \dots, m$ defined recursively by $A_{m+1} = \langle \rangle = 1, A_m = a_m$, and

$$A_n = a_n A_{n+1} + A_{n+2}$$

for all $n = 1, \dots, m - 1$. It is easily seen that then all A_n are positive integers, and $\gcd(A_n, A_{n+1}) = 1$. Moreover, it follows from Theorem 3.1

that
$$[0, a_1, \ldots, a_m] = \frac{A_2}{A_1}.$$

Theorem 3.4. *For fixed $m \geq 2$ let $a_2, \ldots, a_m$ be given positive integers. Then we have for every positive integer k satisfying $(A_2 - 1)k > A_3$ that*

$$[0, (A_2 - 1)k - A_3, a_2, \ldots, a_m] = \frac{1}{A_2 k - A_3} + \frac{1}{(A_2 k - A_3)(A_2 - 1)},$$

where A_n for $n = 2, 3$ is the Muir symbol associated with $a_2, \ldots, a_m$.

Proof. For any a_1, we have
$$[0, a_1, a_2, \ldots, a_m] = \frac{A_2}{A_1} = \frac{A_2}{a_1 A_2 + A_3}.$$

The right-hand side is equal to
$$\frac{1}{A_2 k - A_3} + \frac{1}{(A_2 k - A_3)(A_2 - 1)} = \frac{A_2}{(A_2 k - A_3)(A_2 - 1)}$$

if and only if $a_1 = (A_2 - 1)k - A_3$, provided that $(A_2 - 1)k - A_3 > 0$. This proves the assertion. •

This theorem does not imply the truth of the Erdös–Strauss conjecture since, for example, the number $\frac{4}{13}$ has no representation as a sum of two Egyptian fractions. However, for numbers

$$\frac{4}{n} = [0, a_1, a_2, \ldots, a_m],$$

where a_1 satisfies the condition $a_1 = (A_2 - 1)k - A_3$, the above theorem provides an explicit representation of $\frac{4}{n}$ as a sum of two Egyptian fractions. It is not too difficult to obtain a characterization of those continued fractions which are the sum of two Egyptian fractions.

Theorem 3.5. *Let $a_1, \ldots, a_m$ be positive integers. Then $[0, a_1, \ldots, a_m]$ has a representation as a sum of two Egyptian fractions if and only if*

$$A_1 + \prod_{j=1}^{r} p^{\mu_j} \equiv 0 \bmod A_2$$

for some non-negative integers $\mu_j \leq 2\nu_j$, where $A_1 = \prod_{j=1}^{r} p_j^{\nu_j}$ is the prime factorization of A_1.

We give here only a sketch of the argument. First,

$$[0, a_1, \ldots, a_m] = \frac{1}{x} + \frac{1}{y}$$

for some positive integers x, y if and only if $A_2 x(z - x) = A_1 z$, where $z = x + y$. This gives $z = A_2 \omega$ for some ω. Hence, the identity we started with is equivalent to
$$(A_2 x - A_1)\omega = x^2.$$

The solvability of this equation is left to the reader as Exercise 3.13.

3.4. Infinite continued fractions

We can rewrite algorithm (3.1) for computing the continued fraction of a rational α by applying the iteration

$$(3.4) \qquad \alpha_0 := \alpha, \quad \alpha_n = [\alpha_n] + \frac{1}{\alpha_{n+1}} \quad \text{for} \quad n = 0, 1, \dots .$$

Putting $a_n = [\alpha_n]$ we obtain $\alpha = [a_0, a_1, \dots, a_n, \alpha_{n+1}]$. This algorithm is called the **continued fraction algorithm**. Obviously, if α is rational, the iteration stops after finitely many steps, and this algorithm is nothing but the Euclidean algorithm in disguise.

We illustrate how the continued fraction algorithm works for an irrational number. Let's take $\alpha = \pi = 3.14159\dots$. Then we find

$$a_0 = [\pi] = 3 \quad \text{and} \quad \alpha_1 = \frac{1}{\pi - 3} = 7.06251\dots,$$

$$a_1 = [7.06251\dots] = 7 \quad \text{and} \quad \alpha_2 = \frac{1}{7.06251\dots - 7} = 15.99744\dots,$$

$$a_2 = [15.99744\dots] = 15 \quad \text{and} \quad \alpha_3 = \frac{1}{15.99744\dots - 15}.$$

This gives $\pi = [3, 7, 15, \alpha_3]$.

Now let α be any irrational. Then the iteration does not stop; otherwise α would have a representation as a finite continued fraction, contradicting $\alpha \notin \mathbb{Q}$. Thus by the continued fraction algorithm we get an infinite sequence of finite continued fractions with limit denoted by

$$[a_0, a_1, \dots] := \lim_{m \to \infty} [a_0, a_1, \dots, \alpha_m].$$

This limit $[a_0, a_1, a_2, \dots]$ is an **infinite continued fraction** but the first thing we have to ask is whether the underlying infinite process is convergent, and maybe with limit α.

Theorem 3.6. *Let $\alpha = [a_0, a_1, \dots, a_n, \alpha_{n+1}]$ be irrational with convergents $\frac{p_n}{q_n}$. Then*

$$\alpha - \frac{p_n}{q_n} = \frac{(-1)^n}{q_n(\alpha_{n+1}q_n + q_{n-1})}.$$

In particular,

$$\alpha = \lim_{n \to \infty} \frac{p_n}{q_n} = [a_0, a_1, a_2, \dots].$$

Proof. First, note that all observations from Section 3.2 remain valid if we let $n \to \infty$. Taking into account (3.3) and Theorem 3.1 a short computation shows

$$\alpha - \frac{p_n}{q_n} = \frac{\alpha_{n+1}p_n + p_{n-1}}{\alpha_{n+1}q_n + q_{n-1}} - \frac{p_n}{q_n} = \frac{p_{n-1}q_n - p_n q_{n-1}}{q_n(\alpha_{n+1}q_n + q_{n-1})}.$$

Now Theorem 3.2 implies the first assertion of the theorem.

Since $a_{n+1} \leq \alpha_{n+1}$ we may deduce from Theorem 3.2 that

$$\left| \alpha - \frac{p_n}{q_n} \right| < \frac{1}{q_n(a_{n+1}q_n + q_{n-1})}.$$

In case of irrational α the sequences of the p_n and the q_n are strictly increasing for $n \geq 2$. It thus follows that the sequence of fractions $\frac{p_n}{q_n}$ is spinning around α, those with even index n to the left, the others to the right:

$$\frac{p_0}{q_0} < \frac{p_2}{q_2} < \ldots < \alpha < \ldots < \frac{p_3}{q_3} < \frac{p_1}{q_1}.$$

If α is irrational, the continued fraction algorithm does not terminate, and the sequence q_n of the denominators of the convergents is unbounded. It thus follows from the first statement of the theorem that the distances of consecutive convergents become smaller and smaller. Hence, the convergents $\frac{p_n}{q_n}$ tend to a limit $[a_0, a_1, \ldots]$, and the limit is equal to α. This proves the second assertion. •

It is easily shown that the continued fraction expansion of any irrational number is uniquely determined. This offers an alternative way to construct the set of real numbers $\mathbb{R}$ via continued fractions.

Furthermore, the continued fraction expansion provides an order on the real axis. Given any two real numbers $\alpha = [a_0, \ldots, a_n, \alpha_{n+1}]$ and $\alpha' = [a_0, \ldots, a_n, \alpha'_{n+1}]$ having the same first partial quotients, it follows that any α'' lying in between α and α' has a continued fraction expansion starting with the same partial quotients as α and α', namely,

$$\alpha'' = [a_0, \ldots, a_n, \alpha''_{n+1}]$$

for some α''_{n+1} lying in between α_{n+1} and α'_{n+1}. This is easily seen by induction on n.

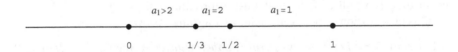

FIGURE 3.1. The location of the first partial quotient in the unit interval.

3.5. Approximating with convergents

Theorem 3.6 indicates the important role continued fractions play in the theory of diophantine approximations. We immediately deduce

Corollary 3.7. *Let* $\alpha = [a_0, a_1, \ldots]$ *be irrational with convergents* $\frac{p_n}{q_n}$. *Then*

(3.5)
$$\left| \alpha - \frac{p_n}{q_n} \right| < \frac{1}{a_{n+1} q_n^2}.$$

This gives another proof of Dirichlet's approximation theorem 2.1 (since any partial quotient is greater than or equal to one). Moreover, it solves our problem of finding explicitly *good* rational approximations. In view of (3.5) the sequence of convergents $\frac{p_n}{q_n}$ is approximating α better and better while n tends to infinity (since the q_n strictly increase), and the increase in accuracy of $\frac{p_n}{q_n}$ upon the previous convergent is proportional to the next partial quotient a_{n+1}.

For instance, we have already noticed that the fraction $\frac{355}{113}$ is an extraordinarily good rational approximation to π. In fact, it is a convergent to

$$\pi = [3, 7, 15, 1, 292, 1, 1, 1, 21, 31, 14, 2, 1, 2, 2, 2, \ldots],$$

namely, the one we obtain by cutting before 292, that is,

$$\frac{355}{113} = [3, 7, 15, 1] = \frac{p_3}{q_3}.$$

Since $a_4 = 292$ is quite large if compared with $q_3 = 113$, formula (3.5) yields the excellent approximation

$$0 < \frac{355}{113} - \pi < \frac{1}{292 \cdot 113^2} = 0.0000002682 \ldots.$$

Furthermore, it follows that the next convergent has a relatively large denominator, namely, $q_4 = a_4 q_3 + q_2 = 292 \cdot 113 + 106 = 33\,102$.

3.6. The law of best approximations

Let's have a closer look at other convergents of π. A short computation shows that the first convergents, located as follows:

$$\frac{3}{1} < \frac{333}{106} < \frac{1\,03993}{33102} < \ldots < \pi < \ldots < \frac{355}{113} < \frac{22}{7},$$

cover exactly the list of the first best approximations (1.9).

This observation is not a miracle, as Lagrange proved in 1770.

Theorem 3.8. *Let α be any real number with convergents $\frac{p_n}{q_n}$. If $n \geq 2$ and p, q are positive integers satisfying $0 < q \leq q_n$ and $\frac{p}{q} \neq \frac{p_n}{q_n}$, then*

$$|q_n \alpha - p_n| < |q\alpha - p|.$$

This so-called law of best approximations shows that we cannot do better than approximating an irrational number by its convergents.

Proof. We may suppose that p and q are coprime. Since

$$|q_n \alpha - p_n| < |q_{n-1}\alpha - p_{n-1}|,$$

it is sufficient to prove the assertion under the assumption that $q_{n-1} < q \leq q_n$; the full statement of the theorem then follows by induction on n.

If $q = q_n$, then $p \neq p_n$ and

$$\left| \frac{p}{q} - \frac{p_n}{q_n} \right| \geq \frac{1}{q_n}.$$

But

$$\left| \alpha - \frac{p_n}{q_n} \right| \leq \frac{1}{q_n q_{n+1}} < \frac{1}{2q_n}$$

by Theorem 3.6 and $q_{n+1} \geq 3$ (since $n \geq 2$). Now the triangle inequality gives

$$\left| \alpha - \frac{p}{q} \right| \geq \left| \frac{p}{q} - \frac{p_n}{q_n} \right| - \left| \alpha - \frac{p_n}{q_n} \right| > \frac{1}{2q_n} > \left| \alpha - \frac{p_n}{q_n} \right|,$$

which implies the inequality in question after multiplication with $q = q_n$.

Now suppose that $q_{n-1} < q < q_n$. By Theorem 3.2, the system of linear equations

$$p_n X + p_{n-1} Y = p \qquad \text{and} \qquad q_n X + q_{n-1} Y = q$$

has the unique solution

$$x = \frac{pq_{n-1} - qp_{n-1}}{p_n q_{n-1} - p_{n-1} q_n} = \pm(pq_{n-1} - qp_{n-1})$$

and

$$y = \frac{pq_n - qp_n}{p_n q_{n-1} - p_{n-1} q_n} = \pm(pq_n - qp_n).$$

Hence, x and y are both non-zero integers. Obviously, x and y have opposite signs. Since $q_n \alpha - p_n$ and $q_{n-1} \alpha - p_{n-1}$ have opposite signs as well, $x(q_n \alpha - p_n)$ and $y(q_{n-1} \alpha - p_{n-1})$ have the same sign. Since

$$q\alpha - p = x(q_n \alpha - p_n) + y(q_{n-1} \alpha - p_{n-1}),$$

we obtain

$$|q\alpha - p| > |q_{n-1} \alpha - p_{n-1}| > |q_n \alpha - p_n|,$$

which was to be shown. •

3.7. Consecutive convergents

By Corollary 3.7 the convergents to an irrational α satisfy the inequality of Dirichlet's approximation theorem. It may happen that a continued fraction contains a long sequence of partial quotients equal to one. Nevertheless, among two consecutive convergents, there is always one which satisfies a stronger inequality. This observation is also attributed to Lagrange.

Theorem 3.9. *Among two consecutive convergents $\frac{p}{q}$ to any real α, there is at least one satisfying*

$$(3.6) \qquad \left| \alpha - \frac{p}{q} \right| < \frac{1}{2q^2}.$$

Furthermore, if $\frac{p}{q}$ is a reduced fraction which solves the latter inequality, then $\frac{p}{q}$ is a convergent to α.

Proof. Since the convergents are alternately less or greater than α,

$$\left| \frac{p_{n+1}}{q_{n+1}} - \frac{p_n}{q_n} \right| = \left| \alpha - \frac{p_{n+1}}{q_{n+1}} \right| + \left| \frac{p_n}{q_n} - \alpha \right|.$$

Now assume that (3.6) is not true for both convergents, $\frac{p_n}{q_n}$ and $\frac{p_{n+1}}{q_{n+1}}$. In view of Theorem 3.2 by combining (3.6) and the previous equation we would then get

$$\frac{1}{q_{n+1}q_n} = \left| \frac{p_{n+1}q_n - p_n q_{n+1}}{q_{n+1}q_n} \right| = \left| \frac{p_{n+1}}{q_{n+1}} - \frac{p_n}{q_n} \right| \geq \frac{1}{2q_{n+1}^2} + \frac{1}{2q_n^2}.$$

This implies $(q_{n+1} - q_n)^2 \leq 0$, a contradiction unless $n = 0$, $a_1 = 1$, and $q_1 = q_0 = 1$. In this case we find

$$0 < \frac{p_1}{q_1} - \alpha = 1 - [1, a_2, a_3, \ldots] < 1 - \frac{a_2}{a_2 + 1} \leq \frac{1}{2},$$

so the assertion is still true. This proves the desired inequality.

Now assume that (3.6) holds. In view of Theorem 3.8, the law of best approximations, it suffices to prove that $\frac{p}{q}$ is a best approximation to α. Let $\frac{P}{Q}$ be a fraction with $\frac{P}{Q} \neq \frac{p}{q}$ such that

$$|Q\alpha - P| \leq |q\alpha - p| < \frac{1}{2q}.$$

It follows that

$$\frac{1}{qQ} \leq \left| \frac{p}{q} - \frac{P}{Q} \right| \leq \left| \alpha - \frac{p}{q} \right| + \left| \frac{P}{Q} - \alpha \right| < \frac{1}{2q^2} + \frac{1}{2qQ} = \frac{q+Q}{2q^2Q}.$$

This implies $q < Q$. Hence $\frac{p}{q}$ is a best approximation to α, which proves the second assertion. •

One can go a bit further. In fact, among three consecutive convergents to any irrational α, there is at least one satisfying the inequality in Hurwitz' theorem 2.8.

3.8. The continued fraction for e

In 1748, Euler found the continued fraction of e but he had no full proof. We shall fill the gaps and start with the following:

Theorem 3.10. *For $k \in \mathbb{N}$,*

$$\frac{\exp\left(\frac{2}{k}\right) + 1}{\exp\left(\frac{2}{k}\right) - 1} = [k, 3k, 5k, 7k, \ldots].$$

The proof relies on fundamental properties of the exponential function.

Proof. For any non-negative integer n define the sequences

$$\alpha_n = \frac{1}{n!} \int_0^1 x^n (1-x)^n \exp\left(\frac{2x}{k}\right) dx,$$

$$\beta_n = \frac{1}{n!} \int_0^1 x^{n+1} (1-x)^n \exp\left(\frac{2x}{k}\right) dx;$$

the integrands should be compared with the functions f_n which we used in the proof of the irrationality of π. A short computation shows

$$(3.7) \qquad \alpha_0 = \frac{k}{2}\left(\exp\left(\frac{2}{k}\right) - 1\right),$$

$$\beta_0 = \frac{k}{2}\exp\left(\frac{2}{k}\right) - \left(\frac{k}{2}\right)^2\left(\exp\left(\frac{2}{k}\right) - 1\right).$$

By induction on n it follows that, for $n \in \mathbb{N}$,

$$(3.8) \qquad \frac{2}{k}\alpha_n + \alpha_{n-1} = 2\beta_{n-1} \qquad \text{and} \qquad k(2n+1)\alpha_n = k\beta_{n-1} - 2\beta_n$$

(we leave the easy proof as Exercise 3.18 to the reader). With regard to (3.8) we may eliminate the β_n's:

$$\frac{2}{k}\alpha_{n+1} + k(2n+1)\alpha_n = \frac{k}{2}\left(\frac{2}{k}\alpha_n + \alpha_{n-1}\right) - \alpha_n = \frac{k}{2}\alpha_{n-1},$$

resp.

$$(3.9) \qquad \frac{2\alpha_{n+1}}{k\alpha_n} + (2n+1)k = \frac{k\alpha_{n-1}}{2\alpha_n}.$$

In view of (3.7) we deduce

$$\alpha_1 = \frac{k}{2}(2\beta_0 - \alpha_0) = \left(\frac{k}{2}\right)^2\left(\exp\left(\frac{2}{k}\right) + 1 - k\left(\exp\left(\frac{2}{k}\right) - 1\right)\right),$$

and

$$\frac{\exp\left(\frac{2}{k}\right) + 1}{\exp\left(\frac{2}{k}\right) - 1} = \frac{\left(\frac{2}{k}\right)^2\alpha_1 + k\left(\exp\left(\frac{2}{k}\right) - 1\right)}{\frac{2}{k}\alpha_0}$$

$$= \frac{\left(\frac{2}{k}\right)^2\alpha_1 + 2\alpha_0}{\frac{2}{k}\alpha_0} = k + \frac{2\alpha_1}{k\alpha_0}.$$

It is easily seen that $2\alpha_{n+1} < k\alpha_n$ for $n \in \mathbb{N}$, which yields in view of (3.9)

$$\frac{\exp\left(\frac{2}{k}\right) + 1}{\exp\left(\frac{2}{k}\right) - 1} = k + \left(\frac{k\alpha_0}{2\alpha_1}\right)^{-1} = \left[k, 3k, \frac{k\alpha_1}{2\alpha_2}\right].$$

Induction on n proves the theorem. •

In the sequel we follow Euler in computing the continued fraction expansion for e. Taking $k = 2$ in Theorem 3.10 we get

$$\frac{e+1}{e-1} = [2, 6, 10, 14, \ldots].$$

Denote the nth convergent of the preceding number by $\frac{p_n}{q_n}$. Further, define the real number E by $E = [A_0, A_1, A_2, \ldots]$, where

$$(3.10) \qquad A_0 = 2, \quad A_{3n-2} = A_{3n} = 1 \quad \text{and} \quad A_{3n-1} = 2n,$$

and let $\frac{P_n}{Q_n}$ be the nth convergent to E. Our next aim is to verify

$$(3.11) \qquad P_{3n+1} = p_n + q_n \qquad \text{and} \qquad Q_{3n+1} = p_n - q_n.$$

We prove (3.11) by induction on n. The formula in question is easily verified for $n = 0, 1$. We find for $n \geq 2$

$$
\begin{aligned}
p_n &= (2 + 4n)p_{n-1} + p_{n-2}, \\
q_n &= (2 + 4n)q_{n-1} + q_{n-2}.
\end{aligned}
$$

On the other side, (3.10) implies

$$
\begin{aligned}
P_{3n-3} &= P_{3n-4} + P_{3n-5}, \\
P_{3n-2} &= P_{3n-3} + P_{3n-4}, \\
P_{3n-1} &= 2nP_{3n-2} + P_{3n-3}, \\
P_{3n} &= P_{3n-1} + P_{3n-2}, \\
P_{3n+1} &= P_{3n} + P_{3n-1}.
\end{aligned}
$$

Analogous formulae hold for the Q_n's. Multiplying these equations with $1, -1, 2, 1$ and 1, respectively, we get

$$
\begin{aligned}
P_{3n+1} &= (2 + 4n)P_{3n-2} + P_{3n-5}, \\
Q_{3n+1} &= (2 + 4n)Q_{3n-2} + Q_{3n-5},
\end{aligned}
$$

similar to the recursion formula for the p_n's and q_n's. This leads to

$$
P_{3n+1} = (2 + 4n)(p_{n-1} + q_{n-1}) + p_{n-2} + q_{n-2} = p_n + q_n,
$$

which is the first formula in (3.11); the other one follows similarly.

Now (3.11) implies

$$
E = \lim_{n \to \infty} \frac{P_{3n+1}}{Q_{3n+1}} = \lim_{n \to \infty} \frac{\dfrac{p_n}{q_n} + 1}{\dfrac{p_n}{q_n} - 1} = \frac{\dfrac{e+1}{e-1} + 1}{\dfrac{e+1}{e-1} - 1} = e.
$$

In view of (3.10) we thus have proved

Theorem 3.11. $e = [2, 1, 2, 1, 1, 4, 1, \ldots]$.

So far no pattern has been found in the first 17 million digits of the continued fraction for π, computed by Gosper in 1985. Curiously, if we switch to another expansion, we may find patterns. Euler found the continued fraction expansion

$$
\frac{\pi}{4} = \cfrac{1}{1 + \cfrac{1}{2 + \cfrac{9}{2 + \cfrac{\cdots}{\quad + \cfrac{(2n+1)^2}{2 + \cdots}}}}} .
$$

This representation follows from a general theorem due to Euler which transforms infinite series into continued fractions. In the case above the underlying infinite series is due to Leibniz,

$$
\frac{\pi}{4} = 1 - \frac{1}{3} + \frac{1}{5} - \frac{1}{7} \pm \cdots .
$$

It is legendary that, faced with this beautiful formula, Leibniz said *God loves the odd integers.*

Exercises

3.1. *Compute the continued fractions of the following numbers:*
$$\frac{57}{75}, \quad \frac{19}{25}, \quad -\frac{23}{97}, \quad \frac{23}{97}, \quad \frac{97}{23}, \quad \frac{99}{70}, \quad -\frac{351}{17}, \quad \frac{13\,254}{53\,412}.$$

3.2. The lunar month has approximately 29.53 days, the solar year 365.24. *Recover the* **Metonic cycle** *of seven leap months every* 19 *years from a sufficiently good approximation to* $\frac{365.24}{29.53}$ *by use of the Euclidean algorithm.*

3.3 *Compute all convergents to the continued fraction expansion of*
$$\frac{77\,708\,431}{2\,640\,858}.$$
Recover Huygens' approximation $\frac{147}{5}$ *from Section 1.8. Propose an approximation which differs by less than* 10^{-5}.

3.4.* The reduced fraction
$$\frac{x}{y} = \frac{70\,226}{40\,545} = 1.73205\,08077 \dots$$
is a fairly good approximation to $\sqrt{3}$.

i) Show that $\frac{x}{y}$ *is a convergent to* $\sqrt{3}$.

ii) Factor numerator x and denominator y into their prime divisors and with respect to these factorizations construct (virtually) a mechanism which approximates $\sqrt{3}$ *with six gears each of which having less than a hundred teeth.*

3.5.* This is a continuation of Exercise 2.3. So far it is not clear when a given rational will appear in the Calkin–Wilf sequence.

i) Compute the continued fraction expansions for the rational numbers appearing in the first four rows of the Calkin–Wilf tree. Is there any pattern? Where will the number $\frac{355}{113}$ *appear? Try to find a rule for how the rationals in the Calkin–Wilf sequence can be enumerated in terms of their continued fraction expansions.*

ii) Prove that the Calkin–Wilf sequence satisfies the following recursion formula:
$$x_1 = \frac{1}{1}, \quad x_{n+1} = \frac{1}{\lfloor x_n \rfloor + 1 - \{x_n\}} \quad \text{for } n \in \mathbb{N}.$$

3.6. Let $a_0, a_1, \dots, a_{m-1}$ and $b_0, b_1, \dots, b_{m-1}$ be positive integers.

i) Show that
$$[a_0, a_1, \dots, a_{m-1}, a_m] = [b_0, b_1, \dots, b_{m-1}, b_m]$$
implies $a_n = b_n$ *for all* $0 \le n \le m$.

ii) What are the conditions on the partial quotients for
$$[a_0, a_1, \dots, a_m] < [b_0, b_1, \dots, b_m] \quad ?$$

3.7. Let $a_0, a_1, \ldots, a_m$ be positive integers. Define the numbers p_m and q_m by

$$\begin{pmatrix} a_0 & 1 \\ 1 & 0 \end{pmatrix} \begin{pmatrix} a_1 & 1 \\ 1 & 0 \end{pmatrix} \cdot \ldots \cdot \begin{pmatrix} a_m & 1 \\ 1 & 0 \end{pmatrix} = \begin{pmatrix} p_m & p_{m-1} \\ q_m & q_{m-1} \end{pmatrix}.$$

Prove that

$$\frac{p_m}{q_m} = [a_0, a_1, \ldots, a_m].$$

3.8. *With the same notation as in Section 3.2 prove for $n \geq 2$ that*

$$\frac{p_n}{p_{n-1}} = [a_n, a_{n-1}, \ldots, a_1, a_0] \quad \text{and} \quad \frac{q_n}{q_{n-1}} = [a_n, a_{n-1}, \ldots, a_2, a_1].$$

3.9. *With the same notation as in Section 3.2 show that*

$$\frac{p_m}{q_m} = a_0 + \sum_{n=1}^{m} \frac{(-1)^{n-1}}{q_n q_{n-1}}.$$

3.10. *Prove all statements on the Muir symbol from Section 3.3.*

3.11.* *Prove that the Greedy algorithm applied to $\frac{4}{n}$ with $n \in \mathbb{N}$ yields a representation as a sum of four Egyptian fractions at most. Show that only in the case $n \equiv 1 \bmod 4$ can it happen that the Greedy algorithm returns more than three Egyptian fractions. Is it possible to restrict n further?*

3.12. *i) Verify the Erdös–Strauss conjecture for $n = 19, 91, 185, 201$ by hand calculation.*
ii) Implement the Greedy algorithm and apply it to

$$\frac{30}{131}, \quad \frac{35}{131}, \quad \frac{65}{131}, \quad \text{and} \quad \frac{31}{311}.$$

iii) Verify the Erdös–Strauss conjecture for all $n \leq 10^4$.

3.13. Let a, b be positive integers.
Prove that there exists an integer $x > \frac{a}{b}$ such that $bx - a$ divides x^2 if and only if there exist non-negative integers $\beta_j \leq 2\nu_j$ such that b divides

$$a + \prod_{j=1}^{r} p^{\beta_j},$$

where $a = \prod_{j=1}^{r} p^{\nu_j}$ is the prime factorization of a. Use this to finish the proof of Theorem 3.5.

3.14.* *For fixed $k \geq 2$, let $a_2, \ldots, a_m$ be positive integers.*
In the notation of Section 3.3, prove that, for every positive integer k satisfying $(A_2 + 1)k + A_3 > 0$,

$$[0, (A_2 + 1)k + A_3, a_2, \ldots, a_m] = \frac{1}{A_2 k + A_3} - \frac{1}{(A_2 k + A_3)(A_2 + 1)}.$$

Hint: Exercise 3.13 does not apply here but one may try something similar.

3.15.* *Replace (3.1) by an algorithm according to the nearest integer algorithm (according to Exercise 1.13). Extend this to the continued fraction algorithm!*

3.16.* *Given an irrational α, prove that among three consecutive convergents to α there is at least one satisfying the inequality in Hurwitz' theorem 2.9.*

3.17. *Use the continued fraction algorithm to find good rational approximations to the logarithm of 3 with base 2.*

From the sequence of convergents of the logarithm of 3 with base 2 one can recover several approximations to a Pythagorean scale, the Chinese scale with five notes to the octave, the scale in Western music, and the modern 12-tone scale. For further reading we refer to [**50**].

3.18. *Fill the gaps in Section 3.8, that are the proof of (3.8) and the verification of the second formula in (3.11).*

3.19.* *Let p and q be positive integers satisfying $|q^2\alpha^2 - p^2| < \alpha$. Show that then $\frac{p}{q}$ is a convergent to α.*

This characterization of convergents dates back to Legendre. We will meet a similar inequality in Theorem 7.1 below.

3.20.* *Show that*

$$\left| e - \frac{p}{q} \right| > \frac{c}{q^2 \log q}$$

for all $\frac{p}{q} \in \mathbb{Q}$ with $q > 1$ and some positive constant c.

CHAPTER 4

The irrationality of $\zeta(3)$

The Riemann zeta-function $\zeta(s)$ is an important object in analytic number theory. The strongest version of the prime number theorem is proved by analyzing the zeta-function as a complex function; in particular, its mysterious zero distribution is an active field of research with plenty of open questions. Also the behavior on the real axis is not yet understood very well. It was a big surprise when Apéry proved in 1978 that $\zeta(3)$ is irrational. In this chapter we prove Apéry's theorem by his original elementary approach. We also sketch Beukers' slightly different proof working with multiple integrals.

4.1. The Riemann zeta-function

The **Riemann zeta-function** is defined by

$$(4.1) \qquad \zeta(s) = \sum_{n=1}^{\infty} \frac{1}{n^s} = \prod_{p} \left(1 - \frac{1}{p^s}\right)^{-1},$$

where the so-called **Euler product** is taken over all prime numbers. The appearing series is the prototype of so-called **Dirichlet series** and is easily seen to be absolutely convergent for $s > 1$ (e.g., by Riemann's integral test), and uniformly convergent in any compact subset. The identity between the series and the product may be regarded as an analytic version of the unique prime factorization of integers. This reformulation was found by Euler in 1737 and gives a first glance on the intimate relation between $\zeta(s)$ and the distribution of prime numbers.

Assume that there are only finitely many primes; then the product converges throughout the complex plane, in contradiction to the divergence of the harmonic series (i.e., the singularity of $\zeta(s)$ at $s = 1$). Thus, there are infinitely many primes. Of course, Euclid gave a simpler proof of this fundamental fact. But Euler's analytic approach is much more powerful. By the outstanding work of Riemann on the zeta-function and the contributions of his successors, this led to a proof of the celebrated prime number theorem A.2. In a weak form this theorem claims that the number $\pi(x)$ of primes $p \leq x$ is asymptotically

$$(4.2) \qquad \pi(x) = (1 + o(1))\frac{x}{\log x}$$

(see Exercise 4.2). There exists an elementary proof due to Erdös and Selberg (independently; see [**76**]), but the nature of prime number distribution is encoded in the distribution of the zeros of the zeta-function.

In order to understand this deep relation one has to study $\zeta(s)$ as a function of a complex variable s. Riemann proved that $\zeta(s)$ can be analytically continued to the whole complex plane, except for a simple pole at $s = 1$ (the divergent harmonic series), and satisfies the functional equation

$$(4.3) \qquad \pi^{-\frac{s}{2}} \Gamma\left(\frac{s}{2}\right) \zeta(s) = \pi^{-\frac{1-s}{2}} \Gamma\left(\frac{1-s}{2}\right) \zeta(1-s),$$

where $\Gamma(s)$ denotes Euler's gamma function. In view of the Euler product it is easily seen that $\zeta(s)$ has no zeros in the half-plane Re $s > 1$. It follows from the functional equation and from basic properties of the gamma function that $\zeta(s)$ vanishes in Re $s < 0$ exactly at the so-called **trivial** zeros $s = -2n$ with $n \in \mathbb{N}$.

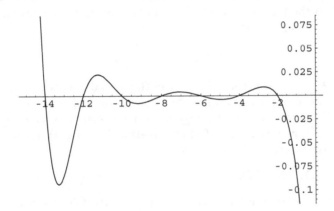

FIGURE 4.1. The trivial zeros of $\zeta(s)$ in the range $-15 \leq s \leq 0$.

All other zeros of $\zeta(s)$ are said to be **nontrivial** and they lie inside the so-called critical strip $0 \leq$ Re $s \leq 1$. It is easily seen that they are non-real; it is a bit more difficult to see that there are infinitely many of them. The functional equation (in addition with the reflection principle $\zeta(s) = \overline{\zeta(\bar{s})}$) shows some symmetries of $\zeta(s)$. In particular, the nontrivial zeros of $\zeta(s)$ have to be distributed symmetrically with respect to the real axis and with respect to the vertical line Re $s = \frac{1}{2}$. It was Riemann's ingenious contribution to number theory that pointed out how the distribution of these nontrivial zeros is linked to the distribution of prime numbers. In fact the error term in the prime number theorem A.2 is *minimal* if $\zeta(s)$ has no zeros in the half-plane Re $s > \frac{1}{2}$. This is the famous yet unproved Riemann hypothesis (for more details we refer the interested reader to Appendix A.3).

4.2. Apéry's theorem

The values of the zeta-function taken at the integers are of special interest. Euler showed for $n \in \mathbb{N}$ that

$$\zeta(2n) = (-1)^{n-1} \frac{(2\pi)^{2n}}{2(2n)!} B_{2n},$$

where the B_k's are the **Bernoulli numbers** defined by the identity

$$(4.4) \qquad \sum_{k=0}^{\infty} B_k \frac{z^k}{k!} = \frac{z}{\exp(z) - 1} = 1 - \frac{1}{2}z + \frac{1}{12}z^2 + \dots .$$

It is not too difficult to see that the B_k's are all rational. Thus it follows from the irrationality of π^2, Theorem 1.5, that $\zeta(2) = \frac{\pi^2}{6}$ is irrational. In Section 9.5 we will prove that π is transcendental, i.e., π is not the root of a polynomial with integer coefficients. This implies that none of the powers π^{2n} is rational, and thus $\zeta(2n) \notin \mathbb{Q}$ for any $n \in \mathbb{N}$.

Nearly nothing is known about the values taken at the odd integers; this is related to the trivial zeros of $\zeta(s)$ and the functional equation. In fact, one can prove, for $m \in \mathbb{N}$,

$$(4.5) \qquad\qquad \zeta(1 - m) = -\frac{B_m}{m}.$$

However, $B_{2n+1} = 0$ for $n \in \mathbb{N}$, so the functional equation does not give any information for the values $\zeta(2n + 1)$. It is widely believed that all values of the zeta-function at integers are irrational. It was a big surprise when Apéry announced a proof of the irrationality of $\zeta(3)$ during the 1978 Journées Arithmétiques in Marseille.

Theorem 4.1. $\zeta(3) \notin \mathbb{Q}$.

Apéry [5] published only a sketch of a proof. We follow van der Poorten [129], which is an interesting source on the exciting story of the proof and reactions upon it, and an unpublished manuscript of Elsner.

4.3. Approximating $\zeta(3)$

Apéry's proof relies on his discovery of a sequence of rational approximations to $\zeta(3)$ with several surprising properties. For integers k, n satisfying $0 \leq k \leq n$, we define

$$c_{n,k} = \sum_{m=1}^{n} \frac{1}{m^3} + \sum_{m=1}^{k} \frac{(-1)^{m-1}}{2m^3 \binom{n}{m} \binom{n+m}{m}},$$

and

$$b_n = \sum_{k=0}^{n} \binom{n}{k}^2 \binom{n+k}{k}^2, \qquad a_n = \sum_{k=0}^{n} c_{n,k} \binom{n}{k}^2 \binom{n+k}{k}^2.$$

These sequences yield an approximation to $\zeta(3)$.

Lemma 4.2.
$$\lim_{n \to \infty} \frac{a_n}{b_n} = \zeta(3).$$

Proof. By definition,

(4.6) $$a_n = b_n \sum_{m=1}^{n} \frac{1}{m^3} + \sum_{k=0}^{n} \binom{n}{k}^2 \binom{n+k}{k}^2 \sum_{m=1}^{k} \frac{(-1)^{m-1}}{2m^3 \binom{n}{m} \binom{n+m}{m}}.$$

First, we shall estimate the double sum on the right. For $1 \le m < n$,

$$\binom{n}{m}\binom{n+m}{m} \ge \binom{n}{1}\binom{n+1}{1} = n(n+1) \ge n^2.$$

The same estimate holds for the binomial coeffcient $\binom{2n}{n}$. Hence,

$$\sum_{m=1}^{k} \frac{(-1)^{m-1}}{2m^3 \binom{n}{m} \binom{n+m}{m}} \ll \frac{1}{n^2} \sum_{m=1}^{k} \frac{1}{m^3} \ll \frac{1}{n^2},$$

where we used the convergence of $\zeta(3) = \sum \frac{1}{m^3}$ for the last estimate. This leads in (4.6) to

(4.7) $$a_n = b_n \sum_{m=1}^{n} \frac{1}{m^3} + O\left(\frac{b_n}{n^2}\right).$$

Next we shall incorporate $\zeta(3)$. Therefore, we write

$$\sum_{m=1}^{n} \frac{1}{m^3} = \zeta(3) - \sum_{m=n+1}^{\infty} \frac{1}{m^3}.$$

To estimate the sum on the right-hand side, note that

(4.8) $$\frac{1}{m^3} \le \int_{m-1}^{m} \frac{du}{u^3} \le \frac{1}{(m-1)^3}$$

for $m \ge 2$. This yields above

$$\sum_{m=n+1}^{\infty} \frac{1}{m^3} \le \int_{n}^{\infty} \frac{du}{u^3} \ll \frac{1}{n^2}.$$

Hence we get from (4.7)

(4.9) $$\frac{a_n}{b_n} - \zeta(3) = \frac{a_n}{b_n} - \sum_{m=1}^{n} \frac{1}{m^3} + O\left(\sum_{m=n+1}^{\infty} \frac{1}{m^3}\right) \ll \frac{1}{n^2},$$

which implies the assertion of the lemma. •

In order to prove the irrationality of $\zeta(3)$ we may think of applying Theorem 2.2 to the sequence $\left(\frac{a_n}{b_n}\right)$. However, this cannot be done straightforward since the a_n are not integers:

$$a_0 = 0, \quad a_1 = 6, \quad a_2 = \frac{351}{4}, \quad \ldots \, .$$

Moreover, the estimate (4.9) is much too poor. Next we shall show that the sequence $\frac{a_n}{b_n}$ converges much faster to its limit than (4.9) indicates. For this purpose we have to look a bit more closely at these sequences.

Our next observation is not necessary for the proof of Apéry's theorem but of interest with respect to attempts to prove the irrationality for other zeta-values along the lines of Apéry's argument. Via

$$c_{n,n} - c_{n-1,n-1} = c_{n,n} \underbrace{-c_{n,n-1} + c_{n,n-1}}_{=0} - c_{n-1,n-1}$$

one can easily compute

$$c_{n,n} - c_{n-1,n-1} = \frac{5}{2} \frac{(-1)^{n-1}}{n^3 \binom{2n}{n}},$$

valid for any $n \in \mathbb{N}$. Since $c_{0,0} = 0$ and $\lim_{n\to\infty} c_{n,n} = \zeta(3)$, we get by summation

(4.10)
$$\zeta(3) = \frac{5}{2} \sum_{m=1}^{\infty} \frac{(-1)^{m-1}}{m^3 \binom{2m}{m}}.$$

This series representation was already known to Hjortnaes [82]. It has some relatives,

(4.11)
$$\zeta(2) = 3 \sum_{n=1}^{\infty} \frac{1}{n^3 \binom{2n}{n}} \quad \text{and} \quad \zeta(4) = \frac{36}{17} \sum_{n=1}^{\infty} \frac{1}{n^4 \binom{2n}{n}}.$$

Extensive computation has suggested that there are no analogous formulae for $\zeta(5)$ and $\zeta(7)$ of this simplicity. If there are coprime integers a, b such that

$$\zeta(5) = \frac{a}{b} \sum_{m=1}^{\infty} \frac{(-1)^{m-1}}{m^5 \binom{2m}{m}},$$

then b has to be *extremely large* (this can be checked by use of a computer algebra package). Nevertheless, Borwein & Bradley [28] succeeded in finding remarkable generalizations of (4.10) for $\zeta(4n + 3)$ empirically determined.

4.4. A recursion formula

Define
$$B(X) = 34X^3 + 51X^2 + 27X + 5.$$
The following property of the sequences of the a_n and of the b_n is the heart of Apery's marvelous proof.

Lemma 4.3. *The sequences $(a_n)_n$ and $(b_n)_n$ satisfy the recursion formula*
$$(n+1)^3 u_{n+1} - B(n) u_n + n^3 u_{n-1} = 0 \quad \text{for} \quad n \in \mathbb{N}.$$

Proof. We write for short
$$\lambda_{n,k} = \binom{n}{k}^2 \binom{n+k}{k}^2 = \frac{(n+k)!^2}{k!^4 (n-k)!^2},$$
and
$$B_{n,k} = 4(2n+1)(2k^2 + k - (2n+1)^2)\lambda_{n,k}.$$
Our first aim is to show that, for $1 \leq k \leq n$,

(4.12) $B_{n,k} - B_{n,k-1} = (n+1)^3 \lambda_{n+1,k} - B(n)\lambda_{n,k} + n^3 \lambda_{n-1,k}.$

Substituting the B's and dividing through $\lambda_{n,k}$, this is equivalent to

(4.13)
$$4(2n+1)(2k^2 + k - (2n+1)^2) - 4(2n+1)(2(k-1)^2$$
$$+ (k-1) - (2n+1)^2)\frac{\lambda_{n,k-1}}{\lambda_{n,k}}$$
$$= (n+1)^3\frac{\lambda_{n+1,k}}{\lambda_{n,k}} - B(n) + n^3\frac{\lambda_{n-1,k}}{\lambda_{n,k}}.$$

We compute
$$\frac{\lambda_{n,k-1}}{\lambda_{n,k}} = \frac{k^4}{(n+k)^2(n+1-k)^2},$$
$$\frac{\lambda_{n+1,k}}{\lambda_{n,k}} = \frac{(n+1+k)^2}{(n+1-k)^2}, \quad \text{and} \quad \frac{\lambda_{n-1,k}}{\lambda_{n,k}} = \frac{(n-k)^2}{(n+k)^2}.$$

Substituting these expressions, we may rewrite (4.13) as
$$4(2n+1)(2k^2 + k - (2n+1)^2) - 4(2n+1)(2(k-1)^2$$
$$+ (k-1) - (2n+1)^2)\frac{k^4}{(n+k)^2(n+1-k)^2}$$
$$= (n+1)^3\frac{(n+1+k)^2}{(n+1-k)^2} - B(n) + n^3\frac{(n-k)^2}{(n+k)^2}.$$

Multiplying this with $(n+k)^2(n+1-k)^2$ gives, equivalently,
$$4(2n+1)(2k^2 + k - (2n+1)^2)(n+k)^2(n+1-k)^2$$
$$- 4(2n+1)(2(k-1)^2 + (k-1) - (2n+1)^2)k^4$$
$$= (n+1)^3(n+1+k)^2(n+k)^2 - B(n)$$
$$+ n^3(n-k)^2(n+1-k)^2.$$

Now multiply out the polynomials in n and k on both sides and compare the coefficients (this boring job can be done by a computer algebra package). This proves (4.12).

Next we use (4.12) with $k = 0, 1, \ldots, n+1$ and sum up. Since the binomial coefficients $\binom{n}{k}$ vanish if $n < k$ or $k < 0$, we get
$$\lambda_{n,n+1} = \lambda_{n-1,n+1} = \lambda_{n-1,n} = \lambda_{n,-1} = 0,$$

which implies $B_{n,n+1} = B_{n,-1} = 0$. From (4.12) it thus follows that
$$0 = (n+1)^3\sum_{k=0}^{n+1}\lambda_{n+1,k} - B(n)\sum_{k=0}^{n}\lambda_{n,k} + n^3\sum_{k=0}^{n-1}\lambda_{n-1,k}.$$

In view of the definition of $\lambda_{n,k}$ the appearing sums are nothing other than $b_{n+1}, b_n,$ and b_{n-1}. Thus the sequence of the b_n satisfies the recursion formula of the lemma.

The same procedure can be applied to the sequence of the a_n. Here we only sketch the idea. It is not difficult to verify, for $1 \le k \le n$,

$$(4.14) \qquad c_{n,k} - c_{n-1,k} \;=\; \frac{(-1)^k k!^2 (n-k-1)!}{n^2 (n+k)!},$$

$$(4.15) \qquad c_{n,k} - c_{n,k-1} \;=\; \frac{(-1)^{k-1}}{2k^3 \binom{n}{k} \binom{n+k}{k}}.$$

Define

$$C_{n,k} = (n+1)^3 \lambda_{n+1,k} c_{n+1,k} - B(n) \lambda_{n,k} c_{n,k} + n^3 \lambda_{n-1,k} c_{n-1,k}.$$

It follows from (4.12) and (4.14) that

$$C_{n,k} \;=\; (B_{n,k} - B_{n,k-1}) c_{n,k} + (-1)^k k!^2 \Big((n+1)^3 \lambda_{n+1,k}$$

$$(4.16) \qquad \times \frac{(n-k)!}{(n+1)^2 (n+k+1)!} - n^3 \lambda_{n-1,k} \frac{(n-k-1)!}{n^2 (n+k)!} \Big).$$

Define

$$A_{n,k} = B_{n,k} c_{n,k} + 5 \frac{(-1)^{k-1} k (2n+1)}{n(n+1)} \binom{n}{k} \binom{n+k}{k}.$$

Then it follows from (4.15) that

$$A_{n,k} - A_{n,k-1} \;=\; (B_{n,k} - B_{n,k-1}) c_{n,k} + B_{n,k-1} \frac{(-1)^{k-1}}{2k^3 3 \binom{n}{k} \binom{n+k}{k}}$$

$$+ 5 \frac{2n+1}{n(n+1)} \Big((-1)^{k-1} k \binom{n}{k} \binom{n+k}{k}$$

$$- (-1)^k (k-1) \binom{n}{k-1} \binom{n+k-1}{k-1} \Big).$$

This and (4.16) give $C_{n,k} = A_{n,k} - A_{n,k-1}$. Hence,

$$\sum_{k=0}^{n+1} C_{n,k} = A_{n,n+1} - A_{n,1} = 0.$$

This yields the recursion formula for the a_n. The lemma is proved. ●

The recursion formula for the sequence of the a_n and of the b_n is the heart of Apéry's proof. The b_n are in fact values of the hypergeometric series and from this point of view the recurrence may be a bit less surprising.

4.5. The speed of convergence

Using the above recursion formula we can improve the estimate (4.9) significantly.

Lemma 4.4. *For* $n \in \mathbb{N}$,

$$\left| \zeta(3) - \frac{a_n}{b_n} \right| \ll \frac{1}{b_n^2}.$$

Proof. From Lemma 4.3 we know that

$$n^3 a_n - B(n-1)a_{n-1} + (n-1)^3 a_{n-2} = 0,$$
$$n^3 b_n - B(n-1)b_{n-1} + (n-1)^3 b_{n-2} = 0,$$

for all $n \geq 2$. Next we shall eliminate the polynomial B. Multiplying these equations with b_{n-1} and a_{n-1}, respectively, and subtracting one from the other, we get

$$n^3(a_n b_{n-1} - a_{n-1}b_n) = (n-1)^3(a_{n-1}b_{n-2} - a_{n-2}b_{n-1}).$$

Substituting this identity with $n-1$ instead of n into the above one, we obtain

$$a_n b_{n-1} - a_{n-1}b_n = \left(\frac{n-1}{n}\right)^3 (a_{n-1}b_{n-2} - a_{n-2}b_{n-1})$$
$$= \left(\frac{n-2}{n}\right)^3 (a_{n-2}b_{n-3} - a_{n-3}b_{n-2}).$$

By induction we can reduce the right-hand side until $n = 2$. Taking into account $a_1 b_0 - a_0 b_1 = 6 \cdot 1 - 0 \cdot 5 = 6$ it follows that

$$a_n b_{n-1} - a_{n-1}b_n = \frac{6}{n^3}$$

for all $n \in \mathbb{N}$. Now define

$$s_n = \zeta(3) - \frac{a_n}{b_n}.$$

Then, for $n \in \mathbb{N}$,

$$s_n - s_{n-1} = \frac{a_{n-1}}{b_{n-1}} - \frac{a_n}{b_n} = \frac{a_n b_{n-1} - a_{n-1}b_n}{b_{n-1}b_n} = \frac{6}{n^3 b_{n-1}b_n}.$$

It follows from Lemma 4.2 that s_n tends to zero as $n \to \infty$. Hence, by induction,

$$\zeta(3) - \frac{a_n}{b_n} = s_n = s_n \underbrace{- s_{n+1} + s_{n+1}}_{=0} = \ldots = \sum_{m=n+1}^{\infty} (s_{m-1} - s_m).$$

Thus

$$\left| \zeta(3) - \frac{a_n}{b_n} \right| = 6 \sum_{m=n+1}^{\infty} \frac{1}{m^3 b_{m-1}b_m}.$$

Since b_n is a strictly increasing sequence of integers, the convergence of $\zeta(3) - \sum_m \frac{1}{m^3}$ implies the assertion of the lemma. •

Next we shall observe the growth of the sequence of the b_n. Rewriting the recursion formula of Lemma 4.3,

$$1 \cdot b_n - \left(34 - \frac{51}{n} + \frac{27}{n^2} - \frac{5}{n^3}\right) b_{n-1} + \left(1 - \frac{3}{n} + \frac{3}{n^2} - \frac{1}{n^3}\right) b_{n-2} = 0,$$

and using the ansatz $b_{n+1} = b_n X$, we have to consider the polynomial

$$Q(X) = 1 \cdot X^2 - 34 X + 1.$$

The positive root of Q is easily seen to be

$$\alpha := 17 + 12\sqrt{2} = (1 + \sqrt{2})^4.$$

This leads to the estimate

(4.17) $$\alpha^n \ll b_n \ll \alpha^n;$$

so b_n increases (quasi) geometrically. Substituting this in Lemma 4.4 gives

(4.18) $$\left| \zeta(3) - \frac{a_n}{b_n} \right| \ll \alpha^{-2n}.$$

4.6. Final steps in the proof

First, we need an upper bound for the least common multiple of the integers $1, 2, \ldots, n$ which we denote by $\operatorname{lcm}[1, 2, \ldots, n]$.

Lemma 4.5. *For $n \in \mathbb{N}$ and any positive ε,*

$$\operatorname{lcm}[1, 2, \ldots, n] \leq \exp((1 + \varepsilon)n).$$

Proof. The prime divisors of $\operatorname{lcm}[1, 2, \ldots, n]$ are exactly the primes $p \leq n$. From the inequality $p^{\nu} \leq n < p^{\nu+1}$ one immediately deduces

$$\nu \leq \frac{\log n}{\log p} < \nu + 1.$$

We write $\nu_p(n)$ for the exponent of the prime p in the unique prime factorization of the integer n. Thus the exponent of any prime p in $\operatorname{lcm}[1, 2, \ldots, n]$ is

(4.19) $$\nu_p(\operatorname{lcm}[1, 2, \ldots, n]) = \left[\frac{\log n}{\log p} \right].$$

Hence we get

$$\operatorname{lcm}[1, 2, \ldots, n] = \prod_{p \leq n} p^{\left[\frac{\log n}{\log p} \right]}.$$

Removing the Gauss brackets we get the upper bound

$$\operatorname{lcm}[1, 2, \ldots, n] \leq \prod_{p \leq n} n = n^{\pi(n)},$$

where $\pi(n)$ counts the number of primes $p \leq n$. It follows from the prime number theorem (4.2) that

$$n^{\pi(n)} = \exp(\pi(n) \log n) \leq \exp((1 + \varepsilon)n).$$

The assertion of the lemma follows. •

Recall that the a_n are not integers, but their denominators are not too long. To see this we consider the numbers $c_{n,k}$.

Lemma 4.6. *For integers $0 \leq k \leq n$,*

$$2 \operatorname{lcm}[1, 2, \ldots, n]^3 \binom{n+k}{k} c_{n,k} \in \mathbb{Z}.$$

Proof. Let $m \leq k \leq n$. By the definition of the $c_{n,k}$ it suffices to show that both

$$2 \operatorname{lcm}[1, 2, \ldots, n]^3 \binom{n}{n+k} \frac{1}{m^3}$$

and

(4.20) $$2 \operatorname{lcm}[1, 2, \ldots, n]^3 \binom{n}{n+k} \frac{1}{2m^3 \binom{n}{m}\binom{n+m}{m}}$$

are integers. The first assertion is obvious. To prove the second one we check the number of times that any given prime p divides the denominator.

First,

$$\nu_p \left(\binom{n}{m} \right) \leq \nu_p(\operatorname{lcm}[1, 2, \ldots, n]) - \nu_p(m).$$

Moreover,

$$\binom{n+m}{m}^{-1} \binom{n+k}{k} = \binom{k}{m}^{-1} \binom{n+k}{k-m}.$$

Taking into account (4.19) it follows that

$$\nu_p \left(m^3 \binom{n}{m} \binom{n+m}{m} \binom{n+k}{k}^{-1} \right) = \nu_p \left(m^3 \binom{n}{m} \binom{k}{m} \binom{n+k}{k-m}^{-1} \right)$$

$$\leq \nu_p(m) + \left\lceil \frac{\log n}{\log p} \right\rceil + \left\lceil \frac{\log k}{\log p} \right\rceil \leq 3 \left\lceil \frac{\log n}{\log p} \right\rceil$$

since $m \leq k \leq n$. In view of (4.19) this implies that the expression in (4.20) is an integer, what we had to prove. •

Now we can do the final step in Apéry's proof of $\zeta(3) \notin \mathbb{Q}$.

Proof of Theorem 4.1. For $n \in \mathbb{N}$, define

$$p_n = 2 \operatorname{lcm}[1, 2, \ldots, n]^3 a_n \quad \text{and} \quad q_n = 2 \operatorname{lcm}[1, 2, \ldots, n]^3 b_n.$$

In view of Lemma 4.6 it follows that $p_n, q_n \in \mathbb{Z}$. Moreover, $\frac{p_n}{q_n} = \frac{a_n}{b_n}$. With respect to Lemma 4.5 and (4.17) we have

(4.21) $$q_n \ll \alpha^n \exp((3 + \varepsilon)n) = \alpha^{n(1 + \frac{3+\varepsilon}{\log \alpha})}.$$

Hence we may rewrite (4.18) as

(4.22) $$\zeta(3) - \frac{p_n}{q_n} \ll \alpha^{-2n} \ll q_n^{-(1+\delta)},$$

where

(4.23) $$\delta = \frac{\log \alpha - 3}{\log \alpha + 3} = 0.08052 \ldots.$$

In view of the irrationality criterion Theorem 2.2 this proves Apéry's theorem. •

4.7. An irrationality measure

Apéry's approach not only proves that $\zeta(3)$ does not lie in $\mathbb{Q}$, but also provides information *how far* this constant is from being rational. For this purpose we attach to an irrational number α the **irrationality exponent** $\mu = \mu(\alpha)$ if μ is the least possible exponent such that for any positive ε the inequality

$$\left| \alpha - \frac{p}{q} \right| < q^{-\mu-\varepsilon}$$

has only finitely many solutions in integers p and q with $q \geq 1$.

Theorem 4.7. *Let $0 < \kappa < \delta$. If there are infinitely many $\frac{p_n}{q_n}$ such that*

$$\left| \alpha - \frac{p_n}{q_n} \right| < q_n^{-1-\delta},$$

and the q_n are monotonic increasing with $q_n < q_{n-1}^{1+\kappa}$, then, for all integers $p, q \geq 1$ sufficiently large with respect to $\varepsilon > 0$,

$$\left| \alpha - \frac{p}{q} \right| > q^{-(1+\frac{1+\kappa}{\delta-\kappa}+\varepsilon)}.$$

Proof. Suppose that

$$\left| \alpha - \frac{p}{q} \right| \leq \frac{1}{q^\tau}$$

for some positive τ. Now select indices n such that $q_{n-1}^{1+\delta} \leq q^\tau < q_n^{1+\delta}$. This implies

$$\frac{1}{qq_n} \leq \left| \frac{p}{q} - \frac{p_n}{q_n} \right| \leq \left| \frac{p}{q} \underbrace{- \alpha + \alpha}_{=0} - \frac{p_n}{q_n} \right|$$

$$\leq \left| \alpha - \frac{p_n}{q_n} \right| + \left| \alpha - \frac{p}{q} \right| \leq \frac{1}{q_n^{1+\delta}} + \frac{1}{q^\tau} < \frac{2}{q^\tau}.$$

It follows that

$$\frac{1}{2} q^\tau < qq_n < qq_{n-1}^{1+\kappa} \leq q^{1+\tau\frac{1+\kappa}{1+\delta}},$$

resp. $\tau \leq \frac{1+\delta}{\delta-\kappa} + \varepsilon$. Note that $\frac{1+\delta}{\delta-\kappa} = 1 + \frac{1+\kappa}{\delta-\kappa}$. This proves the theorem. •

Now we return to our observations concerning $\zeta(3)$. Taking into account (4.17) one can obtain a lower bound for q_n, comparable to the upper bound in (4.21). Hence, the sequence of the q_n increases geometrically, and so we may apply Theorem 4.7 with $\kappa > 0$ arbitrarily small. This leads to an estimate for the irrationality measure of $\zeta(3)$. By (4.23)

$$1 + \frac{1}{\delta} = \frac{2\log\alpha}{\log\alpha - 3} = 13.41782\ldots ,$$

and thus it follows from (4.22) that

Corollary 4.8. $\mu(\zeta(3)) \leq 13.41782\ldots$.

It follows that $\zeta(3)$ is not approximable by rationals of order 13.42. The presently sharpest estimate for the irrationality measure for $\zeta(3)$ is 5.51389 ... given by Rhin & Viola [**130**].

Apéry's method applies to $\zeta(2)$ as well (and this is the aim of Exercise 4.12). Of course, $\zeta(2) \notin \mathbb{Q}$ was already known before, but Apéry's approach improved the sharpest known bound for the irrationality measure of $\pi^2 = 6\zeta(2)$ to

$$\mu(\pi^2) \leq 11.85078 \ldots .$$

4.8. A non-simple continued fraction

As a by-product Apéry's proof yields a beautiful continued fraction expansion for $\zeta(3)$. However, it is not a simple continued fraction but we shall take the chance to get a bit in touch with things like that.

First, we shall prove

Lemma 4.9. *Assume that the sequences (a_n) and (b_n) of positive integers both satisfy the linear recursion formula*

$$A_n u_{n+1} - B_n u_n + C_n u_{n-1} = 0 \quad for \quad n \in \mathbb{N},$$

where A_n, B_n, C_n are positive-valued polynomials in n and $a_0 = 0, b_0 = 1$. Then

$$\frac{a_{n+1}}{b_{n+1}} = \cfrac{a_1}{b_1 - \cfrac{C_1}{B_1 - \cfrac{A_1 C_2}{B_2 - \cfrac{}{\ddots \; B_{n-1} - \cfrac{A_{n-1}C_n}{B_n}}}}}.$$

Proof. Since

$$\frac{B_n a_n - C_n a_{n-1}}{B_n b_n - C_n b_{n-1}} = \frac{A_n a_{n+1}}{A_n b_{n+1}} = \frac{a_{n+1}}{b_{n+1}},$$

we may rewrite the assertion of the theorem as

$$\frac{B_n a_n - C_n a_{n-1}}{B_n b_n - C_n b_{n-1}} = \cfrac{a_1}{b_1 - \cfrac{C_1/A_1}{B_1/A_1 - \cfrac{}{\ddots \; B_{n-1}/A_{n-1} - \cfrac{C_n/A_n}{B_n/A_n}}}}.$$

The proof follows from induction on n. It is easily seen that the assertion holds for $n = 1$. Assume that the formula is true for n. To prove the formula for $n+1$ it suffices to verify the identity

$$\frac{\left(\frac{B_n}{A_n} - \frac{C_{n+1}}{B_{n+1}}\right) a_n - \frac{C_n}{A_n} a_{n-1}}{\left(\frac{B_n}{A_n} - \frac{C_{n+1}}{B_{n+1}}\right) b_n - \frac{C_n}{A_n} b_{n-1}} = \frac{B_{n+1}a_{n+1} - C_{n+1}a_n}{B_{n+1}b_{n+1} - C_{n+1}b_n}.$$

This simple computation is left to the reader. The lemma is proved. •

Put

$$A_n = (n+1)^3, \quad B_n = 34n^3 + 51n^2 + 27n + 5, \quad \text{and} \quad C_n = n^3.$$

Then the recursion formula of the previous lemma is identical with the one of Lemma 4.3. By Lemma 4.2, the convergence of $\frac{a_n}{b_n}$ to $\zeta(3)$, we obtain

$$\zeta(3) = \cfrac{6}{5 - \cfrac{1^6}{117 - \cfrac{}{\cdots \; - \cfrac{n^6}{34n^3 + 51n^2 + 27n + 5 - \; \cdots}}}}.$$

4.9. Beukers' proof

Beukers [19] found a different proof for Apéry's theorem. His proof might be a bit more transparent than Apéry's original one but is not elementary. It relies on improper integrals taken over the so-called **Legendre polynomials**

$$P_n(z) = \frac{1}{n!} \frac{\mathrm{d}^n}{\mathrm{d}z^n} (z^n (1 - z)^n),$$

which we already met in (1.4) when we proved the irrationality of π. We consider

$$\int_0^1 \int_0^1 \frac{-\log(xy)}{1 - xy} P_n(x) P_n(y) \, \mathrm{d}x \, \mathrm{d}y.$$

Here and in the sequel all manipulations of this or similar integrals can be justified by replacing $\int_0^1$ with $\int_\varepsilon^{1-\varepsilon}$ for some small positive ε, and letting $\varepsilon \to 0$.

Let k, ℓ be non-negative integers and z be any non-negative number. In view of the formula for the geometric series,

$$\frac{1}{1 - xy} = \sum_{m=0}^{\infty} (xy)^m,$$

we find

$$(4.24) \qquad \int_0^1 \int_0^1 \frac{x^{k+z} y^{\ell+z}}{1 - xy} \, \mathrm{d}x \, \mathrm{d}y = \sum_{m=0}^{\infty} \frac{1}{(m + k + z + 1)(m + \ell + z + 1)}.$$

Assume that $k > \ell$; then the right-hand side can easily be evaluated as

$$\frac{1}{k - \ell} \left(\frac{1}{\ell + 1 + z} + \ldots + \frac{1}{k + z} \right).$$

Putting $z = 0$, we get

$$(4.25) \qquad \int_0^1 \int_0^1 \frac{x^k y^\ell}{1 - xy} \, \mathrm{d}x \, \mathrm{d}y = \frac{1}{k - \ell} \left(\frac{1}{\ell + 1} + \ldots + \frac{1}{k} \right).$$

Differentiation with respect to z in (4.24) leads in a similar manner to

$$(4.26) \qquad \int_0^1 \int_0^1 \frac{\log(xy)}{1 - xy} x^k y^\ell \, \mathrm{d}x \, \mathrm{d}y = -\frac{1}{k - \ell} \left(\frac{1}{(\ell + 1)^2} + \ldots + \frac{1}{k^2} \right).$$

Note that both integrals (4.25) and (4.26) are rational numbers with denominators dividing $\mathrm{lcm}[1, 2, \ldots, k]^3$. Taking $k = \ell$ in (4.24), differentiating with respect to z, and putting $z = 0$, we get

$$\int_0^1 \int_0^1 \frac{\log(xy)}{1-xy}(xy)^k \,dx\,dy \;=\; -2\sum_{m=0}^{\infty} \frac{1}{(m+k+1)^3}$$

$$(4.27) \hspace{3.5cm} = \; -2\left(\zeta(3) - 1 - \frac{1}{2^3} - \cdots - \frac{1}{k^3}\right).$$

If $k = 0$, the right-hand side is equal to $-2\zeta(3)$. Now the Legendre polynomials come into the game. By (4.25)–(4.27), we find

$$(4.28) \hspace{2cm} \int_0^1 \int_0^1 \frac{-\log(xy)}{1-xy} P_n(x)P_n(y) \,dx\,dy = \frac{A_n + B_n\zeta(3)}{\operatorname{lcm}[1,2,\ldots,n]^3}$$

for some integers A_n, B_n. In view of the identity

$$\frac{-\log(xy)}{1-xy} = \int_0^1 \frac{1}{1-(1-xy)z} \,dz$$

we obtain

$$\int_0^1 \int_0^1 \int_0^1 \frac{P_n(x)P_n(y)}{1-(1-xy)z} \,dx\,dy\,dz = \frac{A_n + B_n\zeta(3)}{\operatorname{lcm}[1,2,\ldots,n]^3}.$$

After an n-fold partial integration with respect to x the integral on the left-hand side can be written as

$$\int_0^1 \int_0^1 \int_0^1 \frac{(xyz)^n(1-x)^n P_n(y)}{(1-(1-xy)z)^{n+1}} \,dx\,dy\,dz.$$

The substitution

$$w = \frac{1-z}{1-(1-xy)z}$$

changes this into

$$\int_0^1 \int_0^1 \int_0^1 (1-x)^n(1-w)^n \frac{P_n(y)}{1-(1-xy)w} \,dx\,dy\,dw.$$

After another n-fold partial integration, now with respect to y, we finally obtain

$$\int_0^1 \int_0^1 \int_0^1 \frac{x^n(1-x)^n y^n(1-y)^n w^n(1-w)^n}{(1-(1-xy)w)^{n+1}} \,dx\,dy\,dw = \frac{A_n + B_n\zeta(3)}{\operatorname{lcm}[1,2,\ldots,n]^3}.$$

It is not too difficult to compute that

$$\frac{x(1-x)y(1-y)w(1-w)}{1-(1-xy)w} \le (\sqrt{2}-1)^4$$

is in the range of integration, and that equality holds if and only if $x = y$. It follows that the integral (4.28) is bounded above by

$$(\sqrt{2}-1)^{4n} \int_0^1 \int_0^1 \int_0^1 \frac{dx\,dy\,dw}{1-(1-xy)w}$$

$$= \; (\sqrt{2}-1)^{4n} \int_0^1 \int_0^1 \frac{-\log(xy)}{1-xy} \,dx\,dy = 2(\sqrt{2}-1)^{4n}\zeta(3),$$

where the last identity comes from (4.27) with $k = 0$. Since integral (4.28) is not zero, it follows that

$$(4.29) \qquad 0 < |A_n + B_n\zeta(3)| < 2(\sqrt{2} - 1)^{4n}\mathrm{lcm}[1, 2, \ldots, n]^3\zeta(3).$$

By Lemma 4.5, the bound $\zeta(3) < 2$, and since $27(\sqrt{2} - 1)^4 < \frac{4}{5}$, the right-hand side is less than $4(\frac{4}{5})^n$.

Now assume that $\zeta(3)$ is rational, say equal to $\frac{a}{b}$ with some positive integer b; then

$$0 < |A_n + B_n\zeta(3)| = \left|A_n + B_n\frac{a}{b}\right| = \frac{|bA_n + aB_n|}{b}.$$

Since b and $|bA_n + aB_n|$ are integers, it follows from (4.29) that

$$\frac{1}{b} \le |A_n + B_n\zeta(3)| \le 4\left(\frac{4}{5}\right)^n,$$

which tends to zero as $n \to \infty$. This gives the desired contradiction. •

Notes on recent results

Apéry's discovery started intensive research on this topic with the aim to obtain similar results for other values of $\zeta(s)$. However, it is still open whether $\zeta(5)$ or any other value taken at positive odd integers is irrational. Recently, Rivoal [135] proved that there are infinitely many irrational numbers in the set of odd zeta-values; more precisely, he showed that for every $\varepsilon > 0$ there exists an integer $N(\varepsilon)$ such that the $\mathbb{Q}$-vector space spanned by the $n + 1$ numbers

$$1, \zeta(3), \zeta(5), \ldots, \zeta(2n - 1), \zeta(2n + 1)$$

has dimension

$$\ge \frac{1 - \varepsilon}{1 + \log 2}\log n,$$

whenever $n > N(\varepsilon)$. The proof relies on a construction based on linear forms in the values of the zeta-function, and estimates upon them. Zudilin [176] used these ideas to show that at least one of the four numbers $\zeta(5), \zeta(7), \zeta(9)$, and $\zeta(11)$ is irrational.*

Exercises

4.1. *Prove the identity between the infinite product and the infinite series in (4.1).*

4.2. *i) Prove that*

$$\int_2^x \frac{du}{\log u} = \frac{x}{\log x} + O\left(\frac{x}{(\log x)^2}\right).$$

Hint: Use partial integration.

---

*The interested reader can find a list of further references and links devoted to the arithmetic study of values of the zeta-function at http://wain.mi.ras.ru/zw/index.html.

ii) Using i) deduce (4.2) from the prime number theorem A.2.
iii) Deduce from (4.2) that for any $\varepsilon > 0$ and sufficiently large x there exists a prime number in the interval $[x, (1+\varepsilon)x]$. Further show for the nth prime number p_n (in ascending order) that $p_n = (1 + o(1))n \log n$.

4.3. *For real $s \geq 2$ prove that $1 < \zeta(s) \leq \zeta(2) < 2$. What is the limit of $\zeta(s)$ as $s \to \infty$?*

4.4. The following evaluation of $\zeta(2)$ by elementary means is due to Calabi. *Verify*

$$\frac{3}{4}\sum_{n=1}^{\infty}\frac{1}{n^2} = \sum_{m=0}^{\infty}\frac{1}{(2m+1)^2} = \sum_{m=0}^{\infty}\int_0^1\int_0^1 x^{2m}y^{2m}\,\mathrm{d}x\,\mathrm{d}y$$

$$= \int_0^1\int_0^1\sum_{m=0}^{\infty}(xy)^{2m}\,\mathrm{d}x\,\mathrm{d}y = \int_0^1\int_0^1\frac{\mathrm{d}x\,\mathrm{d}y}{1-x^2y^2}.$$

Using the transformation

$$x = \frac{\sin u}{\cos v} \quad \text{and} \quad y = \frac{\sin v}{\cos u},$$

compute the appearing double integral above and deduce

$$\zeta(2) = \sum_{n=1}^{\infty}\frac{1}{n^2} = \frac{\pi^2}{6}.$$

Hence, $\zeta(2)$ is irrational. However, nothing is known on the arithmetic nature of the related **Catalan constant**

$$\sum_{n=0}^{\infty}\frac{(-1)^n}{(2n+1)^2}.$$

4.5. This exercise provides some facts about the Bernoulli numbers defined by (4.4).
i) For $n \in \mathbb{N}$, prove $B_{2n+1} = 0$.
Hint: Recall that the odd coefficients of the Taylor series of an even function are all equal to zero.
ii) Show that

$$\sum_{n=0}^{\infty}\sum_{k=0}^{n}\binom{n}{k}B_k\frac{z^n}{n!} = \frac{-z}{\exp(-z)-1},$$

and deduce the recursion formula

$$B_n = (-1)^n\sum_{k=0}^{n}\binom{n}{k}B_k.$$

iii) Deduce from ii) that all Bernoulli numbers are rational.

4.6.* Taking into account the previous exercise we can prove Euler's formula for the values of the zeta-function at even integers.

i) Verify the formulae

$$z\left(\log\frac{\sin(\pi z)}{\pi z}\right)' = \pi z\cot(\pi z) - 1 = \sum_{k=1}^{\infty}(-1)^k\frac{(2\pi)^{2k}}{(2k)!}B_{2k}z^{2k}.$$

The sine function has the following representation as an infinite product:

(4.30)
$$\sin(\pi z) = \pi z\prod_{n=1}^{\infty}\left(1 - \frac{z^2}{n^2}\right).$$

ii) Deduce from i) and the product formula (4.30) Euler's formula

$$\zeta(2n) = (-1)^{n-1}\frac{(2\pi)^{2n}}{2(2n)!}B_{2n}\qquad\text{for}\quad n\in\mathbb{N}.$$

This proves $\zeta(2n)\pi^{-2n}\in\mathbb{Q}$. Euler did not give a proof for (4.30). Any such proof relies on complex numbers.

4.7. *Using Stirling's formula in the form $m! \sim \sqrt{2\pi m}\left(\frac{m}{e}\right)^m$ for $m\in\mathbb{N}$, show that*

$$\lim_{n\to\infty}\left|\frac{B_{2n+2}}{B_{2n}}\right| = \infty.$$

Hint: Use Euler's formula from the previous exercise and Exercise 4.3.

4.8. *Deduce from the functional equation (4.3) and the Exercise 4.6 formula (4.5) for the values of the zeta-function at negative integers.*

4.9.* Using the theory of diophantine approximations Harald Bohr (the brother of the physicist Niels Bohr) & Landau [23] showed that $\zeta(s)$ takes arbitrarily large values in the half-plane of absolute convergence Re $s > 1$ and s not from the neighborhood of the pole at $s = 1$.

i) Given arbitrary real numbers $\alpha_1,\ldots,\alpha_N$, a positive integer q, and a positive number T, prove the existence of a real number $\tau\in[T,q^NT]$ and integers $x_1,\ldots,x_N$ for which

$$|\tau\alpha_n - x_n| \le \frac{1}{q}\qquad\text{for}\quad n\le N.$$

This is a variant of Exercise 2.6.

ii) Writing $s = \sigma + it$ with $\sigma,t\in\mathbb{R}, i = \sqrt{-1}$, show that for $\sigma > 1$

$$\zeta(\sigma) \ge |\zeta(s)| \ge \sum_{n=1}^{N}\frac{\cos(t\log n)}{n^\sigma} - \sum_{n=N+1}^{\infty}\frac{1}{n^\sigma}.$$

iii) Apply i) with $\alpha_n = \frac{\log n}{2\pi}$ to find a real number $\tau\in[T,q^NT]$ such that

$$\cos(\tau\log n) \ge \cos\left(\frac{2\pi}{q}\right)\qquad\text{for}\quad n = 1,\ldots,N.$$

iv) Use the idea in (4.8) to prove for $\sigma > 1$ the estimate

$$\sum_{n=N+1}^{\infty}\frac{1}{n^\sigma} < \int_N^\infty\frac{du}{u^\sigma} = \frac{N^{1-\sigma}}{\sigma - 1}.$$

v) Deduce from ii)–iv) the existence of an infinite sequence of $s = \sigma + it$ with $\sigma \to 1+$ and $t \to \infty$ for which

$$|\zeta(s)| \geq (1 - \varepsilon)\zeta(\sigma),$$

where ε is an y positive constant. Further, show that for arbitrary $T > 0$

$$\limsup_{\sigma>1,t>T} |\zeta(\sigma + it)| = \infty.$$

4.10.* This exercise shows that $|\zeta(s)|$ takes arbitrarily small values in the half-plane of absolute convergence in spite of the fact that $\zeta(s)$ does not vanish in this region.

i) Use Exercises 2.8. and 2.9 to find arbitrarily large values τ such that for any $x > 0$

$$\cos(\tau \log p) = -\cos(\tau \log p - \pi) \leq -\cos\left(\frac{\pi}{3}\right) = -\frac{1}{2} \quad \text{for all primes } p \leq x.$$

In view of (4.1) $\zeta(s)$ is a non-vanishing analytic function in the half-plane $\sigma > 1$. Thus we can define its logarithm (by fixing any of its values).

ii) For $\sigma > 1$, show that

$$\log \zeta(s) = \sum_p \sum_{k=1}^{\infty} \frac{1}{kp^{ks}}$$

and verify the estimate

$$\left| \log \zeta(s) - \sum_p \frac{1}{p^s} \right| \leq \sum_p \sum_{k=2}^{\infty} \frac{1}{p^k} = \sum_p \frac{1}{p(p-1)} \ll 1.$$

iii) Following the ideas of Exercise 4.9, deduce from i)–iii)

$$\liminf_{\sigma>1,t>T} |\zeta(\sigma + it)| = 0.$$

For quantitative versions of the statements of this and the previous exercise we refer to [162]; a larger class of Dirichlet series is treated in [157].

4.11.* *i) For any $a_1, a_2, \ldots$, prove the formal identity*

$$\sum_{m=1}^{M} \frac{a_1 \cdot \ldots \cdot a_{m-1}}{(x + a_1) \cdot \ldots \cdot (x + a_m)} = \frac{1}{x} - \frac{a_1 \cdot \ldots \cdot a_M}{x(x + a_1) \cdot \ldots \cdot (x + a_M)}.$$

Hint: Rewrite the right-hand side as $A_0 - A_M$ and verify that each term on the left is of the form $A_{m-1} - A_m$.

ii) Use i) with $x = n^2, a_m = -m^2$, and $m \leq M \leq n - 1$ to obtain (4.10).

4.12.* This exercise provides a proof of the irrationality of $\zeta(2)$ via Apéry's approach. Let k, n be integers such that $0 \leq k \leq n$. Define

$$\tilde{c}_{n,k} = 2 \sum_{m=1}^{n} \frac{(-1)^{m-1}}{m^2} + \sum_{m=1}^{k} \frac{(-1)^{n+m-1}}{m^2 \binom{n}{m} \binom{n+m}{m}},$$

and
$$\tilde{b}_n = \sum_{k=0}^{n} \binom{n}{k}^2 \binom{n+k}{k}, \quad \tilde{a}_n = \sum_{k=0}^{n} c'_{n,k} \binom{n}{k}^2 \binom{n+k}{k}.$$

i) Prove that
$$\lim_{n\to\infty} \frac{\tilde{a}_n}{\tilde{b}_n} = \zeta(2).$$

ii) Show that the sequences $(\tilde{a}_n)_n$ and $(\tilde{b}_n)_n$ satisfy the recursion formula
$$n^2 u_n = (11n^2 - 11n + 3)u_{n-1} + (n-1)^2 u_{n-2}.$$

iii) Prove that $\zeta(2) \notin \mathbb{Q}$.

4.13. *For the irrationality measure of π^2, show that $\mu(\pi^2) \le 11.86$. Deduce for the irrationality measure of π the estimate $\mu(\pi) \le 23.73$.*

Hint: The first bound follows from the previous exercise along the lines of Section 4.7. For the second bound one may use
$$\left|\pi - \frac{p}{q}\right| = \left(\pi + \frac{p}{q}\right)^{-1} \left|\pi^2 - \frac{p^2}{q^2}\right|.$$

4.14.* *Prove the series representations in (4.11).*

4.15. *Deduce from Exercise 4.12 that*
$$\zeta(2) = \cfrac{5}{3 + \cfrac{1}{25 + \cfrac{}{\cdots + \cfrac{n^4}{11n^2 + 11n + 3 + \cdots}}}}.$$

4.16.* *Use Beukers' approach to prove that $\zeta(2) \notin \mathbb{Q}$.*

4.17.* *In the notation of Section 4.9 show that*
$$\mathcal{J}_n := \int_0^1 \frac{P_n(x)}{1+x}\, \mathrm{d}x = P_n(-1)\log 2 - \frac{p_n}{\mathrm{lcm}[1,2,\dots,n]}$$
for some integer p_n, and
$$(\sqrt{2} - 1 - \varepsilon)^{2n} \le |\mathcal{J}_n| \le (\sqrt{2} - 1)^{2n}$$
for any positive ε and sufficiently large n. Deduce the bound $\mu(\log 2) \le 4.63$ for the irrationality measure of $\log 2$.

This is due to Alladi & Robinson [**4**].

4.18.* *For $\mathrm{Re}\, z > -2$, prove that*
$$\int_0^1 \int_0^1 \frac{(-\log(xy))^z}{1-xy}(1-x)\, \mathrm{d}x\, \mathrm{d}y = \Gamma(z+2)\left(\zeta(z+2) - \frac{1}{z+1}\right).$$

This is due to Chapman [**38**]. The proof requires some knowledge on complex analysis (in particular, the gamma function).

CHAPTER 5

Quadratic irrationals

It is quite easy to compute the continued fraction of an irrational number up to any prescribed length. Usually, such expansions show no pattern; Euler's number e is an exception. In this chapter we investigate other exceptions, namely, quadratic irrationals (that are algebraic numbers of degree two). Their continued fractions have many patterns which imply interesting properties with respect to diophantine approximations. However, this chapter is not only about quadratic irrationals; it also provides a classification of all real numbers with respect to their continued fraction expansion and aspects of diophantine approximations.

5.1. Fibonacci numbers and paper folding

Leonardo da Pisa (1170–1250), known as Fibonacci (in English *block-head*), was one of the most important mathematicians in the Middle Ages, a dark age outside Arabia. He wrote the influential book *Liber abaci* in which he introduced the arabic ciphers and the notion of zero to Europe. Furthermore, he introduced the **Fibonacci numbers**

$$0, 1, 1, 2, 3, 5, 8, 13, 21, 34, 55, 89, \ldots,$$

defined by the recursion

$$F_0 = 0, \; F_1 = 1 \quad \text{and} \quad F_{n+1} = F_n + F_{n-1} \quad \text{for} \quad n \in \mathbb{N}.$$

The Fibonacci sequence describes several growth processes in nature; for instance, they govern the number of leaves or petals of some plants.

In view of (3.3) it follows immediately that

$$(5.1) \qquad \frac{F_{n+1}}{F_n} = \underbrace{[1, \ldots, 1]}_{n} \quad \text{and} \quad \lim_{n \to \infty} \frac{F_{n+1}}{F_n} = [1, 1, 1, \ldots].$$

From Theorem 3.6 we know that the limit $[1, 1, \ldots]$ exists but what is the value of this limit? If something is unknown, it is always good to give it a name. Let's denote the limit above by x; then

$$x = 1 + \frac{1}{x}.$$

This shows that x has to be a root of the quadratic polynomial $X^2 - X - 1 = (X - G)(X + g)$ which we already know from the proof of Hurwitz' theorem

71

2.9, where G is the golden section. It's a small world! Thus,

$$G = \frac{1}{2}(\sqrt{5}+1) = [1,1,\ldots] \quad \text{and} \quad g = \frac{1}{2}(\sqrt{5}-1) = \frac{1}{G} = [0,1,1,\ldots].$$

In view of Theorem 3.6 the golden section has the slowest converging continued fraction expansion that is possible.

The theory of continued fractions gives some information on the Fibonacci sequence. For instance, we immediately deduce from Theorem 3.2

$$F_{n+1}F_{n-1} - F_n^2 = (-1)^n \quad \text{for} \quad n \in \mathbb{N}.$$

The recursive definition of the Fibonacci numbers is often not very useful. Binet obtained an explicit formula for Fibonacci numbers.

Theorem 5.1. *For $n \geq 0$,*

$$F_n = \frac{1}{\sqrt{5}}(G^n - (-g)^n).$$

Proof by induction on n. It is easy to verify Binet's formula for $n = 0, 1$. Now assume that the formula holds for n and $n - 1$. Taking into account the recursion formula for the Fibonacci numbers we find from the induction hypothesis

$$F_n = F_{n-1} + F_{n-2} = \frac{1}{\sqrt{5}}\left(G^{n-1} - (-g)^{n-1} + G^{n-2} - (-g)^{n-2}\right).$$

Since G and $-g$ are the roots of $X^2 - X - 1$, we have $G^2 = G + 1$ and $g^2 = -g + 1$. Thus

$$G^{n-1} + G^{n-2} = G^n \quad \text{and} \quad -g^{n-1} + g^{n-2} = g^n.$$

Substituting these expressions in the equation above leads to the formula of the theorem. •

Fibonacci numbers have plenty of interesting arithmetic and geometric properties; for instance, the geometry of a regular pentagon associated with the golden section appears in some of Penrose's aperiodic tilings of the plane.* The mathematical journal *Fibonacci Quarterly* is exclusively devoted to the Fibonacci sequence.

An irrational number α is said to be a **quadratic irrational** if it is a root of a quadratic polynomial with integers coefficients (in some literature also **quadratic surd**). Since α is irrational, this polynomial is irreducible, and so, in the language of algebra, quadratic irrationals are algebraic numbers of degree two (see Appendix A.7). Any real quadratic irrational α can be written as

(5.2)
$$\alpha = \frac{a + b\sqrt{d}}{c},$$

where $a, b \in \mathbb{Z}$, $c, d \in \mathbb{N}$. This follows easily from solving the underlying quadratic equation. Since we can extract quadratic divisors from d and

---

*Ron Knott's Web page www.mcs.surrey.ac.uk/Personal/R.Knott/Fibonacci/fib.html provides a lot of interesting material on the Fibonacci sequence, e.g., a photograph of the tv tower in Turku showing the first Fibonacci numbers.

put them into b, we further can assume d to be **squarefree**; i.e., all prime divisors of d have multiplicity one.

Hence, the golden section G and its relative g are examples of quadratic irrationals. Our next example is about paper folding. It is not about origami but on the standard paper size in Europe.

The paper size DIN A has a very useful self-similarity property. When folded in half, the proportion remains (nearly) unchanged. It is easy to see that sheets of paper with this property must have the length–width proportion $\sqrt{2} : 1$. Since $\sqrt{2}$ is irrational, for a realization of the DIN A-formate one has to find *good* rational approximations to $\sqrt{2}$. For this purpose we compute the continued fraction of $\sqrt{2}$. We have

$$(5.3) \qquad \sqrt{2} - 1 = \cfrac{1}{1 + \cfrac{1}{\sqrt{2} - 1}}.$$

Multiplying numerator and denominator with the conjugate $\sqrt{2} + 1$, this is equal to

$$\cfrac{1}{1 + \cfrac{\sqrt{2}+1}{\sqrt{2}-1} \cdot \cfrac{1}{\sqrt{2}+1}} = \frac{1}{\sqrt{2}+1} = \frac{1}{2 + \sqrt{2} - 1}.$$

Substituting (5.3) leads to $\sqrt{2} - 1 = [0, 2, 2, \ldots]$. Thus

$$(5.4) \qquad \sqrt{2} = 1 + \sqrt{2} - 1 = [1, 2, 2, 2, \ldots].$$

By definition, DIN A 4 paper has a length of 29.7 and a width of 21 centimeters. This proportion is a convergent to $\sqrt{2}$:

$$\frac{29.7}{21} = \frac{99}{70} = [1, 2, 2, 2, 2, 2] \approx \sqrt{2} = [1, 2, 2, \ldots].$$

Note that the difference of both quantities is $0.00007\ldots$, so $\frac{99}{70}$ provides a sufficiently good approximation to $\sqrt{2}$.

All examples of quadratic irrationals from above share the following property: their continued fraction expansion is eventually periodic! By the same technique as above, for instance, one can show that

$$\sqrt{n^2 + 2} = [n, n, 2n, n, 2n, n, 2n, \ldots]$$

for any $n \in \mathbb{N}$. What might be the reason for this phenomenon?

5.2. Periodic continued fractions

We say that the continued fraction $[a_0, a_1, \ldots]$ is **periodic** if there exists an integer ℓ with $a_{n+\ell} = a_n$ for all sufficiently large n. We write

$$[a_0, a_1, \ldots, a_r, \overline{a_{r+1}, \ldots, a_{r+\ell}}]$$
$$= [a_0, a_1, \ldots, a_r, a_{r+1}, \ldots, a_{r+\ell}, a_{r+1}, \ldots, a_{r+\ell}, \ldots].$$

Here $a_{n+\ell} = a_n$ for all $n \geq r + 1$. The sequence $a_{r+1}, \ldots, a_{r+\ell}$ is called the **period**, and ℓ is its **length**. The period with minimal length is said to be the **primitive period**.

The periodicity of the continued fraction expansion restricts the arithmetic nature.

Theorem 5.2. *A number $\alpha \in \mathbb{R} \setminus \mathbb{Q}$ is quadratic irrational if and only if its continued fraction expansion is eventually periodic.*

This is known as Lagrange's theorem; however, the easy implication that a periodic continued fraction represents a quadratic irrational was already known by Euler. The above characterization of quadratic irrationals can be compared with the irrationality criterion for real numbers which says that a real number is rational if and only if its decimal fraction expansion is eventually periodic.

Proof. First, assume that $\alpha = [\overline{a_0, a_1, \ldots, a_{\ell-1}}]$. Then, by Theorem 3.1,

$$\alpha = \frac{\alpha p_{\ell-1} + p_{\ell-2}}{\alpha q_{\ell-1} + q_{\ell-2}}.$$

Hence,

$$q_{\ell-1}\alpha^2 + (q_{\ell-2} - p_{\ell-1})\alpha - p_{\ell-2} = 0.$$

Since α is irrational, the polynomial $q_{\ell-1}X^2 + (q_{\ell-2} - p_{\ell-1})X - p_{\ell-2}$ is irreducible, and thus its root α is quadratic irrational.

Now suppose that

$$
\begin{aligned}
\alpha &= [a_0, a_1, \ldots, a_r, \overline{a_{r+1}, \ldots, a_{r+\ell}}] \\
&= [a_0, a_1, \ldots, a_r, \beta] \qquad \text{with} \quad \beta = [\overline{a_{r+1}, \ldots, a_{r+\ell}}].
\end{aligned}
$$

We have already proved that β is quadratic irrational. It follows that

(5.5)
$$\alpha = \frac{\beta p_r + p_{r-1}}{\beta q_r + q_{r-1}}$$

is quadratic irrational too since $\mathbb{Q}(\beta)$ is a quadratic number field and α is irrational. There is a more direct way to do this conclusion. Since β is the root of a quadratic equation, there is another quadratic equation having α as root (we leave this explicit argument to the reader).

Conversely, assume that $\alpha = [a_0, a_1, \ldots, a_{n-1}, \alpha_n]$ is quadratic irrational. Then there exists an irreducible polynomial $P(X) = aX^2 + bX + c$ with $a, b, c \in \mathbb{Z}$ such that

$$P(\alpha) = a\alpha^2 + b\alpha + c = 0.$$

Substituting

$$\alpha = \frac{\alpha_n p_{n-1} + p_{n-2}}{\alpha_n q_{n-1} + q_{n-2}}$$

we obtain

$$A_n \alpha_n^2 + B_n \alpha_n + C_n = 0,$$

where

$$
\begin{aligned}
A_n &= ap_{n-1}^2 + bp_{n-1}q_{n-1} + cq_{n-1}^2, \\
B_n &= 2ap_{n-1}p_{n-2} + b(p_{n-1}q_{n-2} + p_{n-2}q_{n-1}) + 2cq_{n-1}q_{n-2}, \\
C_n &= ap_{n-2}^2 + bp_{n-2}q_{n-2} + cq_{n-2}^2.
\end{aligned}
$$

Note that $A_n = 0$ would imply that $\frac{p_{n-2}}{q_{n-2}}$ is a root of $P(X)$, contradicting the irreducibility of P. Thus, $A_n X^2 + B_n X + C_n$ is an irreducible quadratic polynomial with root α_n. A short calculation with regard to Theorem 3.1 shows that, up to the sign, its discriminant coincides with the discriminant of P:

$$(5.6) \quad B_n^2 - 4A_n C_n = (b^2 - 4ac) \underbrace{(p_{n-1}q_{n-2} - p_{n-2}q_{n-1})}_{=\pm 1} = \pm (b^2 - 4ac).$$

In view of Theorem 3.6

$$p_{n-1} = \alpha q_{n-1} + \frac{\delta_{n-1}}{q_{n-1}} \qquad \text{with} \quad |\delta_{n-1}| < 1.$$

Therefore,

$$\begin{aligned} A_n &= a\left(\alpha q_{n-1} + \frac{\delta_{n-1}}{q_{n-1}}\right)^2 + bq_{n-1}\left(\alpha q_{n-1} + \frac{\delta_{n-1}}{q_{n-1}}\right) + cq_{n-1}^2 \\ &= \underbrace{(a\alpha^2 + b\alpha + c)}_{=P(\alpha)=0} q_{n-1}^2 + 2a\alpha\delta_{n-1} + a\frac{\delta_{n-1}^2}{q_{n-1}^2} + b\delta_{n-1} \\ &= 2a\alpha\delta_{n-1} + a\frac{\delta_{n-1}^2}{q_{n-1}^2} + b\delta_{n-1}. \end{aligned}$$

It follows that $|A_n| < 2|a\alpha| + |a| + |b|$. Since $C_n = A_{n-1}$, the same estimate holds for C_n as well. Finally, by (5.6),

$$B_n^2 \le 4|A_n C_n| + |b^2 - 4ac| < 4(2|a\alpha| + |a| + |b|)^2 + |b^2 - 4ac|.$$

Now comes the nice argument which concludes the proof. Since the upper bounds for A_n, B_n and C_n do not depend on n, there are only finitely many different triples (A_n, B_n, C_n). Thus we can find a triple (A, B, C) among them which occurs at least three times, say as $(A_{n_1}, B_{n_1}, C_{n_1})$, $(A_{n_2}, B_{n_2}, C_{n_2})$, and $(A_{n_3}, B_{n_3}, C_{n_3})$. Consequently, the related real numbers $\alpha_{n_1}, \alpha_{n_2}$, and α_{n_3} are all roots of the quadratic polynomial $AX^2 + BX + C$, and at least two of them must be equal. If, for example, $\alpha_{n_1} = \alpha_{n_2}$ then $a_{n_1} = a_{n_2}, a_{n_1+1} = a_{n_2+1}, \ldots$ This proves the theorem. •

From Lagrange's theorem it follows immediately that the partial quotients of quadratic irrationals are bounded. In particular, we find a slight improvement upon Theorem 1.4. From the continued fraction expansion of e (see Theorem 3.11) it follows that e^2 is irrational.

5.3. Galois' theorem

Galois is famous for his group theoretical view on algebraic equations (and his short but intensive life). Less known is his contribution to the theory of continued fractions.

Before we present Galois' theorem, we have to introduce the notion of reducibility. A quadratic polynomial with integer coefficients has two real roots or none. Recall that in case of a quadratic irrational α the other root of the minimal polynomial of α is called the conjugate of α, and is denoted by α' (see Appendix A.7). A quadratic irrational α is called **reduced** if $\alpha > 1$ and if $-1 < \alpha' < 0$. This notion allows a complete characterization of purely periodic continued fractions.

Theorem 5.3. *The continued fraction expansion of a quadratic irrational number α is purely periodic if and only if α is reduced. In this case, if α' denotes the conjugate of $\alpha = [\overline{a_0, a_1, \ldots, a_{\ell-1}}]$, then*

$$-\frac{1}{\alpha'} = [\overline{a_{\ell-1}, \ldots, a_1, a_0}].$$

Proof. Assume that $\alpha = [a_0, a_1, \ldots, a_{n-1}, \alpha_n]$ is reduced. Then, for $n = 0, 1, \ldots,$

(5.7) $$\alpha_n = a_n + \frac{1}{\alpha_{n+1}} \quad \text{and} \quad \alpha_n' = a_n + \frac{1}{\alpha_{n+1}'}$$

(here we have used Exercise 5.12). Consequently, $\alpha_n > 1$. If $\alpha_n' < 0$, it follows that

$$-1 < \alpha_{n+1}' = \frac{1}{\alpha_n' - a_n} < 0;$$

the first inequality follows by definition, the second one from $\alpha_n' < 0 < 1 < a_n$. By induction, all α_n are reduced. In particular,

$$0 < -\alpha_n' = -\frac{1}{\alpha_{n+1}'} - a_n < 1,$$

which leads to

$$a_n = \left[-\frac{1}{\alpha_{n+1}'} \right].$$

Since α is quadratic irrational, by Lagrange's theorem its continued fraction is eventually periodic. Thus there exist $k < m$ for which $\alpha_k = \alpha_m$. It follows that $a_k = a_\ell$ and that $\alpha_k' = \alpha_m'$. Thus,

$$a_{k-1} = \left[-\frac{1}{\alpha_k'} \right] = \left[-\frac{1}{\alpha_m'} \right] = a_{m-1}.$$

We conclude by induction that the continued fraction of α is purely periodic.

For the converse implication we assume that α has a purely periodic continued fraction expansion, say $\alpha = [\overline{a_0, \ldots, a_{\ell-1}}]$. It follows that $a_0 \geq 1$, hence $\alpha > 1$. By the periodicity,

$$\alpha = \frac{\alpha p_{\ell-1} + p_{\ell-2}}{\alpha q_{\ell-1} + q_{\ell-2}},$$

so α is a root of the quadratic polynomial

$$P(X) = q_{\ell-1} X^2 + (q_{\ell-2} - p_{\ell-1}) X - p_{\ell-2}.$$

Since $\alpha > 1$, it suffices to prove that $P(X)$ has a zero in the interval $(-1, 0)$; this zero is α'. Obviously, $P(0) = -p_{\ell-2} < 0$. Further, by (3.3),

$$\begin{aligned} P(-1) &= q_{\ell-1} - q_{\ell-2} + p_{\ell-1} - p_{\ell-2} \\ &= (a_{\ell-1} - 1)(q_{\ell-2} + p_{\ell-2}) + q_{\ell-3} + p_{\ell-3} > 0. \end{aligned}$$

Thus $P(X)$ changes its sign in the interval $(-1,0)$, and hence $-1 < \alpha' < 0$. This shows that α is reduced.

It remains to prove the second assertion. By the periodicity, we deduce from (5.7)

$$\alpha_0' = a_0 + \frac{1}{\alpha_1'}, \quad \alpha_1' = a_1 + \frac{1}{\alpha_2'}, \quad \dots \quad, \quad \alpha_{\ell-1}' = a_{\ell-1} + \frac{1}{\alpha_0'},$$

where we used the periodicity for the last equation. Hence,

$$-\frac{1}{\alpha'} = -\frac{1}{\alpha_0'} = a_{\ell-1} - \alpha_{\ell-1}', \quad -\frac{1}{\alpha_{\ell-1}'} = a_{\ell-2} - \alpha_{\ell-2}', \quad \dots, \quad -\frac{1}{\alpha_1'} = a_0 - \alpha_0'.$$

Putting these equations together gives the desired continued fraction expansion. The theorem is proved. •

5.4. Square roots

Galois' theorem gives a detailed description of the continued fraction expansion of square roots.

Theorem 5.4. *If $d \in \mathbb{N}$ is not a perfect square, then*

$$\sqrt{d} = \left[\left[\sqrt{d} \right], \overline{a_1, a_2, \dots, a_2, a_1, 2\left[\sqrt{d} \right]} \right],$$

where $a_1, a_2, \dots, a_2, a_1$ is a palindrome (i.e., the ends are straddling each other).

For instance,

$$\sqrt{109} = \left[10, \overline{2, 3, 1, 2, 4, 1, 6, 6, 1, 4, 2, 1, 3, 2, 20} \right].$$

One can even show the inverse implication; that is, any continued fraction of this form represents a square root of some rational greater than one (see Perron [**127**]).

Proof. Obviously,

$$-1 < \left[\sqrt{d} \right] - \sqrt{d} < 0 \qquad \text{and} \qquad 1 < \left[\sqrt{d} \right] + \sqrt{d}.$$

Consequently, $\sqrt{d} + \left[\sqrt{d} \right]$ is reduced, and hence by Galois' theorem purely periodic. Thus,

$$\sqrt{d} + \left[\sqrt{d} \right] = \left[\overline{2\left[\sqrt{d} \right], a_1, \dots, a_{\ell-1}} \right]$$

(5.8)
$$= \left[2\left[\sqrt{d} \right], \overline{a_1, \dots, a_{\ell-1}, 2\left[\sqrt{d} \right]} \right].$$

Furthermore,

$$\frac{1}{\sqrt{d} - \left[\sqrt{d} \right]} = -\frac{1}{-\sqrt{d} + \left[\sqrt{d} \right]} = \left[\overline{a_{\ell-1}, \dots, a_1, 2\left[\sqrt{d} \right]} \right].$$

Hence

$$\sqrt{d} = \left[\left[\sqrt{d} \right], \frac{1}{\sqrt{d} - \left[\sqrt{d} \right]} \right] = \left[\left[\sqrt{d} \right], \overline{a_{\ell-1}, \dots, a_1, 2\left[\sqrt{d} \right]} \right].$$

In conjunction with (5.8) the representation of the theorem follows. •

5.5. Equivalent numbers

Two real numbers α and β are said to be **equivalent** if

$$(5.9) \qquad \alpha = \frac{a\beta + b}{c\beta + d},$$

where a, b, c, d are integers with $ad - bc = \pm 1$. In other words, α and β are equivalent, $\alpha \sim \beta$ for short, if there exists a so-called **unimodular transformation** M, given by

$$(5.10) \qquad M : z \mapsto \begin{pmatrix} a & b \\ c & d \end{pmatrix} z := \frac{az + b}{cz + d},$$

such that $\alpha = M\beta$. The restriction $ad - bc = \pm 1$ assures that unimodular transformations form a group. This follows from the fact that the set of all 2×2 matrices with integer entries and determinant ± 1 is a group with respect to matrix multiplication, the **special linear group** $\mathsf{SL}_2(\mathbb{Z})$; for details on the beautiful theory of modular transformations and their geometry see [**32**]. It thus follows that equivalence $\sim$ is symmetric, transitive, and reflexive (we leave the rigorous proof to the reader). This observation enables us to arrange the real numbers in classes of equivalent numbers.

Any real number α is equivalent to any of its tails α_n. This can be seen as follows. In view of Theorem 3.1 we can write

$$\alpha = [a_0, a_1, \dots, a_{n-1}, \alpha_n] = \frac{p_{n-1}\alpha_n + p_{n-2}}{q_{n-1}\alpha_n + q_{n-2}}$$

for some integers a_j, p_j, q_j, and some α_n. Now put

$$M_{n-1} = \begin{pmatrix} p_{n-1} & p_{n-2} \\ q_{n-1} & q_{n-2} \end{pmatrix}.$$

Then

$$\alpha = M_{n-1}\alpha_n.$$

It follows from Theorem 3.2 that $\det(M_{n-1}) = \pm 1$, and so $M_{n-1} \in \mathsf{SL}_2(\mathbb{Z})$. Now we shall try to replace these matrices by simpler ones. If we put

$$A_n = \begin{pmatrix} a_n & 1 \\ 1 & 0 \end{pmatrix},$$

then $\det(A_n) = -1$, and A_n is a unimodular transformation too. Furthermore, it follows by induction on n that

$$M_n = A_0 \cdot \ldots \cdot A_n.$$

Thus, $\alpha = [a_0, a_1, \dots, a_{n-1}, \alpha_n]$ may be regarded as n successive unimodular transformations.

The interpretation of the real line via the actions of unimodular transformations provides a new view on our previous observations on the continued fraction algorithm. Our first aim is to show that any rational number α is equivalent to zero. We may assume that $\alpha = \frac{c}{d}$ with coprime integers c and d. Taking into account Theorem 1.6 there exist integers a, b with $ad - bc = 1$. Now the assertion follows from

$$\frac{c}{d} = \frac{a \cdot 0 + c}{b \cdot 0 + d}.$$

Since equivalence is transitive we have proved

Theorem 5.5. *Any two rational numbers are equivalent.*

It is clear that rational numbers are not equivalent to irrationals. The problem of describing equivalence classes for irrational numbers is more delicate and much more interesting.

5.6. Serret's theorem

In order to investigate equivalence for irrational numbers we start with a technical

Lemma 5.6. *A necessary and sufficient condition for the unimodular transformation (5.9) to be representable as a continued fraction $\alpha = [a_0, a_1, \ldots, a_{n-1}, \beta]$ with $n \geq 2$ is that $c > d > 0$.*

Proof. In view of Theorem 3.1

$$\alpha = [a_0, \ldots, a_{n-1}, \beta] = \frac{p_{n-1}\beta + p_{n-2}}{q_{n-1}\beta + q_{n-2}}.$$

It follows from the definition (3.3) of the q_n that $c = q_{n-1} > d = q_{n-2} > 0$, so the condition is necessary.

The sufficiency can easily be shown by induction on n. The lemma is proved. •

Now we are in the position to prove Serret's theorem which gives a characterization of equivalent numbers with respect to their continued fraction expansions.

Theorem 5.7. *Two irrational numbers α and β are equivalent if and only if their continued fraction expansions $[a_0, a_1, \ldots]$ and $[b_0, b_1, \ldots]$ are eventually identical; more precisely: there exist positive integers m, n and a real number $\gamma > 1$ such that*

$$\alpha = [a_0, \ldots, a_n, \gamma] \qquad and \qquad \beta = [b_0, \ldots, b_m, \gamma].$$

Proof. By Theorem 3.1,

$$\alpha = \frac{p_n\gamma + p_{n-1}}{q_n\gamma + q_{n-1}},$$

where $p_n q_{n-1} - p_{n-1} q_n = \pm 1$. Hence, α and γ are equivalent. Similarly, β and γ are equivalent. Since equivalence is transitive, α and β are equivalent too.

Conversely, assume that α and β are equivalent. Hence, the identity (5.9) holds with $c > d > 0$ by Lemma 5.6. This together with

$$\beta = [b_0, \ldots, b_m, \gamma] = \frac{p_m\gamma + p_{m-1}}{q_m\gamma + q_{m-1}}$$

implies

$$\alpha = \frac{P\gamma + R}{Q\gamma + S},$$

where

$$P = ap_m + bq_m, \qquad R = ap_{m-1} + bq_{m-1},$$
$$Q = cp_m + dq_m, \qquad S = ap_{m-1} + bq_{m-1},$$

$PS - QR = \pm 1$, and a and b are defined by (5.9) (this is easily seen by considering unimodular transformations as matrices). In view of Theorem 3.6

$$p_{m-j} = \alpha q_{m-j} + \frac{\delta_j}{q_{m-j}} \qquad \text{with} \quad |\delta_j| < 1$$

for $j = 0, 1$. Consequently,

$$Q = (c\alpha + d)q_m + \frac{c\delta_0}{q_m} > (c\alpha + d)q_{m-1} + \frac{c\delta_1}{q_{m-1}} = S$$

for sufficiently large m. Obviously, we may assume without loss of generality that $c\alpha + d > 0$, which implies $S > 0$. Because of Lemma 5.6 it follows that $\alpha = [a_0, \ldots, a_n, \gamma]$, which was to be shown. •

For instance, real numbers which differ by an integer are equivalent. In particular, the numbers $G = [\overline{1}]$ and $g = [0, \overline{1}]$ are equivalent. However, adding rational numbers can change the equivalence class.

5.7. The Markoff spectrum

Given a real number x, denote by $\|x\|$ the distance between x and the nearest integer. Whenever we have a sufficiently good rational approximation $\frac{p}{q}$ to α, then

$$|q\alpha - p| = \|q\alpha\|.$$

Then the **Markoff constant** for a real number α is defined by

$$\lambda(\alpha) = \liminf_{q\to\infty} q\|q\alpha\|.$$

Taking into account Corollary 3.7 we have

$$\alpha - \frac{p_n}{q_n} = \frac{1}{\delta_n q_n^2}, \qquad \text{where} \quad \delta_n = (-1)^n \left(\alpha_{n+1} + \frac{q_{n-1}}{q_n} \right)$$

and $\alpha_{n+1} = [a_{n+1}, a_{n+2}, \ldots]$. Note that

$$q_n\|q_n\alpha\| = \frac{1}{|\delta_n|} = \left(\alpha_{n+1} + \frac{q_{n-1}}{q_n} \right)^{-1}.$$

Since

$$\frac{q_{n-1}}{q_n} = \frac{1}{q_n/q_{n-1}} = \frac{1}{a_n + \frac{q_{n-2}}{q_{n-1}}} = [0, a_n, a_{n-1}, \ldots, a_1]$$

(this follows from Exercise 3.8), we get

$$\lambda(\alpha) \;=\; \liminf_{n\to\infty}\frac{1}{|\delta_n|}$$

(5.11)
$$\;=\; \liminf_{n\to\infty}([a_{n+1},a_{n+2},\dots,] + [0,a_n,a_{n-1},\dots,a_1])^{-1}.$$

In view of Theorem 5.7 we obtain

Theorem 5.8. *If α and β are equivalent, then $\lambda(\alpha) = \lambda(\beta)$.*

Note that a non-zero $\lambda(\alpha)$ implies that the inequality

$$\left|\alpha - \frac{p}{q}\right| < \frac{\lambda(\alpha) + \varepsilon}{q^2}$$

has infinitely many solutions in positive integers p, q for any fixed positive ε. In view of Hurwitz' theorem

$$\lambda(\alpha) \le \lambda(G) = \frac{1}{\sqrt{5}}$$

for any real α; by Theorem 5.8 equality holds if and only if α is equivalent to G, for example, $\lambda(g) = \lambda(G)$. This puts new light on Theorem 2.9. Numbers which are not equivalent to G are better approximable. For instance, by (5.4) and (5.11),

$$\lambda(\sqrt{2}) = \lambda\left([\overline{2}]\right) = \left([\overline{2}] + [0,\overline{2}]\right)^{-1} = \frac{1}{\sqrt{8}}.$$

One can show that if α is not equivalent to G (not necessarily equivalent to $\sqrt{2}$), then

$$\lambda(\alpha) \le \lambda(\sqrt{2}) = \frac{1}{\sqrt{8}}.$$

The step from $\lambda(G)$ to $\lambda(\sqrt{2})$ is only the first step as Markoff [**107**] observed. We could continue this process to find further smaller Markoff constants by further restrictions on the continued fraction expansions. The set of all Markoff constants is the **Markoff spectrum**. The beginning of the Markoff spectrum looks as follows:

- $G = [\overline{1}]$ with $\lambda(G) = \frac{1}{\sqrt{5}} = 0.44721\dots$,
- $1 + \sqrt{2} = [\overline{2}]$ with $\lambda(1 + \sqrt{2}) = \frac{1}{\sqrt{8}} = 0.35355\dots$,
- $\frac{9+\sqrt{221}}{10} = [\overline{2,2,1,1}]$ with $\lambda\left(\frac{9+\sqrt{221}}{10}\right) = \frac{5}{\sqrt{221}} = 0.33633\dots$;

for a longer list see Burger [**33**]. There is a deep result due to Markoff [**107**] which states that the Markoff spectrum above $\frac{1}{3}$ consists exactly of numbers of the form

$$\frac{z}{\sqrt{9z^2 - 4}},$$

where z is a positive integer such that there exist $x, y \in \mathbb{N}$ with $\max\{x, y\} \le z$ satisfying the diophantine equation

$$X^2 + Y^2 + Z^2 = 3XYZ.$$

Freiman [**64**] showed that any positive real number less than

$$\frac{153\,640\,040\,533\,216 - 19\,623\,586\,058\sqrt{462}}{693\,746\,111\,282\,512} = 0.22085\ldots$$

is in the Markoff spectrum. Unfortunately, proofs of these remarkable results are far beyond our scope; we refer the interested reader to Cassels [**35**] and Freiman [**64**].

5.8. Badly approximable numbers

We call a number α **badly approximable** if $\lambda(\alpha) > 0$; i.e., there exists a positive constant c, depending only on α, such that for *all* $\frac{p}{q}$

$$\left| \alpha - \frac{p}{q} \right| > \frac{c}{q^2}$$

holds. The following theorem classifies badly approximable numbers.

Theorem 5.9. *An irrational α is badly approximable if and only if its partial quotients are bounded.*

For instance, quadratic irrationals are badly approximable, but not e (see Theorem 3.11).

Proof. In view of Theorem 3.6

$$(5.12) \qquad \frac{1}{(a_{n+1} + 2)q_n^2} \leq \left| \alpha - \frac{p_n}{q_n} \right| \leq \frac{1}{a_{n+1} q_n^2}.$$

By Theorem 3.8, the law of best approximations, other rationals cannot approximate α better than convergents do. This proves the theorem. •

Thus, all real numbers which are equivalent to the golden section G, that are those which have an eventually periodic continued fraction with period $\overline{1}$, are the *worst* approximable irrationals. This explains why consecutive Fibonacci numbers give the worst case for the running time of the Euclidean algorithm (see Exercise 5.8).

Given real numbers α and β, **Littlewood's conjecture** claims that there exists an integer sequence $q_1, q_2, \ldots$ such that

$$\lim_{n \to \infty} q_n \|q_n \alpha\| \cdot \|q_n \beta\| = 0.$$

It is not too difficult to verify Littlewood's conjecture if either α or β is not badly approximable. Recently, Einsiedler, Katok & Lindenstrauss [**56**] showed that the set of exceptions to Littlewood's conjecture has Hausdorff measure zero. But the conjecture is still open.

Notes on the metric theory

Nearly nothing is known about the continued fraction expansions of cubic irrationals or algebraic numbers of higher degree. It is conjectured that their sequence of partial quotients is unbounded. But there is no single cubic irrational with this property known. For some nice discussion we refer to Bombieri & van der Poorten [**27**].

The metric theory of continued fractions investigates the average behavior of the partial quotients. The continued fraction algorithm can be regarded as a discrete dynamical system on the unit interval $[0, 1]$. In fact, the partial remainders α_n in (3.4) satisfy $\alpha_{n+1} = T(\alpha_n)$ with $T : [0, 1] \to [0, 1]$ given by

$$T(x) = \left\{ \frac{1}{x} \right\} \quad \text{mod } 1,$$

where $\{.\}$ stands for the fractional part function. It is known that the set of real numbers $\alpha = [a_0, a_1, \ldots]$ with bounded partial quotients has measure null. On the other side one can prove that if ψ is any positive function, then the inequality

$$a_n = a_n(\alpha) \geq \psi(n)$$

holds for almost all real α if and only if the series

$$\sum_{n=1}^{\infty} \psi(n)$$

diverges (this is a bit in the spirit of our observations in Section 2.3).

One of the highlights is the limit theorem of Gauss–Kusmin. If $\text{meas}_n(x)$ denotes the Lebesgue measure of the set of $\alpha \in (0, 1)$ for which $[a_n, a_{n+1}, \ldots] - a_n < x$, then

$$\lim_{n \to \infty} \text{meas}_n(x) = \frac{\log(1 + x)}{\log 2} \quad \text{for any} \quad x \in (0, 1).$$

It should be noted that Gauss stated this result in a letter to Laplace but never published a proof; the first published proof was given by Kusmin in 1928. For example, it follows that the probability that a positive integer m appears as a partial quotient in the continued fraction expansion of a randomly chosen real number is

$$\frac{1}{\log 2} \int_{\frac{1}{m+1}}^{\frac{1}{m}} \frac{dx}{1 + x} = \frac{1}{\log 2} \log \left(1 + \frac{1}{m(m + 2)} \right).$$

On average, the geometric mean of the partial quotients tends to a limit

$$(5.13) \qquad \lim_{n \to \infty} \sqrt[n]{a_1 \cdot \ldots \cdot a_n} = \prod_{m=1}^{\infty} \left(1 + \frac{1}{m(m + 2)} \right)^{\frac{\log m}{\log 2}} = 2.68545 \ldots.$$

This follows from the theorem of Gauss–Kusmin and ergodicity. For details to this deep result and its applications we refer the interested reader to Khintchine [90] and Flajolet et al. [61].

It is rather difficult to say anything on sums or products of continued fractions. In 1947, Hall [73] showed that any real number can be represented as sum of two irrationals whose continued fractions only contain partial quotients ≤ 4. Hence, any real number is the sum of two badly approximable numbers. This beautiful result is related to the geometry of Cantor sets (see Exercise 13.7). For improvements and further results we refer to Astels [6].

Exercises

5.1. This is Fibonacci's original example of Fibonacci numbers describing processes in nature. A single pair of rabbits is multiplying as follows. It begets a new pair at the end of each month from the second month on and each new pair reproduces itself the same way. So the original pair produces a second pair at the end of the second month. At the end of the third month, the original pair produces a third pair, and at the end of the forth month, the offspring of the two oldest pairs bring the number of pairs up to five.

How many pairs of rabbits will there be at the end of one year?

5.2. *i) For $|z| < g$, prove that*

$$\sum_{m=0}^{\infty} F_m z^m = \frac{z}{1 - z - z^2}.$$

Hint: Use Binet's theorem 5.1 and the formula for the geometric series.

ii) Compute

$$\frac{F_0}{10} + \frac{F_1}{10^2} + \frac{F_2}{10^3} + \ldots + \frac{F_n}{10^{n+1}} + \ldots = \frac{1}{F_{11}} = 0.011235\ldots.$$

Hence, the reciprocal of F_{11} contains the whole Fibonacci sequence in its decimal fraction expansion. This observation is due to de Weger [**171**].

5.3. *For $n \in \mathbb{N}$, show that*

$$\begin{pmatrix} 0 & 1 \\ 1 & 1 \end{pmatrix}^n = \begin{pmatrix} F_{n-1} & F_n \\ F_n & F_{n+1} \end{pmatrix}.$$

5.4.* In 1876, Lucas discovered some beautiful divisibility properties of Fibonacci numbers. Let $m, n \in \mathbb{N}$.

i) Prove that F_n divides F_{mn}.

ii) Show that $\gcd(F_m, F_n) = F_{\gcd(m,n)}$.

Hint: Use the previous exercise.

5.5. *For $n \in \mathbb{N}$, prove that*

$$\sum_{\substack{a,b \geq 0 \\ a+b=n}} \binom{a}{b} = F_{n+1}.$$

Hint: Recall that binomial coefficients satisfy a recursion formula similar to the one for Fibonacci numbers. It might be helpful to have a look at Pascal's triangle.

5.6.* *For $n \in \mathbb{N}$ prove the formulae*

$$\frac{F_{n-1}}{F_n} = \sum_{j=1}^{n-1} \frac{(-1)^{j+1}}{F_j F_{j+1}} \quad \text{and} \quad \frac{\sqrt{5} - 1}{2} = \sum_{j=1}^{\infty} \frac{(-1)^{j+1}}{F_j F_{j+1}}.$$

The value of the latter series has an algebraic value. On the contrary, it is known that

$$\sum_{n=1}^{\infty} \frac{(-1)^n}{F_n^2}$$

does not satisfy any polynomial equation with rational coefficients, i.e., it is *transcendental* in the language of Chapter 9 (cf. [**167**]). There is no conjectural statement that gives a satisfactory description of the arithmetic nature of real numbers given in terms of power series involving the Fibonacci sequence.

5.7. For real x, define the complex-valued functions

$$F_{\pm}(x) = \frac{1}{\sqrt{5}} \left(G^x \pm \exp(\pi i x) g^x \right),$$

where $i = \sqrt{-1}$ denotes the imaginary unit.

i) Show that the function $F_{-}(x)$ interpolates the Fibonacci numbers, i.e., $F_{-}(n) = F_n$ for $n = 0, 1, 2, \ldots$.
ii) Prove the addition formula

$$2F_{\pm}(x+y) = \sqrt{5}(F_{+}(x)F_{\pm}(y) + F_{\mp}(y)F_{-}(x)).$$

iii) For non-negative integers m, n show that

$$2^{m+n} F_{m+n} = 2^m F_m \sum_{0 \le j \le \frac{n}{2}} \binom{n}{2j} 5^j + 2^n F_n \sum_{0 \le j \le \frac{m}{2}} \binom{m}{2j} 5^j.$$

5.8.* Whenever an algorithm is used it is important to know whether it terminates, and in case it does, how fast.

i) For the number of steps m in the Euclidean algorithm (1.11) for the integers $b \le a$ show

$$m \le \left(\log \frac{\sqrt{5}+1}{2} \right)^{-1} (1 + \log a).$$

Hint: Show that the Euclidean algorithm is extraordinarily slow for consecutive Fibonacci numbers. Use Binet's theorem 5.1 to derive the estimate.

ii) Show that any positive integer a has a binary representation

$$a = \sum_{k=0}^{\ell} a_k 2^k, \qquad \text{where} \quad a_k \in \{0,1\}, \ u_\ell = 1.$$

Give an upper bound for the quantity ℓ and deduce that a can be expressed by approximately $\frac{\log a}{\log 2}$ bits.
iii) What does ii) imply for the running time of the Euclidean algorithm?

The Euclidean algorithm terminates in polynomial time in the input data. The estimate for the running time of the Euclidean algorithm is due to Lamé in 1845, long before the computer age. It is remarkable that the average case does not fall much behind the bound for the worst case. Heilbronn [**79**] showed that the average length of the Euclidean algorithm is

$$\frac{12}{\pi^2}\log 2 \log a.$$

5.9. *i) For $n \in \mathbb{N}$ prove that*

$$\sqrt{n^2+1} = \left[n, \overline{2n}\right] \qquad and \qquad \sqrt{n^2+2} = \left[n, \overline{n, 2n}\right].$$

ii) Find for each $n \in \mathbb{N}$ the continued fraction expansion for the positive root of the polynomial

$$(6n^2+1)X^2 + 3n(8n^2+1)X - (12n^2+1).$$

5.10. *For arbitrary $a_1, a_2 \in \mathbb{Z}$, show that*

$$\sqrt{a_0^2 + k(a_1 a_2 + 1) - a_2^2} = \left[a_0, \overline{a_1, a_2, a_1, 2a_0}\right],$$

where $a_0 = \frac{1}{2}(k(a_1^2 a_2 + 2a_1) - (a_1 a_2 + 1)a_2)$.

5.11. *Let α be a root of the quadratic equation $AX^2 + BX + C = 0$.*
i) Show that α is irrational if and only if $B^2 - 4AC \neq 0$.
ii) Prove that any real quadratic irrational α has a representation of the form (5.2), that is,

$$\alpha = \frac{a + b\sqrt{d}}{c},$$

where $a, b \in \mathbb{Z}$, $c, d \in \mathbb{N}$ with squarefree d.

5.12. Recall the definition of the conjugate α' of a given quadratic irrational α from Section 5.3.

i) For any $a, b, c, d \in \mathbb{Z}$ with positive c and squarefree d show that

$$\left(\frac{a + b\sqrt{d}}{c}\right)' = \frac{a - b\sqrt{d}}{c}.$$

ii) If α and β are quadratic irrationals with conjugates α' and β', prove that

$$(\alpha + \beta)' = \alpha' + \beta' \qquad and \qquad (\alpha\beta)' = \alpha'\beta'.$$

iii) If F is a rational function with rational coefficients and d is not a perfect square, show that $F(\sqrt{d}) = 0$ implies $F(-\sqrt{d}) = 0$.

5.13. *i) Prove that if α is a quadratic irrational, so is*

$$\frac{a\alpha + b}{c\alpha + d},$$

where a, b, c, d are integers, and a and c are not both zero.
ii) In the proof of Theorem 5.2, construct a quadratic polynomial with root α, where α is given by (5.5). Deduce that α is quadratic irrational.

5.14.* *Give an estimate for the length of the period of the continued fraction expansion of a quadratic irrational.*

5.15. *Prove that the relation ∼ introduced in Section 5.5 is an equivalence relation; i.e., it is reflexive, symmetric, and transitive.*

5.16.* We may also consider a unimodular transformation M of the form (5.10) for $z \in \mathbb{C} \cup \{\infty\}$ by setting $M(\infty) = \frac{a}{c}$ if $c \neq 0$, $M(\infty) = \infty$ otherwise, and $M(-\frac{d}{c}) = \infty$.

i) Show that a unimodular transformation maps straight lines and circles onto straight lines and circles.

ii) Prove that a unimodular transformation is determined by the mapping of three points.

ii) Show that a unimodular transformation M has at most three fixed points, and that it is the identity if it has three.

5.17. *Show that a real number α is equivalent to $\alpha + q$ for any integer q. If q is rational, prove that this can only hold for quadratic irrational α. Can it happen that a quadratic irrational α is equivalent to $\alpha + q$ for any rational number q?*

5.18. We have already seen that a positive real $\alpha = [a_0, a_1, a_2, \ldots]$ is equivalent to its reciprocal. The same holds true for the additive inverse of α. *Show that*

$$-\alpha = [-a_0 - 1, 1, a_1 - 1, a_2, a_3, \ldots] \quad \text{if} \quad a_1 > 1,$$

and

$$-\alpha = [-a_0 - 1, a_2 + 1, a_3, a_4, \ldots] \qquad \text{if} \quad a_1 = 1.$$

5.19. *For $n \in \mathbb{N}$, compute the Markoff constants for $[\overline{n}]$ and $\sqrt{n^2 + 2}$.*

5.20. *For the denominators q_n of the nth convergent to the continued fraction expansion of $\alpha = [a_0, a_1, \ldots]$, prove that $q_n \geq F_n$. When does equality hold?*

5.21.* *Compute the geometric mean of the partial quotients of* e *and compare the result with the average case (5.13).*

5.22.* The so-called $3x + 1$-**conjecture** (or **Collatz problem**) claims that for any positive integer x_0 the sequence (x_n), defined by the recursion

$$x_{n+1} = \begin{cases} \frac{x_n}{2} & \text{if} \quad x_n \text{ is even,} \\ 3x_n + 1 & \text{if} \quad x_n \text{ is odd,} \end{cases}$$

eventually runs into the cycle $4, 2, 1, 4, 2, 1, \ldots$

Prove that the $3x+1$-conjecture is equivalent to the fact that the real numbers $[x_0, x_1, x_2, \ldots]$ with $x_0 \in \mathbb{N}$, and $x_1, x_2, \ldots$ defined by the recursion above, all lie in the equivalence class of $\overline{[4, 2, 1]}$. Compute the Markoff constant of $\overline{[4, 2, 1]}$.

For the state of art of the $3x + 1$-conjecture see Lagarias [**97**].

CHAPTER 6

The Pell equation

This chapter is devoted to one of the oldest diophantine equations, the Pell equation. John Pell was an English mathematician who lived in the seventeenth century but he had nothing to do with this equation; it was Euler who mistakenly attributed a solution method to Pell which in fact was found by Pell's contemporaries Wallis and Lord Brouncker. The history of the Pell equation dates back at least to the ancient Greeks. Its main importance lies in the role it plays in the arithmetic of quadratic number fields. The solution of the Pell equation relies on the theory of diophantine approximation, and so it is our first non-trivial example for the close relation between diophantine approximations and diophantine equations. Nevertheless, the Pell equation still contains some unsolved interesting questions.

6.1. The cattle problem

In a letter to Eratosthenes $(267-197$ B.C.), Archimedes posed the so-called **cattle problem** in which he asks for the number of bulls and cows belonging to the Sun god, subject to certain arithmetic restrictions. This problem was forgotten over the centuries until it was rediscovered by G.E. Lessing in the Wolfenbüttel library in 1773. A nice English version goes as follows:

> *The Sun god's cattle, apply thy care*
> *to count their numbers, hast thou wisdom's share.*
> *They grazed of old on the Thrinacian floor*
> *of Sic'ly's island, herded in to four,*
> *colour by colour: one herd white as cream,*
> *the next in coats glowing with ebon gleam,*
> *brown-skinned the third, and stained with spots the last.*
> *Each herds saw bulls in power unsurpassed,*
> *in ratios these: count half the ebon-hued,*
> *add one third more, then all the brown include;*
> *thus, friend, canst thou the white bulls' number tell.*
> *The ebon did the brown exceed as well,*
> *now by a fourth and fifth part of the stained.*
> *To know the spotted - all bulls that remained -*
> *reckon again the brown bulls, and unite*
> *these with a sixth and seventh of the white.*
> *Among the cows, the tale of silver-haired*

was, when with bulls and cows of black compared,
exactly one in three plus one in four.
The black cows counted one in four once more,
plus now a fifth, of the bespeckled breed
when, bulls withal, they wandered out to feed.
The speckled cows tallied a fifth and sixth of all the brown-
haired, males and females mixed.
Lastly, the brown cows numbered half a third
and one in seven of the silver herd.
Tellst thou unfailingly how many head
the Sun possessed, o friend, both bulls well-fed
and cows of ev'ry colour - no-one will
deny thou hast numbers' art and skill,
though not yet dost thou rank among the wise.
But come! also the foll'wing recognise.
Whene'er the Sun God's white bulls joined the black,
their multitude would gather in a pack
of equal length and breadth, and squarely throng
Thrinacia's territory broad and long.
But when the brown bulls mingled with the flecked,
in rows growing from one would they collect,
forming a perfect triangle, with ne'er
a diff'rent-coloured bull, and none to spare.
Friend, canst thou analyse this in thy mind,
and of these masses all the measures find,
go forth in glory! be assured all deem
thy wisdom in this discipline supreme!
(cf. [**103**])

Denoting by w, x, y, and z the numbers of the white, black, dappled, and brown bulls, and by $\mathcal{W}, \mathcal{X}, \mathcal{Y}$, and $\mathcal{Z}$ the numbers of the white, black, dappled, and brown cows, respectively, one has to solve the system of linear diophantine equations

(6.1)
$$\begin{cases} w = \left(\frac{1}{2} + \frac{1}{3}\right) x + z, \quad x = \left(\frac{1}{4} + \frac{1}{5}\right) y + z, \\ \\ y = \left(\frac{1}{6} + \frac{1}{7}\right) w + z, \end{cases}$$

(6.2)
$$\begin{cases} \mathcal{W} = \left(\frac{1}{3} + \frac{1}{4}\right) (x + \mathcal{X}), \quad \mathcal{X} = \left(\frac{1}{4} + \frac{1}{5}\right) (y + \mathcal{Y}), \\ \\ \mathcal{Y} = \left(\frac{1}{5} + \frac{1}{6}\right) (z + \mathcal{Z}), \quad \mathcal{Z} = \left(\frac{1}{6} + \frac{1}{7}\right) (w + \mathcal{W}). \end{cases}$$

Due to Archimedes everyone who can solve this problem is *merely competent*; but to win the prize for *supreme wisdom* one has to meet the additional condition that $w+x$ has to be a square, and that $y+z$ has to be a **triangular**

number, i.e., a number of the form

$$1 + 2 + \ldots + n = \frac{1}{2}n(n+1)$$

(imagine the numbered balls in pool billiard put in triangle before starting). Hence, the additional condition can be rewritten as

(6.3) $w + x = u^2$ and $y + z = \frac{1}{2}v(v+1)$

for some positive integers u and v.

The system of linear equations (6.1) and (6.2) consists of eight unknowns, and so it can be solved by standard linear algebra techniques. It is easily seen that the general solution to (6.1) is given by

$$(w, x, y, z) = m \cdot (2226, 1602, 1580, 891), \qquad \text{where} \quad m \in \mathbb{N}.$$

It turns out that the system (6.2) is solvable if and only if m is a multiple of 4657. Setting $m = 4657 \cdot M$ we obtain for the general solution of (6.2)

$$(\mathcal{W}, X, Y, Z) = M \cdot (7\,206\,360, 4\,893\,246, 3\,515\,820, 5\,439\,213),$$

where $M \in \mathbb{N}$. This solves the first part of the cattle problem. To solve also the second part (6.3) one has to find an M such that $w + x = 4657 \cdot 3828 \cdot M$ is a square and $y + z = 4657 \cdot 2471 \cdot M$ is triangular. By the prime factorization $4657 \cdot 3828 = 2^2 \cdot 3 \cdot 11 \cdot 29 \cdot 4657$ it follows that $w + x$ is a square if and only if

$$M = 3 \cdot 11 \cdot 29 \cdot 4657 \cdot Y^2,$$

where $Y \in \mathbb{N}$. Since $y + z$ is triangular if and only if $8(y+z) + 1$ is a square:

$$8(y + z) + 1 = 4v^2 + 4v + 1 = (2v + 1)^2 =: X^2,$$

one has to solve the quadratic equation

$$X^2 - 410\,286\,423\,278\,424\,Y^2 = 1,$$

where $410\,286\,423\,278\,424 = 2 \cdot 3 \cdot 7 \cdot 11 \cdot 29 \cdot 353 \cdot (2 \cdot 4657)^2$, which does not look too easy. It seems that the ancient Greeks were unable to solve this equation. We know that Archimedes was aware of the solution $x = 1351$ and $y = 780$ of the related equation

$$X^2 - 3Y^2 = 1,$$

but there is no belief that Archimedes could have had any solution to the cattle problem. We refer to Vardi [163] and Lenstra [103] for nicely written surveys on the cattle problem, its history, and the first prize winners for supreme wisdom.

6.2. Lattice points on hyperbolas

We consider a more general equation. The **Pell equation** is defined by

(6.4) $X^2 - dY^2 = 1,$ where $d \in \mathbb{N}.$

We are interested in integral solutions. Looking at this problem with the eyes of a geometer, we have to find the set of intersections of a hyperbola with the lattice $\mathbb{Z}^2$. Since both the hyperbola and the lattice are rather *thin* sets in the Euclidean plane, it is not clear what we should expect to find. The result will be that the set of solutions to the Pell equation depends intimately on the arithmetic nature of $\sqrt{d}$.

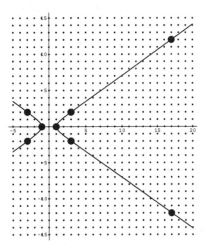

FIGURE 6.1. Integer lattice points and the hyperbola associated with the Pell equation $X^2 - 2Y^2 = 1$.

Our first observation is simple but rather important. If d is a perfect square, say $d = m^2$, we can factor the left-hand side:

$$X^2 - dY^2 = (X - mY)(X + mY).$$

If now x, y is a solution of (6.4) in integers, the factors on the right have to be integers too. Since their product is equal to 1, it follows that both factors are either $+1$ or -1. Thus there are only finitely many integral solutions in this case (which is boring). What are these solutions? Adding both factors, $x \pm my$ we get $2x = \pm 2$ which leads to two solutions $(x, y) = (\pm 1, 0)$. Obviously, these are always solutions, independent of d, but are there more?

By symmetry, if (x, y) is a solution of (6.4), then the tuples $(\pm x, -y)$ and $(-x, y)$ are solutions too. So it suffices to look for solutions in positive integers. Furthermore, in the sequel we may assume that d is not a perfect square, resp. $\sqrt{d} \notin \mathbb{Q}$. Sometimes the expression $x + y\sqrt{d}$, where (x, y) is a solution of (6.4), is also called a solution.

Our first deeper observation is due to Euler and Lagrange. At first glance one might think that knowledge of the integers alone should be sufficient for the study of diophantine equations. Now we shall see what a good idea it can be to look on a diophantine problem from a broader perspective. Assume that x, y is a solution. Having the case of d being a perfect square in mind, we may factor the left-hand side of (6.4) in the number field $\mathbb{Q}(\sqrt{d})$. So

$$(x - y\sqrt{d})(x + y\sqrt{d}) = x^2 - dy^2 = 1,$$

resp.

$$\left| \sqrt{d} - \frac{x}{y} \right| = \frac{1}{y^2(\sqrt{d} + \frac{x}{y})} < \frac{1}{2y^2}.$$

In view of Theorem 3.9 all solutions of (6.4) can be found among the convergents to $\sqrt{d}$. For instance, the sequence of convergents $\frac{p_n}{q_n}$ to $\sqrt{2}$ starts with

$$\frac{1}{1}, \frac{\mathbf{3}}{\mathbf{2}}, \frac{7}{5}, \frac{\mathbf{17}}{\mathbf{12}}, \frac{41}{29}, \frac{\mathbf{99}}{\mathbf{70}}, \cdots \quad \rightarrow \quad \sqrt{2} = [1, \overline{2}],$$

and in fact we get for $p_n^2 - 2q_n^2$ the values

$$1^2 - 2 \cdot 1^2 = -1, \quad \mathbf{3}^2 - 2 \cdot \mathbf{2}^2 = +1, \quad 7^2 - 2 \cdot 5^2 = -1,$$

$$\mathbf{17}^2 - 2 \cdot \mathbf{12}^2 = +1, \quad 41^2 - 2 \cdot 29^2 = -1, \quad \mathbf{99}^2 - 2 \cdot \mathbf{70}^2 = +1.$$

This gives us the first non-trivial solutions to (6.4) with $d = 2$. The regularity is astonishing! Furthermore, it suggests that instead of (6.4) we consider the more general quadratic equation

$$(6.5) \qquad\qquad X^2 - dY^2 = \pm 1;$$

according to the sign we speak about the **plus** and the **minus** equation, respectively.

6.3. An infinitude of solutions

For d being not a perfect square, the continued fraction expansion of $\sqrt{d}$ is periodic by Lagrange's theorem 5.2. In what follows $\ell = \ell(d)$ denotes the length of the primitive period of the continued fraction expansion of $\sqrt{d}$. By Theorem 5.4, we write

$$(6.6) \quad \sqrt{d} = \left[\left[\sqrt{d} \right], \overline{a_1, \ldots, a_{\ell-1}, 2\left[\sqrt{d} \right]} \right] = \left[\left[\sqrt{d} \right], a_1, \ldots, a_{n-1}, \alpha_n \right].$$

Our first aim is to show that, for $n = 0, 1, \ldots$, there exist some integers P_n, Q_n for which

$$(6.7) \qquad \alpha_n = \frac{P_n + \sqrt{d}}{Q_n}, \qquad \text{where} \quad d - P_n^2 \equiv 0 \bmod Q_n.$$

Clearly, $\alpha_0 = \sqrt{d}$, so we have $Q_0 = 1$ and $P_0 = 0$. Furthermore,

$$\alpha_1 = \frac{1}{\sqrt{d} - \left[\sqrt{d} \right]} = \frac{\sqrt{d} + \left[\sqrt{d} \right]}{d - \left[\sqrt{d} \right]^2},$$

which gives $Q_1 = d - [\sqrt{d}]^2$ and $P_1 = [\sqrt{d}]$. Thus, (6.7) holds for $n = 0$ and $n = 1$. Now assume that (6.7) is true for n. Then it follows that

$$\alpha_{n+1} = \frac{1}{\alpha_n - a_n} = \frac{Q_n}{P_n - a_n Q_n + \sqrt{d}}$$

$$= \frac{Q_n(P_n - a_n Q_n - \sqrt{d})}{(P_n - a_n Q_n)^2 - d} =: \frac{P_{n+1} + \sqrt{d}}{Q_{n+1}},$$

where $P_{n+1} = a_n Q_n - P_n$ and

$$Q_{n+1} = \frac{d - (P_n - a_n Q_n)^2}{Q_n} = \underbrace{\frac{d - P_n^2}{Q_n}}_{\in \mathbb{Z}} + 2a_n P_n - a_n^2 Q_n$$

are integral. Since

$$Q_n = \frac{d - (P_n - a_n Q_n)^2}{Q_{n+1}} = \frac{d - P_{n+1}^2}{Q_{n+1}},$$

it follows that Q_{n+1} divides $d - P_{n+1}^2$. Hence, by induction on n we see that each α_n has a representation of the form (6.7).

Now (6.7) yields

$$\sqrt{d} = \frac{\alpha_n p_{n-1} + p_{n-2}}{\alpha_n q_{n-1} + q_{n-2}} = \frac{(P_n + \sqrt{d})p_{n-1} + p_{n-2}Q_n}{(P_n + \sqrt{d})q_{n-1} + q_{n-2}Q_n},$$

resp.

$$\sqrt{d}((P_n + \sqrt{d})q_{n-1} + q_{n-2}Q_n) = (P_n + \sqrt{d})p_{n-1} + p_{n-2}Q_n.$$

Since $\sqrt{d} \notin \mathbb{Q}$, splitting the latter identity into its rational and its irrational parts yields

$$dq_{n-1} = P_n p_{n-1} + p_{n-2}Q_n \quad \text{and} \quad p_{n-1} = P_n q_{n-1} + q_{n-2}Q_n.$$

Multiplying the first one by q_{n-1} and the second one by p_{n-1}, subtraction of both equations gives

$$p_{n-1}^2 - dq_{n-1}^2 = Q_n(p_{n-1}q_{n-2} - p_{n-2}q_{n-1}).$$

With regard to Theorem 3.1 we get

Lemma 6.1. *For $n \in \mathbb{N}$,*

$$p_{n-1}^2 - dq_{n-1}^2 = (-1)^n Q_n.$$

If n is a multiple of the length of the primitive period ℓ, $n = k\ell$ say, then

$$\frac{P_{k\ell} + \sqrt{d}}{Q_{k\ell}} = \alpha_{k\ell} = [0, \overline{a_1, \ldots, a_{\ell-1}}] = \sqrt{d} - [\sqrt{d}],$$

by (6.7). Consequently, $Q_{k\ell} = 1$. In view of Lemma 6.1

$$(6.8) \qquad p_{k\ell-1}^2 - dq_{k\ell-1}^2 = (-1)^{k\ell}.$$

We thus have proved

Theorem 6.2. *If $\sqrt{d} \notin \mathbb{Q}$, then the Pell equation (6.4) has infinitely many solutions in positive integers x_k, y_k given by*

$$(6.9) \qquad (x_k, y_k) = \begin{cases} (p_{k\ell-1}, q_{k\ell-1}) & \text{if } \ell \equiv 0 \bmod 2, \\ (p_{2k\ell-1}, q_{2k\ell-1}) & \text{if } \ell \equiv 1 \bmod 2. \end{cases}$$

This might be compared with the example $d = 2$ discussed at the end of the previous section.

6.4. The minimal solution

Now we have to investigate whether there are more solutions than those given by Theorem 6.2. The solution $x, y \in \mathbb{N}$ of (6.4) with minimal x is called **minimal solution**. It is clear that $x \leq p_{2\ell-1}$ in the notation of the previous section, and so finding the minimal solution to the Pell equation is reduced to a finite problem.

Lemma 6.3. *With the notation from Section 6.3, $Q_n \neq -1$ for all $n \in \mathbb{N}$, and $Q_n = +1$ if and only if n is a multiple of ℓ. In particular, the minimal solution to the Pell equation (6.4) is given by $(x_1, y_1) = (p_{\ell-1}, q_{\ell-1})$ or $(p_{2\ell-1}, q_{2\ell-1})$ according to ℓ even or odd.*

Proof. If α_n is defined by (6.6), the numbers

$$\alpha_0, \; \alpha_1, \; \ldots, \; \alpha_\ell$$

are all distinct; otherwise we would get a contradiction with ℓ being minimal. Hence, we have $\alpha_1 = \alpha_{n+1}$ if and only if n is a multiple of ℓ. In the previous section we have already seen that the numbers P_n and Q_n appearing in the representation (6.7) are integers, and, in particular, $Q_{k\ell} = 1$. In view of Lemma 6.1 a solution x, y of the Pell equation (6.4) corresponds to p_{n-1} and q_{n-1} with $Q_n = 1$ or $Q_n = -1$. Thus it suffices to show that $Q_n \neq \pm 1$ for $1 \leq n < \ell$.

Assume that $Q_n = 1$; then we get $\alpha_n = P_n + \sqrt{d}$ from (6.7). For $n \geq 1$, the numbers α_n have a purely periodic continued fraction expansion,

$$\alpha_n = [\overline{a_n, \ldots, a_\ell, a_1, \ldots, a_{n-1}}].$$

By Galois' theorem 5.3, they are reduced; i.e.,

$$\sqrt{d} - 1 < P_n < \sqrt{d}.$$

This implies $P_n = [\sqrt{d}]$, so $\alpha_n = [\sqrt{d}] + \sqrt{d}$, and thus $\alpha_{n+1} = \alpha_1$. It follows that n must be a multiple of ℓ.

Now assume that $Q_n = -1$. Then we get $\alpha_n = -P_n - \sqrt{d}$. Now Galois' theorem 5.3 yields

$$-1 < -P_n + \sqrt{d} < 0 \qquad \text{and} \qquad 1 < -P_n - \sqrt{d},$$

which leads to $\sqrt{d} < P_n < -\sqrt{d} + 1$, a contradiction. Thus, $Q_n \neq -1$ for all $n \in \mathbb{N}$. The lemma is proved. •

In the following table some continued fraction expansions of $\sqrt{d}$ and the related minimal solutions (x_1, y_1) are listed:

$\sqrt{2} =$	$[1, \overline{2}]$	$(3, 2)$
$\sqrt{3} =$	$[1, \overline{1, 2}]$	$(2, 1)$
$\sqrt{5} =$	$[2, \overline{4}]$	$(9, 4)$
$\sqrt{19} =$	$[4, \overline{2, 1, 3, 1, 2, 8}]$	$(170, 39)$
$\sqrt{61} =$	$[7, \overline{1, 4, 3, 1, 2, 2, 1, 3, 4, 1, 14}]$	$(1\,766\,319\,049, 226\,153\,980)$
$\sqrt{99} =$	$[9, \overline{1, 18}]$	$(10, 1)$
$\sqrt{2002} =$	$[44, \overline{1, 2, 1, 9, 5, 6, 9, 1, 2, 1, 88}]$	$(11\,325\,887, 253\,128)$

It can be shown that the minimal solution of the Pell equation related to the cattle problem leads to a herd consisting of $77602\ldots81\,800$ many bulls and cows, a number having $206\,545$ digits.

The size of the minimal solution (x_1, y_1) of the Pell equation (6.4) behaves quite irregularly with increasing d.* Hua [85] proved that

$$x_1 + y_1\sqrt{d} < 2d^{\sqrt{d}}.$$

Of course, this depends on the length of the primitive period $\ell(\sqrt{d})$ of the continued fraction expansion of $\sqrt{d}$. Kraitchik [94] proved that the minimal period length satisfies

$$\ell(\sqrt{d}) \leq 0.72\,\sqrt{d}\log d,$$

where d is any integer greater than 7; it is conjectured that $\log d$ can be replaced by $\log\log d$ (cf. [134]). This is an interesting topic related to several deep and important questions, for example, the class number problem.

6.5. The group of solutions

We return to our example with $d = 2$. We observe that

$$(3 + 2\sqrt{2})^2 = 17 + 12\sqrt{2} \qquad \text{and} \qquad 17^2 - 2\cdot 12^2 = (3^2 - 2\cdot 2^2)^2 = 1.$$

This suggests that we can find *new* solutions of the Pell equation by taking powers of *old* ones. What's going on here?

The Pell equation is very important with respect to its role in quadratic number fields (see Appendix A.7). For instance, an integer $\alpha = x + y\sqrt{d}$ in $\mathbb{Q}(\sqrt{d})$, where d is a squarefree integer with $d \equiv 3 \bmod 4$, is a unit if and only if its norm is equal to one, that is,

$$\mathrm{N}(x + y\sqrt{d}) = (x - y\sqrt{d})(x + y\sqrt{d}) = x^2 - dy^2 = 1.$$

Hence, solutions to the Pell equation correspond one-to-one to units in the ring of integers of $\mathbb{Q}(\sqrt{d})$. It is a well-known fact that the units of a ring form a multiplicative group. Consequently, the solutions of the Pell equation form an infinite group. We shall show that this group is cyclic, generated by the minimal solution.

Theorem 6.4. *The Pell equation (6.4) has infinitely many solutions* $x, y \in \mathbb{Z}$; *all of them are up to the sign given by powers of the minimal solution:*

$$x + y\sqrt{d} := \pm(x_1 \mid y_1\sqrt{d})^{\pm n}\,, \qquad \text{where} \quad n = 0, 1, 2, \ldots.$$

Proof. In view of

$$(6.10) \qquad \frac{\pm 1}{x + y\sqrt{d}} = \frac{\pm(x - y\sqrt{d})}{x^2 - dy^2} = \pm x \mp y\sqrt{d}$$

---

*The lazy reader can find an *online* Pell solver on the Web page www.bioinfo.rpi.edu/ zukerm/cgi − bin/dq.html.

it suffices to show that all solutions of (6.4) in positive integers are given by $x + y\sqrt{d} = \varepsilon^n$, where $\varepsilon := x_1 + y_1\sqrt{d}$ and $n \in \mathbb{N}$. For any positive solution we can find an $n \in \mathbb{N}$ such that

$$\varepsilon^n \le x + y\sqrt{d} < \varepsilon^{n+1}.$$

Define

$$\mathcal{X} + \mathcal{Y}\sqrt{d} = \varepsilon^{-n}(x + y\sqrt{d});$$

then we have to prove that the latter expression is equal to one. Since $\sqrt{d}$ is irrational, conjugation (6.10) leads to

$$\mathcal{X} - \mathcal{Y}\sqrt{d} = \varepsilon^n(x - y\sqrt{d}).$$

Multiplying the latter equation with the previous one, we deduce

$$\mathcal{X}^2 - d\mathcal{Y}^2 = \varepsilon^{-n+n}\underbrace{(x - y\sqrt{d})(x + y\sqrt{d})}_{=x^2-dy^2=1} = 1.$$

Suppose now that $1 < \mathcal{X} + \mathcal{Y}\sqrt{d} < \varepsilon$, then, again with (6.10),

$$0 < \varepsilon^{-1} < (\mathcal{X} + \mathcal{Y}\sqrt{d})^{-1} = \mathcal{X} - \mathcal{Y}\sqrt{d} < 1.$$

It follows that

$$\begin{aligned}
2\mathcal{X} &= (\mathcal{X} + \mathcal{Y}\sqrt{d}) + (\mathcal{X} - \mathcal{Y}\sqrt{d}) > 1 + \varepsilon^{-1} > 0, \\
2\mathcal{Y}\sqrt{d} &= (\mathcal{X} + \mathcal{Y}\sqrt{d}) - (\mathcal{X} - \mathcal{Y}\sqrt{d}) > 1 - 1 = 0.
\end{aligned}$$

Thus, $\mathcal{X}$ and $\mathcal{Y}$ are positive integral solutions of the Pell equation with $1 < \mathcal{X} + \mathcal{Y}\sqrt{d} < \varepsilon$. Since $x + y\sqrt{d}$ increases with y, we get $\mathcal{Y} < y_1$ and $\mathcal{X} < x_1$, contradicting the fact that (x_1, y_1) is the minimal solution. It follows that $\mathcal{X} + \mathcal{Y}\sqrt{d} = 1$. The assertion of the theorem follows. •

This result, together with Theorem 6.2, gives a detailed description of the set of solutions to the Pell equation. In view of Lemmas 6.1 and 6.3, powers of the minimal solution correspond bijectively to solutions coming from the convergents to $\sqrt{d}$:

$$x_k + y_k\sqrt{d} = (x_1 + y_1\sqrt{d})^k.$$

This complete solution of the Pell equation was first proved by Legendre; however, partial results were already known by Euler and Lagrange.

6.6. The minus equation

The minus equation is not much more difficult than the plus equation. In view of (6.8) the minus equation has infinitely many solutions if the length ℓ of the primitive period of the continued fraction expansion of $\sqrt{d}$ is odd. Taking into account Lemma 6.3 it follows that the converse implication holds too.

Hence, in the notation of Section 6.3,

Theorem 6.5. *The minus equation*

$$X^2 - dY^2 = -1$$

has infinitely many solutions if and only if ℓ is odd. In this case, all positive solutions are given by

$$(x_k, y_k) = (p_{(2k-1)\ell-1}, q_{(2k-1)\ell-1}).$$

The set of solutions to the minus equation do not form a group (since there simply does not exist a neutral element, $(x_0, y_0) = (1, 0)$ is a solution of the plus equation). Nevertheless, combining a solution of the minus equation with a solution of the plus equation leads to a further solution of the minus equation. We illustrate this simple fact by an example.

The minimal period length of the continued fraction $\sqrt{5} = \left[2, \overline{4}\right]$ is odd; thus the minus equation is solvable. The minimal solutions of the minus and of the plus equation are given by $(2, 1)$ and $(9, 4)$, respectively. We compute

$$(2 + 1\sqrt{5}) \cdot (9 + 4\sqrt{5}) = 38 + 17\sqrt{5}.$$

Indeed, this leads to a solution of the minus equation:

$$38^2 - 5 \cdot 17^2 = -1.$$

Squares modulo 4 are congruent to 0 or 1. Thus, if $d \equiv 3 \bmod 4$,

$$x^2 - dy^2 \equiv \quad 0 \quad \text{or} \quad 1 \quad \bmod 4.$$

Hence, the minus equation is unsolvable if $d \equiv 3 \bmod 4$. In combination with the theorem above, it follows that the length of the primitive period of the continued fractions to $\sqrt{d}$ is automatically even whenever $d \equiv 3 \bmod 4$.

6.7. The polynomial Pell equation

Given a diophantine equation, it is often interesting to ask for polynomial solutions. With regard to the Pell equation we shall investigate now whether there are polynomial solutions of the Pell equation which we rewrite as

(6.11) $$P^2 - D \cdot Q^2 = 1,$$

where D is a fixed polynomial, and P and Q are polynomials in the same variables as D and with coefficients in the same field or ring as those of D. We shall refer to this as the **polynomial Pell equation**. Many aspects of the Pell equation do not depend on the type of solutions we ask for. Following Dubickas & Steuding [48] we shall study (6.11) by purely algebraic methods, so all results hold for polynomials in several variables over any field of characteristic $\neq 2$.

Again, the solutions $(P, Q) = (1, 0)$ and $(-1, 0)$ are called **trivial**, and all other solutions are said to be **non-trivial**. Clearly, for $\deg D = 0$, we have $\deg P = \deg Q = 0$. This is the case we discussed previously. In the sequel we may assume that $\deg D > 0$. Obviously, the polynomial Pell equation has no solutions if $\deg D$ is an odd number. Therefore, we assume that $\deg D$ is even, so that $\deg D \geq 2$. Also if (P, Q) is a non-trivial solution, then so are $(P,-Q)$ and $(-P,\pm Q)$. Again $P+Q\sqrt{D}$, where (P,Q) is a solution of (6.11), is also called a solution.

Let R be a subring of $\mathbb{C}$, and let $D \in \mathbb{C}[X]$ (or $D \in R[X]$). We are interested in polynomials $P,Q \in \mathbb{C}[X]$ (or in $R[X]$, respectively) satisfying the Pell equation (6.11). We call a solution (p,q) **minimal** if $p,q \in \mathbb{C}[X]$ and p has the smallest degree among all non-trivial solutions of (6.11) (with fixed D). Since

$$2\deg p = \deg D + 2\deg q,$$

the polynomial q has the smallest non-negative degree among all Q in non-trivial solutions (P,Q).

The set of polynomial solutions of the polynomial Pell equation forms an abelian group generated by one of the minimal solutions and $(-1,0)$. The multiplication in the abelian group corresponds to the multiplication of two expressions of the form $P+Q\sqrt{D}$. In case there is at least one non-trivial solution of (6.11), there exists a minimal solution, and all solutions are obtained as powers of the minimal solution.

Theorem 6.6. *Suppose that $D \in \mathbb{C}[X]$. If (6.11) has a non-trivial solution in $\mathbb{C}[X]$, then it has precisely four minimal solutions. If (p,q) is one of these, then all non-trivial solutions of (6.11) are obtained as $(\pm P, \pm Q)$, where*

$$P+Q\sqrt{D} = (p+q\sqrt{D})^n$$

for some $n \in \mathbb{N}$.

Furthermore, if $D \in R[X]$, where R is a subring of $\mathbb{C}$, is such that (6.11) has a non-trivial solution in $R[X]$, then there is a positive integer m such that every non-trivial solution of (6.11) in $R[X]$ is of the form $(\pm P, \pm Q)$, where

$$P+Q\sqrt{D} = (p+q\sqrt{D})^{mn}$$

for some $n \in \mathbb{N}$.

Proof. We start with the formal identity

$$(Pp \pm DQq)^2 - D(Pq \pm Qp)^2 = (P^2 - DQ^2)(p^2 - Dq^2).$$

Thus, if (p,q) and (P,Q) are two solutions of (6.11), then so are

(6.12) $(Pp + DQq, Pq + Qp)$ and $(Pp - DQq, Qp - Pq)$.

Further, it follows that

$$(Pq + Qp)(Qp - Pq) = \det\begin{pmatrix} p^2 & q^2 \\ P^2 & Q^2 \end{pmatrix} = \det\begin{pmatrix} p^2 - Dq^2 & q^2 \\ P^2 - DQ^2 & Q^2 \end{pmatrix}$$

(6.13) $= \det\begin{pmatrix} 1 & q^2 \\ 1 & Q^2 \end{pmatrix} = Q^2 - q^2;$

here we do the step from the first determinant to the second one simply by adding a multiple of the second column.

Assume now that (p,q) and (P,Q) are two minimal solutions. Let $k = \deg Q = \deg q$. Then (6.13) implies that

$$\deg(Pq + Qp) + \deg(Qp - Pq) \le 2k.$$

If both degrees on the left-hand side would be equal to k, then

$$Pq + Qp + Qp - Pq = 2Qp$$

would be of degree $\leq k$, so $\deg p = 0$, a contradiction. Hence, for instance, $\deg(Qp - Pq) < k$; the other case can be treated in the same way. Then the second solution in (6.12) must be trivial, so $Qp = Pq$ and $Pp = \pm 1 + DQq$. Setting

$$\lambda = \lambda(X) = \frac{P(X)}{p(X)} = \frac{Q(X)}{q(X)},$$

we have

$$\lambda p^2 = pP = \pm 1 + DQq = \pm 1 + \lambda Dq^2.$$

Thus,

$$\lambda = \lambda(p^2 - Dq^2) = \pm 1,$$

giving $(P, Q) = (p, q)$ or $(-p, -q)$; the other case yields $(P, Q) = (-p, q)$ or $(p, -q)$. This proves the first assertion of the theorem.

Now assume that all solutions of (6.11) with Q of degree $< \ell$ are of the form $(\pm P, \pm Q)$, where

$$P + Q\sqrt{D} = (p + q\sqrt{D})^n$$

with some $n \in \mathbb{N}$. Consider a solution (P, Q) of (6.11) with $\deg Q = \ell$. Again (6.13) shows, by the same reasoning as above, that either $\deg(Pq + Qp) < \ell$ or $\deg(Qp - Pq) < \ell$. In the first case, using (6.12) we get by induction that (up to a sign)

$$Pp + DQq + (Pq + Qp)\sqrt{D} = (p + q\sqrt{D})^k$$

for some positive integer k. This yields

$$P + Q\sqrt{D} = (p + q\sqrt{D})^{k-1},$$

and so (P, Q) is of the form required. The other case can be treated in the same way. This proves the "complex" case of the theorem.

For the "subring" part note that every solution in $R[X]$ must be a certain positive power of a minimal complex solution. Take the smallest power m for which $(p + q\sqrt{D})^m$ gives the expression $P + Q\sqrt{D}$ with $P, Q \in R[X]$. It follows that $(p + q\sqrt{D})^{mn}$ are solutions of (6.11) for every $n \in \mathbb{N}$. If it would be a solution in $R[X]$ different from $(p + q\sqrt{D})^{mn}$, say $(p + q\sqrt{D})^{mn+r}$ with $0 < r < m$, then

$$(p + q\sqrt{D})^r = (p + q\sqrt{D})^{mn+r}(p - q\sqrt{D})^{mn}$$

would give a solution of (6.11) in $R[X]$, a contradiction with the minimality of m. This completes the proof of the theorem. •

Note that the equation

$$P + Q\sqrt{D} = (p + q\sqrt{D})^n$$

can be rewritten as

$$P = \frac{1}{2}\left((p+q\sqrt{D})^n + (p-q\sqrt{D})^n\right),$$

$$Q = \frac{1}{2\sqrt{D}}\left((p+q\sqrt{D})^n - (p-q\sqrt{D})^n\right).$$

Corollary 6.7. *A solution* (P,Q) *of (6.11) satisfying* $\deg P = \frac{1}{2}\deg D$ *and* $\deg Q = 0$ *is minimal.*

6.8. Nathanson's theorem

The main difficulty in solving polynomial Pell equations is to determine whether non-trivial solutions exist or not. Chowla asked for solutions of (6.11) in $\mathbb{Z}[X]$ when

$$D(X) = X^2 + d \in \mathbb{Z}[X].$$

Nathanson [121] proved that there are no non-trivial solutions if $d \neq \pm 1, \pm 2$. For $d = 1$ or $d = \pm 2$, he found sequences of polynomial solutions, given by $P_0 = 1, Q_0 = 0$, and, for $n \in \mathbb{N}$,

(6.14)
$$\begin{cases} P_n(X) = \left(\frac{2}{d}X^2 + 1\right)P_{n-1}(X) + \frac{2}{d}X(X^2+d)Q_{n-1}(X), \\[2mm] Q_n(X) = \frac{2}{d}XP_{n-1}(X) + \left(\frac{2}{d}X^2 + 1\right)Q_{n-1}(X); \end{cases}$$

furthermore, he showed that the only integer polynomials which satisfy (6.11) are of the form $(\pm P_n, \pm Q_n)$. Note that the case $d = -2$ includes an old result of Brahmagupta (see Exercise 6.3). For $d = -1$ Nathanson gave another family of solutions (see Exercise 6.19).

As an application of our results from the previous section we can prove now Nathanson's theorem. Our proof is rather different from his approach.

Theorem 6.8. *Let* $D(X) = X^{2k} + d$ *with* $k \in \mathbb{N}$ *and* $0 \neq d \in \mathbb{Z}$. *Then the polynomial Pell equation (6.11) has a non-trivial solution* $P, Q \in \mathbb{Z}[X]$ *if and only if* $d \in \{\pm 1, \pm 2\}$. *In the case of solvability, the set of all non-trivial solutions can be deduced from (6.15) below.*

In Section 11.4 we shall prove that there are no solutions for $d = 0$.

Proof. First, we look for solutions in complex polynomials. By Corollary 6.7,

$$\left(\frac{X^k}{\sqrt{-d}}, \frac{1}{\sqrt{-d}}\right)$$

is the minimal solution of (6.11). By Theorem 6.6 all non-trivial solutions (P,Q) of (6.11) in complex polynomials are given by

(6.15)
$$\left(\frac{X^k + \sqrt{X^{2k}+d}}{\sqrt{-d}}\right)^n \qquad \text{for} \quad n \in \mathbb{N}.$$

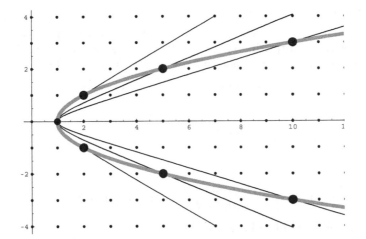

FIGURE 6.2. The hyperbolas $x^2 - Dy^2 = 1$ with $D = D(X) = X^2 + 1$ for $X = 1, 2, 3$ in the range $0 < x \leq 11$ and the curve $x = D(y)$ (resp., $Q_1(x) = P_1(y)$ with $d = 2$ in terms of Nathanson's polynomials). Polynomial solutions with integer coefficients generate minimal solutions in integers.

Let d be a non-zero integer. By the binomial theorem, the coefficient for X^{kn} in P is equal to

$$(-d)^{-\frac{n}{2}} \left(1 + \binom{n}{2} + \binom{n}{4} + \ldots \right) = 2^{n-1}(-d)^{-\frac{n}{2}}.$$

Consequently, if (6.11) has a non-trivial solution in $\mathbb{Z}[X]$, then there is some $n \in \mathbb{N}$ for which

$$2^{n-1}(-d)^{-\frac{n}{2}} \in \mathbb{Z}.$$

This is true if and only if $d \in \{\pm 1, \pm 2\}$. The theorem is proved. $\bullet$

We shall shortly discuss how to deduce Nathanson's families of solutions for $d \in \{1, \pm 2\}$. Taking into account (6.15)

(6.16) $-\left(\dfrac{X^k + \sqrt{X^{2k} + d}}{\sqrt{-d}} \right)^2 = \dfrac{2}{d}X^{2k} + 1 + \dfrac{2}{d}X^k \sqrt{X^{2k} + d},$

we are led to the polynomials

$$\frac{2}{d}X^{2k} + 1 \qquad \text{and} \qquad \frac{2}{d}X^k$$

with integer coefficients. For $k = 1$, these are exactly the polynomials P_1, Q_1 in (6.14).

However, since we had to raise the minimal *complex* solution to the power two for finding a minimal *integer* solution (6.16) (that is, $m = 2$ in Theorem 6.6), we lost some complex solutions. For example, for $d = k = 1$, that is, $D(X) = X^2 + 1$, we have two series of solutions of (6.11) in $\mathbb{C}[X]$

(corresponding to $m = 2$ in the second part of Theorem 6.6). The integer solutions of the Pell equation are integral lattice points on a hyperbola, and so it is not at all surprising that the polynomial solutions of (6.11) can be parametrized by the hyperbolic sine and cosine functions. Let

$$t = \operatorname{arsinh}(X) = \log(X + \sqrt{X^2 + 1}).$$

Then these series are

(6.17) $\left(\pm \cosh(nt), \pm \dfrac{\sinh(nt)}{\cosh(t)} \right)$ for $n = 0, 2, 4, \ldots,$

and

(6.18) $\left(\pm i \sinh(nt), \pm i \dfrac{\cosh(nt)}{\cosh(t)} \right)$ for $n = 1, 3, 5, \ldots.$

These are solutions of (6.11) since

$$\cosh(nt)^2 - \sinh(nt)^2 = 1.$$

It is easily seen that the expressions (6.17) and (6.18) are in fact polynomials in X for the corresponding parity of n; and by Theorem 6.6, for every $n \in \mathbb{N}$, there are at most four solutions (P, Q) with $\deg P = n$. Thus (6.17) and (6.18) give all complex solutions. In fact, the series (6.17) must be identical to the polynomials found by Nathanson (6.14). They consist of polynomials with integer coefficients. Similarly, the series (6.18) consist of integer polynomials multiplied by $\pm i$. These correspond to the solutions of the equation

$$P(X)^2 - (X^2 + 1)Q(X)^2 = -1.$$

Notes for further reading

The quest for polynomial solutions of the Pell equation is rather old. In 1826, Abel [1] observed that any non-trivial solution (P, Q) of (6.11) produces a surprising integral of the shape

(6.19) $\displaystyle \int \frac{P'(t)\,dt}{Q(t)\sqrt{D(t)}} = \log\left(P(t) + Q(t)\sqrt{D(t)} \right).$

Here, Q divides P' which follows immediately by differentiation of (6.11).

Avanzi & Zannier [7] proved that if $D \in \mathbb{K}[X]$ is a separable polynomial of degree four, where $\mathbb{K}$ is any field of characteristic zero, then the polynomial Pell equation (6.11) has a solution (P, Q) if and only if the two points at infinity of the associated elliptic curve

(6.20) $Y^2 = D(X)$

differ by a torsion point. If this is the case, the solutions (P, Q) of (6.11) have a parametrization in terms of Chebyshev polynomials. For arbitrary D, Hazama [78] described how the polynomial solutions of (6.11) may be analyzed via the generalized Jacobian of the curve (6.20). For $\deg D = 2$, this gives a concrete characterization by treating (6.11) as a twist of a conic by a conic. Recently, Webb & Yokota [170] and McLaughlin [112] studied the polynomial Pell equation via the continued fraction expansions of polynomials. Their analytical approach seems to be rather technical.

Exercises

6.1. *Show that there are no integer solutions to $X^2 - 2Y^2 = 0$ apart from $x = y = 0$.*

6.2. The following exercise is only interesting if the reader does not know more than the content of the first two sections.

For $d = 11$ $(d = 13)$, write the convergents $\frac{p_n}{q_n}$ to $\sqrt{d}$ for the first two periods and compute the values $p_n^2 - dq_n^2$. Give a conjecture on how the set of solutions of the Pell equation looks.

6.3. *For $2 \leq t \in \mathbb{N}$, prove that $(x, y) = (t^2 - 1, t)$ and $(x, y) = (2(t^2 - 1)^2 - 1, 2t(t^2 - 1))$ are solutions of the Pell equation*

$$X^2 - (t^2 - 2)Y^2 = 1.$$

This observation is due to Brahmagupta (598–630).

6.4. Let $a \geq 2$ be an integer. For $n \in \mathbb{N}$, define

$$x_0 = 1, \ x_1 = a, \quad \text{and} \quad x_{n+2} = 2ax_{n+1} - x_n,$$
$$y_0 = 0, \ y_1 = 1, \quad \text{and} \quad y_{n+2} = 2ay_{n+1} - y_n.$$

Prove that the sequence $(x_n, y_n)_{n \geq 0}$ yields all positive integer solutions of the Pell equation

$$X^2 - (a^2 - 1)Y^2 = 1.$$

6.5. Let d be a positive integer such that $\sqrt{d} \notin \mathbb{Q}$ and let D be an integer satisfying $|D| < \sqrt{d}$.

Prove that if there exists a solution to

$$X^2 - dY^2 = D$$

in coprime integers x, y, then there exists a convergent $\frac{p_n}{q_n}$ to $\sqrt{d}$ such that $p_n = x$ and $q_n = y$. Furthermore, show that whenever there exists one integer solution, then there are infinitely many integer solutions.

6.6. *Let A and B be positive integers. Show that if the Pell equations*

$$X^2 - AY^2 = +1 \quad \text{and} \quad X^2 - BY^2 = -1$$

have a common solution (x, y), then $y = \pm 1$.

6.7. How can rational solutions of the Pell equation be found? Geometry helps!

Prove that the Pell equation (6.4) with $\sqrt{d} \notin \mathbb{Q}$ has an infinity of rational solutions, generated by

$$x = \frac{dt^2 + 1}{dt^2 - 1} \quad \text{and} \quad y = \frac{2t}{dt^2 - 1} \quad \text{for} \quad t \in \mathbb{Q}.$$

Hint: To show that this gives a parametrization of all points on (x, y) on the conic $X^2 - dY^2 = 1$ with rational coordinates (up to the sign) compute the intersection point of the straight line through $(1, 0)$ and slope t with the hyperbola $X^2 - dY^2 = 1$.

6.8. It is clear that all points (x, y) on the unit circle $X^2 + Y^2 = 1$ with integer coordinates are given by $x = \pm 1, y = \pm 1$. However, there are infinitely many with rational coordinates.

Show that

$$x = \frac{1 - t^2}{1 + t^2} \qquad and \qquad y = \frac{2t}{1 + t^2}$$

provide parametrization of all points on the circle with rational coordinates. Deduce Theorem 1.1 on a parametrization of the Pythagorean triples.

This construction of rational points is called **Bachet's method**.

6.9.* *Popeye has a lot of spinach tins, in between 10^9 and 10^{10}. It is known that the tins can be arranged by a square and also a triangle with equal sides. What is the number of his tins?*

6.10.* *Give a rigorous solution of the first and the second part of the cattle problem. How large is approximately the second least solution of the second part of the cattle problem?*

Hint: We recommend using a computer; it might be helpful to look at [29] or [163].

6.11.* **Newton's square root method** provides fairly good approximations to square roots. Given a real number d and a first approximation $x_0 \neq 0$ to $\sqrt{d}$, then the recursion defined by

$$x_n = \frac{1}{2}\left(x_{n-1} + \frac{d}{x_{n-1}}\right) \qquad for \quad n \in \mathbb{N},$$

is rapidly convergent to $\sqrt{d}$ (in fact, it is a second order process). Mc Bride [111] investigated this iteration with respect to diophantine approximations and the Pell equation.

i) Apply Newton's square root method to $d = 2$ with initial value $x_0 = \frac{3}{2}$ and compute the next 5 values of the iteration process. How good is the approximation x_5?

ii) Compute the first 8 convergents to $\sqrt{2}$ and compare them with the approximations to $\sqrt{2}$ obtained in i).

iii) Let d be any positive integer which is not a perfect square. Prove that if x_0 is the convergent to $\sqrt{d}$ according to the minimal solution of the Pell equation $X^2 - dY^2 = 1$, all following x_n are solutions (and, in particular, convergents) too. Characterize the sequence x_n!

6.12. The binomial series, given by

$$(x + a)^n = a^n + na^{n-1}x + \frac{n(n-1)}{2}a^{n-2}x^2 + \dots,$$

is convergent for any real numbers n, $a > 0$, and x such that $|x| < a$.

Use the binomial series expansion for finding good rational approximations to

$$\sqrt{2} = \frac{1}{5}(49+1)^{\frac{1}{2}}.$$

How good are these approximations?

6.13. *Compute, if possible, the minimal solutions to the following Pell equations*

$$X^2 - dY^2 = \pm 1 \qquad for \quad d = 3,\ 17,\ 20,\ 35\,109.$$

6.14. *Prove that if $x_1^2 - Dy_1^2 = q_1$ and $x_2^2 - Dy_2^2 = q_2$, then*

$$(x_1 x_2 + Dy_1 y_2)^2 - D(x_1 y_2 + x_2 y_1)^2 = q_1 q_2.$$

6.15.* *If d is squarefree and $\sqrt{d} = [a_0, \overline{a_1, \ldots, a_\ell}]$ is the continued fraction expansion of $\sqrt{d}$, show that $a_j < 2a_\ell = 2[\sqrt{d}]$ for $1 \leq j < \ell$.*

Hint: Recall Theorem 5.4 and the proof of Lemma 6.1.

6.16. *For $1 \leq d \leq 200$ and d being not a perfect square, compute the record holders among $\sqrt{d}$ having a longer primitive period in the continued fraction expansion than any one before.*

6.17. *Solve the Pell equation*

$$X^2 - 433Y^2 = 1.$$

Give an estimate for the size of powers of the minimal solution.

The Pell equation was rediscovered several times in history. The above problem was posed by Fermat in 1657.

6.18.* *Show that the negative Pell equation (6.5) is solvable if and only if there exists a primitive Pythagorean triple (a, b, c) and positive integers m, n such that*

$$d = m^2 + n^2 \qquad and \qquad |ma - nb| = 1.$$

Hint: The "if"-part is easy; for the "only if"-part one may have a look into Grytczuk et al. [**71**].

6.19. *Give a description of the set of solutions of the polynomial Pell equation (6.11) with $D(X) = X^2 - 1$. Recover Nathanson's family of all polynomial solutions, up to the sign given recursively by $P_0 = 1, Q_0 = 0$, and, for $n \in \mathbb{N}$,*

$$P_n(X) \quad = \quad X P_{n-1}(X) + (X^2 - 1)Q_{n-1}(X),$$
$$Q_n(X) \quad = \quad P_{n-1}(X) + X Q_{n-1}(X).$$

6.20.* *For $n \in \mathbb{N}$, the* **Chebyshev polynomials of the first kind** *and* **of the second kind** *are defined by*

$$T_n(\cos \theta) = \cos(n\theta) \qquad and \qquad U_n(\cos \theta) = \frac{\sin((n+1)\theta)}{\sin \theta},$$

respectively; note that U_n has removable singularities at $\theta = \pi m, m \in \mathbb{Z}$.

Rewrite the polynomials of the families (6.17) and (6.18) in terms of Chebyshev polynomials.

6.21.* Let P, D, and Q be polynomials with complex coefficients satisfying $P^2 - DQ^2 = 1$.

Deduce Abel's identity (6.19).

6.22.* *Show that if* $P, Q \in \mathbb{C}[X]$ *satisfy the identity*

$$P(X)^2 - (X^2 + k)Q(X)^2 = 1,$$

then $P' = (\pm \deg P) \cdot Q$.

This exercise is from the final round of the 1978 Swedish Mathematical Olympiad.

6.23. Let z and d be positive integers, and d not a perfect cube.

Prove that if (x, y) *is a solution in positive integers to the equation*

$$X^3 - dY^3 = z$$

with $y > 2zd^{-\frac{2}{3}}$, *then* $\frac{x}{y}$ *is a convergent to* $\sqrt[3]{d}$.

6.24.* Since it has nothing to do with Pell or the Pell equation, we conclude with a nice task due to Ramanujan.

For $n \in \mathbb{N}$, *prove that*

$$n + 1 = \sqrt{1 + n\sqrt{1 + (n+1)\sqrt{1 + (n+2)\sqrt{\cdots}}}}.$$

CHAPTER 7

Factoring with continued fractions

Sending secret messages is no any longer the business of spies. We are living in times with exponentially increasing electronic information exchange, and almost every day we make use of modern encryption systems (e.g., smart cards). Many of these cryptosystems, like the famous RSA, rely on the observation that it is simple to multiply integers but, conversely, that it is rather difficult to find the prime factorization of a given large integer. Factoring methods for large integers can be used to test how secure such cryptosystems are. In this chapter we shall consider the continued fraction method CFRAC.

7.1. The RSA cryptosystem

Most of the modern cryptosystems in practice are public-key cryptosystems. The idea of public-key cryptography dates at least back to Diffie & Hellman [46] in 1976; however, there is some evidence that some units of the British secret service already used similar principles in the late 1960s (cf. [154]). We may imagine a large group of people, each of whom wants to send secret messages to any one of the others. For such systems it is very convenient to have a **public-key cryptosystem** where any one of its users has a *public* key, known by all users, and a *private* key which is the secret of its owner. Such a cryptosystem can be realized by use of a so-called **trapdoor function**, that is, roughly speaking, a simple operation which has an inverse that can hardly be computed. (For more information on the fascinating subject of cryptography we refer to [113].)

The first public-key cryptosystem in practice was the famous RSA **cryptosystem**, invented by Rivest, Shamir & Adleman [136] in 1978 (named after the initials of their inventors). It depends on the difficulty in factoring large integers. It is easy to multiply integers, e.g.,

$$75\,658\,934\,651 \; \cdot \; 15\,643\,985\,657,$$

but, conversely, it is rather difficult to find the prime factorization of a given large integer (with few comparatively large prime factors), e.g.,

$$1\,183\,607\,288\,504\,144\,300\,707.$$

There are fast primality tests known which may be used to generate primes; here *fast* means that they succeed in testing a given integer N on primality in polynomial time with respect to the binary input length $O(\log N)$ (see Exercise 5.8.*ii*). On the contrary, there is no comparably fast factoring algorithm known, and it is conjectured that there is no such polynomial time algorithm at all. Note that this depends very much on the hardware which is used; if quantum computers are constructed, Shor's factoring algorithm could split integers in polynomial time (see [150]). This would in theory be the end of RSA and many others cryptosystems.

How does RSA work? A user, say Bob, chooses two distinct *large* prime numbers p and q (having about a hundred digits or a bit more) and computes $N = pq$. Then Bob computes

(7.1) $$\varphi(N) = (p-1)(q-1) = N + 1 - p - q,$$

where $\varphi(N)$ is Euler's totient, the number of integers $1 \leq a \leq N$ coprime with N; for (7.1) we used that φ is a multiplicative function (see Exercises 2.11 and 2.12) but note that Bob can perform this step without this knowledge just by using formula (7.1). Next Bob randomly chooses an integer e such that $1 < e < \varphi(N)$ and e is coprime with $\varphi(N)$. Testing coprimality is easily done by use of the Euclidean algorithm. Then Bob computes the multiplicative inverse d of e mod $\varphi(N)$, i.e.,

$$d \equiv e^{-1} \bmod \varphi(N);$$

by Theorem 1.6, this is equivalent to the existence of an integer f such that

(7.2) $$de = 1 + f\varphi(N).$$

Then Bob's **public key** is the pair (N, e); the value d, the factorizations of N, and also $\varphi(N)$ are secret; in fact, these factorizations are not needed any more. The publishing of the public key may take the form of a telephone book. A key never has to be sent from one user to another — all the keys are public knowledge! The main problem in the history of espionage, the transmission of keys, is circumvented.

Now imagine Alice wants to send a secret love letter to Bob. What has Alice to do? First, she has to translate the content of her letter into a sequence of numbers. This procedure must be known to all users in the system, in particular, Bob. We may think of coding like this

$$\text{blank} \mapsto 99, \ \mathsf{A} \mapsto 10, \ \mathsf{B} \mapsto 11, \ \ldots, \ \mathsf{Z} \mapsto 35,$$

and a message is coded simply by putting the numbers according to the plain text in sequence, building blocks of size $< N$ (however, in practice one should be a bit more cautious). Thus the secret message is given in its digital form as a large integer, say M. Alice knows Bob's public key (N, e) but not his secret d or the factorization of N. Alice has to raise the message M to the power e and reduces it modulo N; let us call this coded message C, i.e.,

$$C = M^e \bmod N,$$

where here and in the sequel $c = a \bmod N$ stands for the least positive residue c of the residue class a modulo N (to fix one of the infinitely many representatives of $a \bmod N$). Then C is sent to Bob, and nothing else. Now

Bob can decode C simply by computing the dth power of C modulo N. In fact, by Fermat's little theorem A.4,

$$(7.3) \quad C^d \equiv (M^e)^d = M^{de} = M^{1+f\varphi(N)} = M \cdot (M^f)^{\varphi(N)} \equiv M \bmod N,$$

Bob simply recovers Alice's love letter in its digital form M (after finding the appropriate member in the corresponding residue class $\bmod N$). The letters d and e stand for decryption and encryption, respectively.

The main point in the RSA cryptosystem is that, according to present knowledge, the decryption exponent d cannot be easily derived from the publicly known N and e, but only from the factorization of N or the knowledge of $\varphi(N)$ — if some care was taken in the choice of e and d (more on this topic in the preceding section). In fact, if $\varphi(N)$ is known, one can easily compute the inverse d of e modulo $\varphi(N)$ by use of the Euclidean algorithm, that is, solving the linear diophantine equation (7.2) in the unknowns f and d. If the factorization of N is known, then one gets $\varphi(N)$ via (7.1).

We shall give an example. The word LOVE becomes $M = 21\,243\,114$. Let's assume that Bob has chosen the primes $p = 15\,373$ and $q = 12\,373$, so $N = 190\,210\,129$ and $\varphi(N) = 190\,182\,384$, and further $e = 154\,201\,933$ which gives $d = 37$. However, in practice prime numbers with approximately 100 digits are used. Using Bob's public key (N, e) Alice then encrypts her message M as

$$C = 64\,353\,547 \equiv 21\,243\,114^{154\,201\,933} \bmod 190\,210\,129.$$

Now imagine that jealous Jill wants to know what stupid Alice wrote to her dear Bob. Jill may try to factor $N = 190\,210\,129$ to find the secret decryption exponent d. For small integers like this example this can be done by so-called **trial division**, that is, trying all primes up to the square root of N (one must not go further since a composite number has at least one divisor less than or equal to its square root, otherwise we would get a contradiction to the compositeness). For large numbers, trial division is not a good factoring algorithm since there are *too many* prime numbers. Given a hundred digit number N, by the prime number theorem A.2, there are about 10^{48} possible prime divisors for N to check. Imagine a million processors in parallel, each of them performing 10^{12} trial divisions per second for distinct subsets of the possible divisors. Then, in the worst case, it takes more than 10^{22} years to factor the hundred-digit number N, which is much longer than the age of our universe.

7.2. A diophantine attack on RSA

To speed up RSA encryption and decryption it is reasonable to use a small public or secret decryption exponent. Such a choice of small d or e is, in particular, of interest whenever there is a large difference in the computing power between two devices, for example, in communication between a smart card and a larger computer. Now we shall briefly discuss an interesting diophantine attack on those RSA cryptosystems due to Wiener [**173**]. It provides candidates for the secret decryption exponent d without trying to factor N. For further refinements of Wiener's attack see Dujella [**49**].

Assume that $N = pq$ with $q < p < 2q$ and that the private exponent d is rather small, say

$$(7.4) \qquad\qquad d < \frac{1}{3}N^{\frac{1}{4}}.$$

As we shall show now, d is then the denominator of a convergent to the continued fraction of $\frac{e}{N}$. For any integer f,

$$\left| \frac{f}{d} - \frac{e}{N} \right| = \frac{|fN - de|}{dN}.$$

It follows from (7.1) and $q < p < 2q$ that

$$N - 3\sqrt{N} < \varphi(N) < N.$$

Now we take f satisfying (7.2) with $1 \le f < d$. Then

$$\left| \frac{f}{d} - \frac{e}{N} \right| = \frac{|f\varphi(N) - de + f(N - \varphi(N))|}{dN} < \frac{3f\sqrt{N}}{dN} = \frac{3f}{d\sqrt{N}} < \frac{1}{2d^2},$$

where the last inequality is a consequence of our assumption (7.4). It thus follows from Theorem 3.9 that the fraction $\frac{f}{d}$ can be found among the convergents of the continued fraction for $\frac{e}{N}$. So we can get the secret information d from the public information e and N. It can be shown that one can succeed to do so in polynomial time (i.e., operation count bounded by a power of $\log N$).

To illustrate this attack we return to the example from the previous section. First, we notice that Alice's choice of p, q, and d satisfies (7.4). The convergents to the continued fraction

$$\frac{e}{N} = \frac{154\,201\,933}{190\,210\,129} = [0, 1, 4, 3, 1, 1, 5, 1, 1, 1, 1, 3, 1, 201, 2, 1, 1, 3, 1, 7]$$

are

$$0,\ 1,\ \frac{4}{5},\ \frac{13}{16},\ \frac{17}{21},\ \mathbf{\frac{30}{37}},\ \frac{167}{206},\ \ldots\ .$$

To find among these convergents the fraction $\frac{f}{d}$ we check which d satisfies $(M^e)^d \equiv M \bmod N$ for some randomly chosen M. This test leads here to the convergent $\frac{30}{37}$. Having found the right d, jealous Jill can recover Alice's secret message from the coded message C simply by performing (7.3). In the case of $C = 64\,353\,547$ this leads to $M = 21\,243\,114$. Converting M into letters yields Alice's LOVE.

Another possible attack on RSA-type cryptosystems is to try to factor the publicly known N. This is the theme we shall consider in the sequel of this chapter.

7.3. An old idea of Fermat

The quest for a fast factoring algorithm is much older than the computer age. Gauss wrote in his famous *Disquisitiones Arithmeticae* [**65**] in 1801

> *The problem of distinguishing prime numbers from com-*
> *posite numbers and of resolving the latter into their prime*
> *factors is known to be one of the most important and use-*
> *ful in arithmetic (...) Nevertheless we must confess that*
> *all methods that have been proposed thus far are either re-*
> *stricted to very special cases or are so laborious and prolix*
> *that (...) these methods do not apply at all to larger num-*
> *bers.*

We start with an old idea from Fermat. Suppose that we are interested in the prime factorization of a *large* integer N. If there are integers $\mathcal{X}, \mathcal{Y}$ for which

$$\mathcal{X}^2 \equiv \mathcal{Y}^2 \bmod N \qquad \text{and} \qquad \mathcal{X} \not\equiv \pm\mathcal{Y} \bmod N,$$

then the greatest common divisor $\gcd(N, \mathcal{X} + \mathcal{Y})$ is a non-trivial factor of N. This can be seen as follows. Consider the identity

$$\mathcal{X}^2 - \mathcal{Y}^2 = (\mathcal{X} - \mathcal{Y})(\mathcal{X} + \mathcal{Y}).$$

The left-hand side is a multiple of N. If neither of the factors on the right is a multiple of N, both of them have a common divisor with N. The question is how to find such pairs $\mathcal{X}, \mathcal{Y}$. To look randomly for pairs of squares which satisfy these conditions is hopeless.

In 1926, Kraitchik proposed to search randomly for sufficiently many squares which lie in the same residue class mod N, such that certain combinations among them lead to non-trivial divisors of N. More precisely, having sufficiently many congruences

$$x_j^2 \equiv (-1)^{\varepsilon_{0j}} \ell_1^{\varepsilon_{1j}} \ell_2^{\varepsilon_{2j}} \cdot \ldots \cdot \ell_m^{\varepsilon_{mj}} \bmod N,$$

where the ℓ_i are *small* prime numbers and the ε_{ij} are the related exponents, by Gaussian elimination modulo 2 one may hope to find a relation of the form

$$(7.5) \qquad \sum_{j \leq k} \delta_j(\varepsilon_{0j}, \ldots, \varepsilon_{mj}) \equiv (0, \ldots, 0) \quad \bmod 2,$$

where $\delta_j \in \{0, 1\}$. Then, setting

$$(7.6) \qquad \mathcal{X} = \prod_{j \leq k} x_j^{\delta_j} \qquad \text{and} \qquad \mathcal{Y} = (-1)^{\nu_0} \ell_1^{\nu_1} \ell_2^{\nu_2} \cdot \ldots \cdot \ell_m^{\nu_m},$$

where the ν_i are given by

$$\sum_{j \leq k} \delta_j(\varepsilon_{0j}, \ldots, \varepsilon_{mj}) = 2(\nu_0, \nu_1, \ldots, \nu_m),$$

we get $\mathcal{X}^2 \equiv \mathcal{Y}^2 \bmod N$. This splits N if $\mathcal{X} \not\equiv \pm\mathcal{Y} \bmod N$. The set of prime numbers ℓ_i chosen to find the congruences, in addition with -1, is called **factor basis** for N.

Kraitchik proposed to generate the squares x_j^2 by the nearest integers x to $\sqrt{N}$; however, this is not very convenient in practice.

7.4. CFRAC

One of the first modern factorization methods is the continued fraction method CFRAC due to Lehmer & Powers [101] from 1931. The first implementation was realized by Brillhart & Morrison [31] with which on September 13, 1970 they factored the 39-digit seventh Fermat number*:

$$\mathcal{F}_7 = 2^{2^7} + 1 \quad = \quad 340\,282\,366\,920\,938\,463\,463\,374\,607\,431\,768\,211\,457$$
$$= \quad 59\,649\,589\,127\,497\,217 \cdot 5\,704\,689\,200\,685\,129\,054\,721.$$

Soon after, CFRAC became the main factoring algorithm in practice; actually it was the first algorithm of *expected* subexponential running time. Until the 1980s it was the method of choice for factoring large integers but it has a limit at around 50 digits.

In the sequel of this chapter we let $m \bmod N$ denote the least value of m modulo N in absolute value. Then CFRAC has the following form:

For $j = 0, 1, 2, \ldots$ successively

1. Compute the jth convergent $\frac{p_j}{q_j}$ of the continued fraction expansion to $\sqrt{N}$.

2. Compute $p_j^2 \bmod N$. After doing this for several j, look at the numbers $\pm p_j^2 \bmod N$ which factor into a product of small primes. Define the factor base $\mathcal{B}$ to consist of -1, the primes which either occur in more than one of the $p_j^2 \bmod N$ or in just one $p_j^2 \bmod N$ to an even power.

3. List all of the numbers $p_j^2 \bmod N$ which can be expressed as a product of numbers in the factor base $\mathcal{B}$. If possible, find a subset of numbers ℓ's of $\mathcal{B}$ for which the exponents ε according to the prime numbers in $\mathcal{B}$ sum to zero modulo two as in (7.5), and define $\mathcal{X}, \mathcal{Y}$ by (7.6). If $\mathcal{X} \not\equiv \pm \mathcal{Y} \bmod N$, then $\gcd(\mathcal{X}+\mathcal{Y}, N)$ is a non-trivial factor of N. If this is impossible, then compute more p_j and $p_j^2 \bmod N$, enlarging the factor base $\mathcal{B}$ if necessary.

Of course, to speed up the algorithm one should reduce $\bmod N$ whenever it is possible. Once the number of completely factored integers exceeds the size of the factor base, we can find a product of them which is a perfect square. With a little luck this yields a non-trivial divisor of our given number (by the observations from the previous section). The crucial property of the values p_j is, as we shall show now, that their squares have *small* residues modulo N. Otherwise, CFRAC would hinge on the problem of finding an appropriate factor base $\mathcal{B}$.

Theorem 7.1. *Let $\alpha > 1$ be irrational. Then the convergents $\frac{p_j}{q_j}$ to α satisfy the inequality*

$$|q_j^2 \alpha^2 - p_j^2| < 2\alpha.$$

---

*See www.prothsearch.net/fermat.html#Prime for the current knowledge on Fermat numbers.

In particular, if $\alpha = \sqrt{N}$, where $N \in \mathbb{N}$ is not a perfect square, the residue $p_j^2 \bmod N$ is of modulus less than $2\sqrt{N}$.

Proof. By the triangle inequality,

$$|q_j^2 \alpha^2 - p_j^2| = q_j^2 \left|\alpha - \frac{p_j}{q_j}\right| \cdot \left|\alpha + \frac{p_j}{q_j}\right| \leq q_j^2 \left|\alpha - \frac{p_j}{q_j}\right| \left(2\alpha + \left|\alpha - \frac{p_j}{q_j}\right|\right).$$

It follows from Theorem 3.6 that

$$|q_j^2 \alpha^2 - p_j^2| < \frac{q_j^2}{q_j q_{j+1}} \left(2\alpha + \frac{1}{q_j q_{j+1}}\right).$$

Hence,

$$|q_j^2 \alpha^2 - p_j^2| - 2\alpha < 2\alpha \left(-1 + \frac{q_j}{q_{j+1}} + \frac{1}{2\alpha q_{j+1}^2}\right) < 2\alpha \left(-1 + \frac{q_j + 1}{q_{j+1}}\right),$$

which is less or equal to zero. This proves the first assertion of the theorem. The claim on $p_j^2 \bmod N$ for $\alpha = \sqrt{N}$ is an immediate consequence. •

Therefore, the sequence of the numerators of the convergents of $\sqrt{N}$ provides a sequence of integers whose squares have *small* residues mod N. This property of continued fractions is essential for the running time of CFRAC. If the squares are generated by the nearest integers x to $\sqrt{N}$, as proposed by Kraitchik, one observes that $|x^2 - N|$ grows fairly quick; more precisely, it is equal to $2\sqrt{N}|x - \sqrt{N}|$, which reduces the probability that $x^2 - N$ splits completely using only primes from the factor basis. The so-called quadratic sieve overcomes this difficulty by a sieving process (as in the sieve of Eratosthenes from elementary number theory).

We illustrate CFRAC by two examples. First, consider $N = 8777$. It is easy to compute the beginning of the continued fraction expansion to $\sqrt{N}$ and its first convergents:

$$\frac{93}{1}, \frac{94}{1}, \frac{281}{3}, \frac{1499}{16}, \ldots \quad \rightarrow \quad \sqrt{8777} = [93, 1, 2, 5, 1, \ldots].$$

This leads to

j	0	1	2	3
$p_j \bmod 8777$	93	94	281	1499
$p_j^2 \bmod 8777$	-128	59	-32	89

It is not too difficult to factor the *small* numbers $p_j^2 \bmod 8777$. As a result we may choose the factor base $\mathcal{B} = \{-1, 2\}$, so that $p_j^2 \bmod N$ splits over the factor base $\mathcal{B}$ for $j = 0$ and $j = 2$:

$$-128 = (-1) \cdot 2^7 \quad \text{and} \quad -32 = (-1) \cdot 2^5.$$

The corresponding vectors of exponents ε_{ij} are $\{1, 7\}$ and $\{1, 5\}$. Their sum adds up to zero modulo two, which yields

$$\mathcal{X} = 93 \cdot 281 \equiv -198 \bmod 8777 \quad \text{and} \quad \mathcal{Y} = -2^6 = -64.$$

It follows that $198^2 \equiv 64^2 \bmod 8777$. Since $198 \not\equiv \pm 64 \bmod 8777$, we see that $\gcd(198 + 64, 8777) = 131$ is a non-trivial divisor of 8777 (the computation of the greatest common divisor is quickly done by the Euclidean algorithms). This leads to the prime factorization

$$8777 = 67 \cdot 131.$$

Now we shall consider a more advanced example, namely, the one from the section on the Wiener attack on RSA. But this time we shall factor $N = 190\,210\,129$. For this purpose we first compute the first, say 40 partial quotients of the continued fraction expansion of $\sqrt{N}$:

$$\sqrt{190\,210\,129} = [13\,791, 1, 2, 51, 7, 1, 4, 6, 1, 30, \ldots, 5, 1, 1, \ldots],$$

and the numerators of the corresponding convergents modulo N:

$$13\,791, \; 13\,792, \; 41\,375, \; 2\,123\,917, \; 14\,908\,794, \; \ldots .$$

Next we square these numbers and factor them modulo N. This leads to a list of 40 relatively small integers; among them we find

$$
\begin{array}{rcrcl}
(-1) \cdot 31 \cdot 643 & = & -19\,933 & \equiv & 14\,908\,794^2, \\
2^4 \cdot 3^3 \cdot 31 & = & 13\,392 & \equiv & 90\,815\,340^2, \\
7 \cdot 41^2 & = & 11\,767 & \equiv & 35\,820\,798^2, \\
(-1) \cdot 59^2 & = & -3\,481 & \equiv & (-47\,468\,570)^2, \\
3 \cdot 7 \cdot 643 & = & 13\,503 & \equiv & 33\,354\,409^2,
\end{array}
$$

all modulo $190\,210\,129$; precisely, that are the fifth, the 12th, the 28th, the 37th, and the 40th convergent. We note that the exponents in the prime factorization of the product of the terms on the left-hand sides are all even. Thus we put

$$
\begin{aligned}
\mathcal{X} & = & 14\,908\,794 \cdot 90\,815\,340 \cdot 35\,820\,798 \cdot 47\,468\,570 \cdot 33\,354\,409 \\
& = & -76\,788\,535\,071\,124\,709\,298\,162\,037\,310\,419\,850\,400 \\
& \equiv & 154\,612\,427 \quad \bmod 190\,210\,129,
\end{aligned}
$$

and

$$
\begin{aligned}
\mathcal{Y} = -2^2 \cdot 3^2 \cdot 7 \cdot 31 \cdot 41 \cdot 59 \cdot 643 & = & -12\,150\,917\,604 \\
& \equiv & 22\,530\,652 \quad \bmod 190\,210\,129.
\end{aligned}
$$

Finally, we compute

$$\gcd(\mathcal{X} + \mathcal{Y}, N) = \gcd(154\,612\,427 + 22\,530\,652, 190\,210\,129) = 15\,373.$$

This gives a proper divisor of $N = 190\,210\,129$ and leads to the complete factorization

$$N = 190\,210\,129 = 12\,373 \cdot 15\,373.$$

Alice's secret is not safe!

7.5. Examples of failures

It is a well-known fact that CFRAC does not work for prime powers $N = p^k$. This causes no difficulties since it is quite easy to check whether a given N is a prime power or not. Now we will explicitly construct another infinite family of examples for failures, which are less harmless.

Recall our study of the Pell equation. With regard to Lemma 6.1 the sequence of the numerators p_j of the convergents to the continued fraction of $\sqrt{N}$ is periodic mod N, the length of the period bounded by the length of the primitive period of the continued fraction to $\sqrt{N}$. There are many integers N for which the continued fraction $\sqrt{N}$ has a rather short period. But if the period of the continued fraction expansion of $\sqrt{N}$ is too *short*, then CFRAC can only produce a *small* factor basis, which reduces the chances for factoring N.

For instance, if $N = k^2 + 2$ for some integer k, then by Lemma 6.1

$$(-1)^j Q_j = p_{j-1}^2 - (k^2 + 2)q_{j-1}^2 \equiv p_{j-1}^2 \bmod (k^2 + 2).$$

It turns out that the sequence Q_j takes only the values 1 and 2. Alternatively, with regard to Exercise 5.9 and (3.3), we can compute directly

$$p_j \bmod (k^2 + 2) = \begin{cases} \pm k & \text{if } j \text{ is even,} \\ \pm 1 & \text{if } j \text{ is odd.} \end{cases}$$

Again it follows that $p_j^2 \bmod (k^2 + 2)$ is either equal to 1 or -2. With the notation used in the CFRAC algorithm this gives $\mathcal{X} \equiv +\mathcal{Y}$ or $\equiv -\mathcal{Y} \bmod N$. CFRAC does not work for numbers $N = k^2 + 2$.

One strategy to overcome this problem is to replace N by some multiple cN, where c is some suitably chosen integer, hoping that the continued fraction expansion of $\sqrt{cN}$ has a longer period. However, it can happen that if a proper divisor is found, it is a divisor of c; see Riesel [134] for further details. We conclude with another refinement of CFRAC based on weighted mediants from Šleževičienė & Steuding [156].

7.6. Weighted mediants and a refinement

For two distinct positive reduced fractions $\frac{a}{b}, \frac{c}{d}$ we define their **weighted mediant** with positive integral weights λ, μ by

$$\frac{a\lambda + c\mu}{b\lambda + d\mu};$$

for $\lambda = \mu$ this is the mediant of $\frac{a}{b}, \frac{c}{d}$ we already know from our study on Farey fractions. It is easily seen that the weighted mediant lies in between $\frac{a}{b}$ and $\frac{c}{d}$. One can show that each rational number in the interval with limits $\frac{a}{b}, \frac{c}{d}$ is a mediant of the upper and lower bound for some weight λ, μ.

If we have two excellent rational approximations $\frac{a}{b}, \frac{c}{d}$ to an irrational α, then the weighted mediant of $\frac{a}{b}, \frac{c}{d}$ is also a good approximation if the weights are sufficiently small.

Theorem 7.2. *Let $\alpha > 1$ be irrational. If $\frac{a}{b}, \frac{c}{d}$ are two consecutive convergents to α with $d < b$, then*

$$|(b\lambda + d\mu)^2 \alpha^2 - (a\lambda + c\mu)^2| < 2(\lambda^2 + 3\lambda\mu + 2\mu^2)(\alpha + 1)$$

for any positive coprime integers λ, μ. In particular, if $\alpha = \sqrt{N}$, where $N \in \mathbb{N}$ is not a perfect square, then

$$(a\lambda + c\mu)^2 \bmod N < 2(\lambda^2 + 3\lambda\mu + 2\mu^2)(\sqrt{N} + 1).$$

Proof. The proof follows the argument in the proof of Theorem 7.1. First,

$$\left| \alpha - \frac{a\lambda + c\mu}{b\lambda + d\mu} \right| \leq \left| \alpha - \frac{a}{b} \right| + \left| \frac{a}{b} - \frac{a\lambda + c\mu}{b\lambda + d\mu} \right| \leq \left| \alpha - \frac{a}{b} \right| + \frac{\mu|bc - ad|}{b(b\lambda + d\mu)}.$$

Now let $\frac{a}{b}, \frac{c}{d}$ be two convergents to an irrational $\alpha > 1$. Then, similar to the proof of Theorem 6.1,

$$\begin{aligned}
&|(b\lambda + d\mu)^2 \alpha^2 - (a\lambda + c\mu)^2| \\
&= \quad (b\lambda + d\mu)^2 \left| \alpha - \frac{a\lambda + c\mu}{b\lambda + d\mu} \right| \cdot \left| \alpha + \frac{a\lambda + c\mu}{b\lambda + d\mu} \right| \\
&\leq \quad (b\lambda + d\mu)^2 \left(\left| \alpha - \frac{a}{b} \right| + \frac{\mu|bc - ad|}{b(b\lambda + d\mu)} \right) \left(2\alpha + \left| \alpha - \frac{a}{b} \right| + \frac{\mu|bc - ad|}{b(b\lambda + d\mu)} \right).
\end{aligned}$$

In view of Theorem 3.6 the latter quantity is less than

$$\left(\lambda^2 + 2\lambda\mu\frac{d}{b} + \mu^2\frac{d^2}{b^2} + \mu|bc - ad| \left(\lambda + \mu\frac{d}{b} \right) \right) \left(2\alpha + \frac{1}{b^2} + \frac{\mu|bc - ad|}{b(b\lambda + d\mu)} \right).$$

Since $d < b$ and $|bc - dy| = 1$ for two consecutive convergents $\frac{a}{b}, \frac{c}{d}$, we find in this case

$$\begin{aligned}
&|(b\lambda + d\mu)^2 \alpha^2 - (a\lambda + c\mu)^2| \\
&< \quad (\lambda^2 + 2\lambda\mu + \mu^2 + \mu(\lambda + \mu)) \left(2\alpha + \frac{1}{b^2} + \frac{\mu}{bd(\lambda + \mu)} \right).
\end{aligned}$$

This yields the estimate of the theorem. •

By the estimate of the theorem, the squares $(a\lambda + c\mu)^2 \bmod N$ of numerators of weighted mediants to consecutive convergents with weights $1 \leq \lambda, \mu \leq C$, where C is any constant, are bounded by $\sqrt{N}$ as in the case of the ordinary convergents.

Consequently, we thus can use weighted mediants for generating squares; the effort for factoring the squares into an appropriate factor base is approximately the same as if one works with convergents only. It is obvious how we can use this simple idea for a refinement of CFRAC. If the period $\ell(\sqrt{N})$ of the continued fraction expansion of $\sqrt{N}$ is too short, i.e., if we cannot factor N by the congruences coming from the squares of the numerators of the convergents to $\sqrt{N}$, then one can work with weighted mediants of the convergents additionally. Thus we add to the CFRAC algorithm as a fourth step:

4. If the full period did not lead to a factorization of N, compute for $1 \le j \le \ell(\sqrt{N})$ and coprime non-negative integers λ, μ the numbers $p_j(\lambda, \mu) := \lambda p_{j-1} + \mu p_j \bmod N$ and return to step 2 by replacing $p_j \bmod N$ with $p_j(\lambda, \mu)$.

We shall give an example. In the case of numbers $N = k^2 + 2$ these weighted mediants are

$$\frac{\lambda k + \mu(k^2 + 1)}{\lambda + \mu k} \quad \text{and} \quad \frac{\lambda(k^2 + 1) + \mu(2k^3 + 3k)}{\lambda k + \mu(2k^2 + 1)}.$$

We need only the squares of the numerators modulo N:

$$\lambda k^2 - 2\lambda \mu k + \mu^2 \quad \text{and} \quad \lambda^2 + 2\lambda \mu k + \mu^2 k^2.$$

Varying the weights $\lambda, \mu = 0, 1, \ldots$ gives plenty of good candidates for building up an appropriate factor base. Note that the case $\lambda = 1$ and $\mu = 0$ yields the *old* CFRAC algorithm. For instance, this refinement was used for the factorization of the 45-digit number

$$
\begin{aligned}
N &= 12\,345\,678\,901\,234\,567\,890\,123^2 + 2 \\
&= 152\,415\,787\,532\,388\,367\,504\,942\,236\,884\,722\,755\,800\,955\,131 \\
&= 19 \cdot 8\,021\,883\,554\,336\,229\,868\,681\,170\,362\,353\,829\,252\,681\,849.
\end{aligned}
$$

We do not analyze the speed of CFRAC but it is much behind the recent number field sieve of Pollard and its improvements due to A. Lenstra & H. Lenstra and others (see [42]). Nevertheless, there are still no algorithms known to date that enable the factorization of *large* integers in a *reasonable* amount of time, even if the most powerful computers available are used.

Notes on primality testing

In August 2002, the computer scientist Agrawal together with his students Kayal and Saxena [2] succeeded in finding a first deterministic primality test in polynomial time without assuming any unproved hypothesis. This recent breakthrough in a longtime search for the solution of a fundamental problem shows that the decision problem of testing a given integer on primality is in the class P of, roughly speaking, problems solvable in polynomial time (in the input data). Factoring a given integer N seems to be harder than deciding whether N is prime or composite. It is expected that there exists no polynomial time algorithm for factoring integers. It is known that the factoring problem lies in the class NP, roughly speaking, the class of decision problems having solutions that, once given, can be verified in polynomial time. By definition the classes P and NP seem to be quite different: solving a problem seems to be harder than verifying a given solution. It is widely expected that factoring marks a difference between these classes of decision problems. However, it is an open problem to prove (or disprove) $P \ne NP$. This fundamental conjecture in theoretical computer science is one of the seven millennium problems.[†]

[†]For more information we refer to http://www.claymath.org/millennium/.

Continued fractions can also be used for testing numbers on primality. These tests are based on so-called Lucas sequences, generalizations of the Fibonacci sequence. As a particular case, they include the first efficient primality test for numbers of a special shape. In 1644, the monk Mersenne investigated the numbers defined by

$$M_p := 2^p - 1,$$

where p is prime. It is easily seen that these numbers cannot be prime if the exponent is composite. Prime M_p are called **Mersenne primes**. Mersenne computed an erroneous list of Mersenne primes for $p \leq 257$. Euler found the following simple criterion: if $p \equiv 3 \bmod 4$ is prime, then $2p + 1$ divides M_p if and only if $2p + 1$ is prime too; the first proof was found by Lagrange in 1775. By this primality test Euler found $M_{11} = 2047$ to be composite, contradicting Mersenne's list. In 1878, Lucas discovered the following criterion: given an odd prime p, the number M_p is prime if and only if M_p divides the integer S_{p-1}, where the sequence of the S_k is defined by the recursion

$$S_1 = 4 , \qquad S_{k+1} = S_k^2 - 2 \qquad \text{for} \quad k = 1, 2, \ldots;$$

the first proof is attributed to Lehmer in 1935. The sequence of the S_k starts with

$$S_1 = 4 , \quad S_2 = 14 = 2 \cdot 7 , \quad S_3 = 194 , \quad S_4 = 37634 = 2 \cdot \mathbf{31} \cdot 607 , \quad \ldots ,$$

which yields the first Mersenne primes $M_3 = 7$ and $M_5 = 31$. From an algorithmic point of view it is rather useful to reduce S_k modulo M_p. This is the so-called **Lucas–Lehmer test**. This primality test is very fast and so it is not too surprising that for more than a century the world records for the largest known prime numbers are always Mersenne primes.[‡]

More details on topics we touched in this chapter can be found in the excellent monograph Cohen [**40**]; for applications of continued fraction algorithms to primality tests we refer to Riesel [**134**].

Exercises

7.1. *Show that a positive integer $N \geq 2$ is prime if and only if $\varphi(N) > N - \sqrt{N}$.*

7.2. *For $N = pq$, where p and q are distinct prime numbers, show that knowing $\varphi(N)$ is equivalent to knowing the factorization of N.*
Hint: Verify $p^2 - p(N - \varphi(N) + 1) + N = 0$ and consider this as a quadratic equation in the unknown p.

7.3. *Using the prime numbers $p = 1447$ and $q = 1451$ encrypt the message RSA along the RSA scheme with the encryption exponent $e = 1\,497\,643$. Then do the test: decrypt the coded message.*

[‡]The current largest prime is $2^{25\,964\,951} - 1$, found by Nowak in 2005 within the GIMPS project, the Great Internet Mersenne Prime Search, initiated by Woltman; for more information see www.mersenne.org.

7.4. *Given* $N = 399\,400\,189$ *and* $e = 1\,497\,643$, *try to find the secret decryption exponent* d *in the* RSA *scheme. Decrypt the coded message* $C = 2\,006\,746$. *Use the ideas from Section 7.2 to play the role of Jill!*

7.5. Here is another attack on RSA, the **Low exponent attack**.

i) Suppose that $m, e \in \mathbb{N}$ *and* $N_1, \ldots, N_e$ *are pairwise coprime integers such that* $0 \leq m < N_j$ *for* $1 \leq j \leq e$. *Assume that*

$$C \equiv M^e \bmod N_j \quad \text{for} \quad 1 \leq j \leq e$$

and $0 \leq C < N_1 \cdot \ldots \cdot N_e$. *Show that* $C = M^e$.

ii) Assume that a bank is using the same encryption exponent e *for all clients. If* (N_j, e) *is the public key of client* j, *how can one use i) to decrypt coded messages* C *without knowing the secret key* d?

Hint: Use the Chinese remainder theorem.

7.6. *Let* a, b, c, *and* m *be integers,* m *odd and* a *coprime with* m. *Find a criterion for the solvability of the quadratic congruence*

$$aX^2 + bX + c \equiv 0 \bmod m$$

similar to the one for solving quadratic equations over the real numbers.

7.7.* *Classify the set of integers which can be written as a difference of two squares.*

7.8. *Use Kraitchik's method from Section 7.3 to factor the number* $N = 2041$, *i.e., factor the integers* $x^2 \bmod N$ *for* $x = 46, 47, \ldots$.

7.9. *Use* CFRAC *to factor the integers* $N = 1969, 9073, 17\,873$, *and* $25\,511$.

7.10.* *Show that* CFRAC *fails for prime powers.*

7.11.* *Implement* CFRAC, *including the strategies to overcome failures. Try to factor some big integers!*

7.12. *Given rational numbers* $\frac{a}{b} < \frac{c}{d}$, *show that any rational number* $\frac{p}{q} \in \left(\frac{a}{b}, \frac{c}{d}\right)$ *has a representation of the form*

$$\frac{p}{q} = \frac{a\lambda + c\mu}{b\lambda + d\mu}$$

for some positive integers λ, μ.

7.13. *For* $n \in \mathbb{N}$, *show that* $2^n - 1$ *is composite if* n *is composite.*

7.14. A positive integer n is said to be **perfect** if

$$2n = \sum_{d \mid n} d,$$

For instance, 6 and 28 are perfect numbers.

Show that if $M_p = 2^p - 1$ *is a Mersenne prime, then* $2^{p-1}(2^p - 1)$ *is perfect. Prove that all even perfect numbers are of this form.*

The first statement is due to Euclid; the second one is due to Euler. Note that it is yet unknown whether odd perfect numbers exist or not.

CHAPTER 8

Geometry of numbers

In this chapter we present a further approach toward diophantine analysis. Geometry of numbers was created by Minkowski [**117**] at the turn to the twentieth century. In this theory arithmetic results are proved by using geometric arguments; in this sense, *geometry rules arithmetic*. Originated as a tool in number theory, nowadays geometry of numbers also plays a significant role in several other fields, e.g., cryptography, coding theory, and numerical integration. However, we will only sketch its applications to diophantine analysis.

8.1. Minkowski's convex body theorem

Consider the diophantine equation

(8.1)
$$2X^2 + 10XY + 13Y^2 = 1;$$

are there integer solutions or not? We may reformulate this question as a problem about lattice points, namely, whether the intersection of $\mathbb{Z}^2$ with the ellipse given by the equation

(8.2)
$$2x^2 + 10xy + 13y^2 < 2$$

contains more than the origin. It is easy to see that any $(x,y) \neq (0,0)$ in the intersection is a solution to the diophantine equation (8.1) (this requires a bit linear algebra and we will disclose the secret later). An answer to the latter question can be found by Minkowski's famous convex body theorem. However, before we can state this simple but far-reaching theorem we have to introduce some notions.

A subset $\mathcal{C}$ of the vector space $\mathbb{R}^n$ is said to be **convex** if it contains every point on the line segment connecting any two of its points. It is not too difficult to show that the volume of a bounded convex set $\mathcal{C}$ always exists, and we shall denote it by $\mathrm{vol}(\mathcal{C})$; for a rigorous proof one has to define $\mathrm{vol}(\mathcal{C})$ as the supremum of the sums of the volumes of networks of small n-dimensional cubes whose vertices lie inside $\mathcal{C}$. If $\mathcal{C}$ is convex and not contained in a hyperplane, then it is called a **convex body**. A set $\mathcal{C}$ is said to be **symmetric** (around the origin) if $\mathcal{C} = -\mathcal{C}$; that is, with any vector $\mathbf{x} \in \mathcal{C}$ we also have $-\mathbf{x} \in \mathcal{C}$ (where the minus sign has to be understood with respect to each coordinate). Examples of symmetric convex bodies are spheres, ellipsoids, and cubes centered at the origin.

In this first section we shall only deal with integer lattices $\mathbb{Z}^n$, and so a point in $\mathbb{R}^n$ is said to be a lattice point if its coordinates are integers. A rigorous definition of the fundamental notion of a lattice will be given in the following section.

Theorem 8.1. *Let* $\mathcal{C} \subset \mathbb{R}^n$ *be symmetric, convex, and bounded with* $\mathrm{vol}(\mathcal{C}) > 2^n$. *Then* $\mathcal{C}$ *contains at least one lattice point different from the origin.*

This is Minkowski's convex body theorem from 1891. There are many interesting proofs of this result. Here we follow Mordell's rather simple approach. For the first reading, we recommend considering only the two-dimensional case, the Euclidean plane.

Proof. Let t be a fixed positive integer. The hyperplanes

$$x_j = \frac{2z_j}{t} \qquad (1 \le j \le n; z_1, \ldots, z_n \in \mathbb{Z})$$

divide the space into cubes of volume $(\frac{2}{t})^n$. Denote by $N(t)$ the number of corners of these cubes in $\mathcal{C}$. Then

$$\lim_{t \to \infty} \left(\frac{2}{t}\right)^n N(t) = \mathrm{vol}(\mathcal{C}).$$

In view of the assumption, $\mathrm{vol}(\mathcal{C}) > 2^n$, it follows that $N(t) > t^n$ for sufficiently large t. However, the $(z_1, \ldots, z_n)$ give at most t^n different tuples of remainders when the z_j are divided by t. Consequently, $\mathcal{C}$ contains two distinct points

$$\mathsf{P}_1 := \left(\frac{2z_1^{(1)}}{t}, \ldots, \frac{2z_n^{(1)}}{t}\right), \quad \mathsf{P}_2 := \left(\frac{2z_1^{(2)}}{t}, \ldots, \frac{2z_n^{(2)}}{t}\right),$$

such that $z_j^{(1)} - z_j^{(2)}$ for $1 \le j \le n$ all are divisible by t. Since $\mathcal{C}$ is symmetric with respect to the origin, we have $-\mathsf{P}_2 \in \mathcal{C}$. Thus, the midpoint M of P_1 and $-\mathsf{P}_2$, explicitly given by

$$\mathsf{M} = \left(\frac{z_1^{(1)} - z_1^{(2)}}{t}, \ldots, \frac{z_n^{(1)} - z_n^{(2)}}{t}\right),$$

has integral coordinates and lies inside $\mathcal{C}$. Since $\mathsf{P}_1 \ne \mathsf{P}_2$, at least one of the coordinates of M is non-zero. The theorem is proved. •

It is not difficult to skip the condition on $\mathcal{C}$ to be bounded.

We return to the diophantine equation (8.1). By principal axis transformation (8.2) can be transformed to

$$\frac{15 - \sqrt{221}}{4} \mathcal{X}^2 + \frac{15 + \sqrt{221}}{4} \mathcal{Y}^2 < 1.$$

Since this ellipse has area $2\pi > 4 = 2^2$, Theorem 8.1 implies the existence of a solution to (8.1).

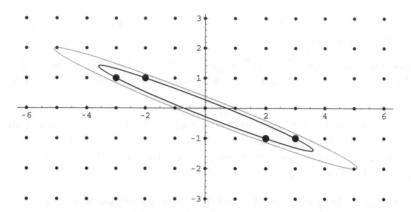

FIGURE 8.1. An example for Minkowski's convex body theorem: the ellipse given by (8.2) and inside the ellipse (8.1) with non-zero lattice points.

8.2. General lattices

In this section we shall prove Minkowski's first theorem, that is, a generalization of Theorem 8.1 to arbitrary lattices. First, we give a rigorous definition of a lattice and introduce some important notions.

Let $\mathbf{z}_1, \ldots, \mathbf{z}_r$ be an r-tuple of linearly independent vectors in $\mathbb{R}^n$. Then the **lattice** Λ with basis $\mathbf{z}_1, \ldots, \mathbf{z}_r$ is defined to be the set of all linear combinations of the $\mathbf{z}_j$ with integer coefficients, i.e.,

$$\Lambda = \mathbf{z}_1 \mathbb{Z} + \ldots + \mathbf{z}_r \mathbb{Z} = \{ m_1 \mathbf{z}_1 + \ldots + m_r \mathbf{z}_r : (m_1, \ldots, m_r) \in \mathbb{Z}^r \}.$$

The number r of linearly independent vectors in a lattice is called the **rank** of the lattice Λ; implicitly, we have $r \leq n$ because of the linear independence (by Theorem 2.5, for example, the linear combinations of the numbers 1 and $\sqrt{2}$ form a dense set on the real line). Algebraically speaking, a lattice $\Lambda \subset \mathbb{R}^n$ is a free $\mathbb{Z}$-module whose rank r as a $\mathbb{Z}$-module is equal to the dimension of the $\mathbb{R}$-vector space generated by Λ. A lattice $\Lambda \subset \mathbb{R}^n$ is called a **full** lattice if its rank is equal to n. Thus, a full lattice is a free $\mathbb{Z}$-module of rank n that generates $\mathbb{R}^n$ as an $\mathbb{R}$-vector space. Note that a lattice $\Lambda \subset \mathbb{R}^n$ of rank $r < n$ becomes a full lattice in a suitable r-dimensional subspace of $\mathbb{R}^n$.

In general, a lattice can have many different bases; nevertheless, the rank is obviously an invariant. Our first aim is to tidy up a bit among all these bases.

Lemma 8.2. *Let* $\{\mathbf{z}_1, \ldots, \mathbf{z}_r\}$ *and* $\{\mathbf{w}_1, \ldots, \mathbf{w}_r\}$ *be any two bases of a lattice* Λ. *Then there exists an* $r \times r$ *matrix* M *with integral entries and* $\det(M) = \pm 1$ *such that* $\mathbf{z}_j = M\mathbf{w}_j$ *for* $1 \leq j \leq r$.

In the language of linear algebra the matrix in the lemma is an element of the general linear group $\mathsf{SL}_r(\mathbb{Z})$ and represents a unimodular transformation.

Proof. Since both, $\{\mathbf{z}_j\}_j$ and $\{\mathbf{w}_j\}_j$ are bases of Λ, there are $r \times r$ matrices M, N having integral entries such that

$$\mathbf{z}_j = M\mathbf{w}_j \qquad \text{and} \qquad \mathbf{w}_j = N\mathbf{z}_j$$

for $1 \leq j \leq r$. It follows that MN is the identity matrix and so $\det(M)\det(N) = 1$. Since both determinants are integers, they have to be $= \pm 1$. The lemma is proved. •

Let $\Lambda \subset \mathbb{R}^n$ be a lattice with basis $\mathbf{z}_1, \dots, \mathbf{z}_r$. Then we define the **determinant** of Λ by

$$\det(\Lambda) = \sqrt{\det((\mathbf{z}_i^t \mathbf{z}_j)_{1 \leq i, j \leq r})},$$

where $\mathbf{x}^t$ denotes the transpose of the vector $\mathbf{x}$, and $\mathbf{x}^t \mathbf{y}$ is the standard (Euclidean) **inner product** of the vectors $\mathbf{x}$ and $\mathbf{y}$ in $\mathbb{R}^n$. The appearing matrix $(\mathbf{z}_i^t \mathbf{z}_j)_{1 \leq i, j \leq r}$ is called **Gram-matrix**. Since it is symmetric with positive eigenvalues, the determinant is positive and we may take its square root. It is not difficult to show, by Lemma 8.2, that the determinant of a lattice does not depend on the chosen basis.

In the case of a full lattice $\Lambda \subset \mathbb{R}^n$, i.e., Λ has a basis of length n, $\mathbf{z}_1, \dots, \mathbf{z}_n$ say, one can alternatively define the determinant of Λ by

$$\det(\Lambda) = |\det(Z)|,$$

where Z is the $n \times n$ matrix with the vectors $\mathbf{z}_j$ as columns. To see this note that $Z^t Z$ is the matrix whose entries consist of the inner products $\mathbf{z}_i^t \mathbf{z}_j$. Thus,

$$\det(\Lambda)^2 = \det((\mathbf{z}_i^t \mathbf{z}_j)_{1 \leq i, j \leq n}) = \det(Z^t)\det(Z),$$

which implies the claim. It follows that, geometrically speaking, $\det(\Lambda)$ is the volume of the parallelepiped

$$\{\lambda_1 \mathbf{z}_1 + \dots + \lambda_n \mathbf{z}_n : 0 \leq \lambda_1, \dots, \lambda_n \leq 1\}.$$

Now we are in the position to generalize Minkowski's convex body theorem 8.1 to arbitrary lattices. There are several possibilities how this can be realized. For instance, given a full lattice Λ with basis $\mathbf{z}_1, \dots, \mathbf{z}_n$, we define a linear mapping $f : \mathbb{R}^n \to \mathbb{R}^n$ by setting

$$f(x_1, \dots, x_n) = x_1 \mathbf{z}_1 + \dots + x_n \mathbf{z}_n.$$

Obviously, f is bijective and $\Lambda = f(\mathbb{Z}^n)$. For any convex set $\mathcal{C}$, we have

(8.3) $\text{vol}(f(\mathcal{C})) = \det(\Lambda)\text{vol}(\mathcal{C})$

This identity is clear for arbitrary cubes $\mathcal{C}$, and since any convex set can be approximated by a disjoint union of sufficiently small cubes as good as we please, (8.3) holds also for arbitrary convex $\mathcal{C}$ (alternatively, one can use the transformation formula for multiple integrals). Now consider $f^{-1}(\mathcal{C})$. This is a convex body with volume

$$\text{vol}(f^{-1}(\mathcal{C})) = \frac{\text{vol}(\mathcal{C})}{\det(\Lambda)}.$$

If $\mathcal{C}$ is symmetric, $f^{-1}(\mathcal{C})$ is symmetric too, and an application of Theorem 8.1 yields the existence of a non-zero lattice point $\mathbf{x} \in f^{-1}(\mathcal{C}) \cap \mathbb{Z}^n$. Obviously, $\mathbf{0} \neq f(\mathbf{x}) \in \mathcal{C} \cap \Lambda$. Thus we have proved

Theorem 8.3. *Let Λ be a full lattice in $\mathbb{R}^n$ and let $\mathcal{C} \subset \mathbb{R}^n$ be symmetric and convex with $\operatorname{vol}(\mathcal{C}) > 2^n \det(\Lambda)$. Then $\mathcal{C}$ contains a lattice point of Λ different from the origin.*

8.3. The lattice basis theorem

Now we give a characterization of lattices as discrete subgroups.

Theorem 8.4. *$\Lambda \subset \mathbb{R}^n$ is a lattice if and only if Λ is a discrete subgroup of the additive group $\mathbb{R}^n$.*

Proof. First, we show that a lattice Λ is a discrete subgroup of $\mathbb{R}^n$. Obviously, for any two elements $\mathbf{x}_1, \mathbf{x}_2$ in Λ, also their difference $\mathbf{x}_1 - \mathbf{x}_2$ is an element of Λ. Thus, Λ is an additive group. Now assume that $\mathbf{z}_1, \ldots, \mathbf{z}_r$ is a basis of Λ and let Z be the matrix with the vectors $\mathbf{z}_j$ as columns. Then the lattice Λ has no point $\neq \mathbf{0}$ inside

$$Z\mathcal{C} := \{ Z\mathbf{c} : \mathbf{c} \in \mathcal{C} \},$$

where $\mathcal{C}$ denotes the open unit cube $\mathcal{C}$ in $\mathbb{R}^r$. More generally, for any $\mathbf{x} \in \Lambda$, there is no lattice point $\neq \mathbf{x}$ in the interior of $\mathbf{x} + Z\mathcal{C}$. Since $Z\mathcal{C}$ contains some sphere of sufficiently small positive radius ϱ, any two lattice points of Λ have distance at least ϱ. Hence, Λ is discrete.

Next we have to show that any discrete subgroup which contains $r \leq n$ linearly independent vectors is a lattice. For this aim we choose inductively r points $\mathbf{z}_1, \ldots, \mathbf{z}_r$ and then prove that Λ is a lattice with basis $\{ \mathbf{z}_1, \ldots, \mathbf{z}_r \}$. Let $\mathbf{z}_1 \neq \mathbf{0}$ be an arbitrary point of Λ with the property that the segment joining $\mathbf{0}$ and $\mathbf{z}_1$ does not contain any other lattice point; this is possible since Λ is discrete. Now assume that $\mathbf{z}_1, \ldots, \mathbf{z}_{j-1} \in \Lambda$ have been chosen. Denote by L_j the $(j-1)$-dimensional linear subspace of $\mathbb{R}^n$ generated by $\mathbf{z}_1, \ldots, \mathbf{z}_{j-1}$,

$$L_j = \mathbb{R}\mathbf{z}_1 + \ldots + \mathbb{R}\mathbf{z}_{j-1},$$

and choose an arbitrary point $\mathbf{x}$ in Λ which does not lie in L_j. Let P_j be the j-dimensional parallelepiped spanned by the vectors $\mathbf{z}_1, \ldots, \mathbf{z}_{j-1}, \mathbf{x}$, i.e.,

$$P_j = \{ \lambda_1 \mathbf{z}_1 + \ldots + \lambda_{j-1} \mathbf{z}_{j-1} + \lambda_j \mathbf{x} : 0 \leq \lambda_1, \ldots, \lambda_j < 1 \}.$$

Since Λ is discrete, the number of points lying in the intersection of Λ and P_j is finite but greater than or equal to one (since $\mathbf{x} \in \Lambda \cap P_j$). Now let $\mathbf{z}_j$ be a point in $\Lambda \cap P_j$ with minimum distance to L_j. Now suppose that r points $\mathbf{z}_1, \ldots, \mathbf{z}_r$ have been chosen according to this procedure. Then, $\mathbf{z}_j \notin L_j$ for $2 \leq j \leq r$. It thus follows that $\mathbf{z}_1, \ldots, \mathbf{z}_r$ are linearly independent. Since all these points lie in Λ and Λ is a group, also all linear combinations

$$m_1 \mathbf{z}_1 + \ldots + m_r \mathbf{z}_r \qquad \text{for} \quad m_j \in \mathbb{Z}$$

lie in Λ. Hence, it remains to be shown that there are no other points in Λ.

Let $\mathbf{x}$ be an arbitrary point of Λ. Then

$$\mathbf{x} = \xi_1 \mathbf{z}_1 + \ldots + \xi_r \mathbf{z}_r$$

for some $\xi_j \in \mathbb{R}$. Now let ξ_j' be the fractional part of ξ_j, i.e., $\xi_j' = \{\xi_j\} = \xi_j - [\xi_j]$ for $1 \le j \le r$. Let

$$\mathbf{x}' = \xi_1' \mathbf{z}_1 + \ldots + \xi_r' \mathbf{z}_r \in \Lambda.$$

In view of the choice of $\mathbf{z}_r$ it follows that $\xi_r' = 0$, and by induction, $\xi_j' = 0$ for all j. Hence, $\xi_j \in \mathbb{Z}$ and $\mathbf{x}' = \mathbf{x}$. This shows that Λ is a lattice and $\{\mathbf{z}_1, \ldots, \mathbf{z}_r\}$ a basis. The theorem is proved. •

For our later purpose we shall investigate integer lattices which are given by congruences.

Lemma 8.5. *Let $k, r \in \mathbb{N}$ and $a_{ij} \in \mathbb{Z}$ for $1 \le i \le k$ and $1 \le j \le r$. Then*

$$\Lambda := \left\{ \mathbf{z} = (z_1, \ldots, z_r) \in \mathbb{Z}^r : \sum_{j=1}^{r} a_{ij} z_j \equiv 0 \mod m_i \quad \text{for} \quad 1 \le i \le k \right\}$$

is a lattice with determinant $\det(\Lambda) \le m_1 \cdot \ldots \cdot m_k$.

Proof. It follows from Theorem 8.4 that Λ is a lattice; more precisely, it is a sublattice of $\mathbb{Z}^r$ and $\mathbf{z}^{(1)}, \mathbf{z}^{(2)} \in \mathbb{Z}^r$ lie in the same residue class modulo Λ if and only if

$$\sum_{j=1}^{r} a_{ij} z_j^{(1)} \equiv \sum_{j=1}^{r} a_{ij} z_j^{(2)} \mod m_i$$

for $1 \le i \le k$. Hence, the residue classes $\mathbf{z} + \Lambda$ in $\mathbb{Z}^r / \Lambda$ are characterized by the residue classes $\sum_j a_{ij} z_j \mod m_i$ for $i = 1, 2, \ldots, k$. The number of these residue classes is at most $m_1 \cdot \ldots \cdot m_k$. This proves the lemma. •

8.4. Sums of squares

Before we dive deeper into the geometry of numbers we give a nice number theoretical application. The squares of integers form a rather *thin* subset of $\mathbb{N}$, so how many squares do we expect are needed such that any positive integer can be written as a sum of squares? At first glance it is not clear whether the number of squares needed can be bounded or not. As Lagrange showed in 1770 (by an elementary but lengthy proof), any positive integer can be written as a sum of four integer squares; here terms of the form $0 = 0^2$ are allowed.

We start with a result attributed to Fermat on prime numbers which have a representation as a sum of two squares.

Theorem 8.6. *Let $p \equiv 1 \mod 4$ be prime. Then there exist positive integers a, b such that $p = a^2 + b^2$.*

Proof. From elementary number theory we know that -1 is a quadratic residue modulo any prime $p \equiv 1 \mod 4$ (see Theorem A.6). Hence there exists an integer q such that

(8.4) $$q^2 \equiv -1 \mod p.$$

We consider

$$\Lambda := \begin{pmatrix} 1 \\ q \end{pmatrix} \mathbb{Z} + \begin{pmatrix} 0 \\ p \end{pmatrix} \mathbb{Z}.$$

By Lemma 8.5, Λ is a lattice with determinant $\det(\Lambda) = p$. Now let

$$\mathcal{C} = \{(x, y) \in \mathbb{R}^2 : x^2 + y^2 < 2p\}.$$

This is a disk with center in the origin, so it fulfills the assumptions of Minkowski's theorem 8.3. Since

$$\mathrm{vol}(\mathcal{C}) = 2\pi p > 4p = 2^2 \det(\Lambda),$$

there exists a non-zero lattice point $(a, b) \in \mathcal{C}$. By construction, $0 < a^2 + b^2 < 2p$. Besides, we have

$$\begin{pmatrix} a \\ b \end{pmatrix} = j \begin{pmatrix} 1 \\ q \end{pmatrix} + k \begin{pmatrix} 0 \\ p \end{pmatrix}$$

for some integers j, k. This implies $a = j$ and $b = jq + kp$. A simple computation shows that

$$a^2 + b^2 = j^2 + (jq + kp)^2 = j^2(1 + q^2) + \underbrace{2jkpq + k^2p^2}_{\equiv 0} \equiv j^2(1 + q^2) \equiv 0 \bmod p,$$

where we used (8.4) in the final step. The only remaining possibility for $a^2 + b^2$, being a multiple of p, is to be equal to p. The theorem is proved. $\bullet$

We can interpret representations as sums of two squares as a statement about complex numbers. A complex number $z = x + iy$ having an integral real- and integer-part is called a **Gaussian integer**. The squared modulus of $z = x + iy$ is

$$|z|^2 = (x + iy)(x - iy) = x^2 + y^2.$$

Since the set $\mathbb{Z}[i]$ of Gaussian integers forms a ring, it follows that the set of integers which can be represented as a sum of two squares is multiplicatively closed. For any complex numbers we have

$$(x_1 + iy_1) \cdot (x_2 + iy_2) = x_1x_2 - y_1y_2 + i(x_1y_2 + x_2y_1).$$

Hence, the norm equation of the complex numbers implies the identity

$$(x_1^2 + y_1^2) \cdot (x_2^2 + y_2^2) = (x_1x_2 - y_1y_2)^2 + (x_1y_2 + x_2y_1)^2.$$

For instance,

$$97 \cdot 30\,449 = (4^2 + 9^2) \cdot (100^2 + 143^2) = 887^2 + 1\,472^2.$$

On the contrary, the prime 7 cannot be written as a sum of two squares. Gauss showed that a positive integer n has a representation as a sum of three squares if and only if $n \not\equiv 7 \bmod 8$. This yields a proof of Lagrange's theorem; however, Gauss' reasoning is rather difficult. It is more surprising that Minkowski's convex body theorem also provides quite a tricky and short proof of Lagrange's theorem, found by Davenport.

Theorem 8.7. *Each positive integer can be written as a sum of four squares.*

Proof. Similar to the norm equation for complex numbers we have the norm equation for quaternions (see Exercise 8.12):

$$
\left(x_1^2 + x_2^2 + x_3^2 + x_4^2\right) \cdot \left(y_1^2 + y_2^2 + y_3^2 + y_4^2\right)
$$
$$
= \left(x_1y_1 + x_2y_2 + x_3y_3 + x_4y_4\right)^2 + \left(-x_1y_2 + x_2y_1 - x_3y_4 + x_4y_3\right)^2
$$
$$
(8.5) \quad + \left(-x_1y_3 + x_2y_4 + x_3y_1 - x_4y_2\right)^2 + \left(-x_1y_4 - x_2y_3 + x_3y_2 + x_4y_1\right)^2 .
$$

In view of this identity it suffices to show the assertion for prime numbers p. Obviously, $2 = 1^2 + 1^2 + 0^2 + 0^2$. Thus we may assume that p is odd (with regard to Theorem 8.6 we could even restrict to $p \equiv 3 \bmod 4$).

We start with the observation that for any given prime p there are integers a and b such that

$$(8.6) \qquad\qquad\qquad a^2 + b^2 + 1 \equiv 0 \bmod p.$$

In fact, there are as many quadratic residues as non-residues modulo p (see Exercise 8.10). Hence, both quantities a^2 and $-(b^2+1)$ run through exactly $\frac{p+1}{2}$ residue classes as a and b run through a full residue system modulo p. Consequently, they must have at least one common residue class, which implies the solvability of (8.6).

With the numbers a, b satisfying (8.6) consider the system of congruences given by

$$(8.7) \qquad\qquad X_3 \equiv aX_1 + bX_2, \quad X_4 \equiv bX_1 - aX_2 \ \bmod p.$$

By Lemma 8.5 this defines a lattice Λ in $\mathbb{R}^4$ with determinant

$$
\det(\Lambda) = \det \begin{pmatrix} 1 & 0 & 0 & 0 \\ 0 & 1 & 0 & 0 \\ a & b & p & 0 \\ b & -a & 0 & p \end{pmatrix} = p^2.
$$

Next, consider the convex body

$$
\mathcal{C} := \{(x_1, \ldots, x_4) \in \mathbb{R}^4 \ : \ x_1^2 + \ldots + x_4^2 < 2p\}.
$$

The volume of a four-dimensional sphere of radius r is $\frac{\pi^2}{2}r^4$ (see (8.8) resp. Exercise 8.15). Since

$$
\mathrm{vol}(\mathcal{C}) = 2\pi^2 p^2 > 16p^2 = 2^4 \det(\Lambda),
$$

by Minkowski's first theorem 8.3, there exists a non-zero lattice point $\mathbf{x} := (x_1, \ldots, x_4) \in \Lambda \cap \mathcal{C}$. In particular, it follows from (8.7) that

$$
x_1^2 + \ldots + x_4^2 \equiv (a^2 + b^2 + 1)(x_1^2 + x_2^2) \equiv 0 \bmod p.
$$

Since $\mathbf{0} \neq \mathbf{x} \in \mathcal{C}$, it follows that the left-hand side is equal to p. This proves the theorem. •

8.5. Applications to linear and quadratic forms

Minkowski's theorem provides also a direct application to the theory of diophantine approximations, namely, another proof of Dirichlet's approximation theorem. Let $\alpha \in (0,1)$ and Q be a positive integer. Consider the set

$$\mathcal{C} := \left\{ (x,y) \in \mathbb{R}^2 : -Q - \frac{1}{2} \le x \le Q + \frac{1}{2}, |x\alpha - y| < \frac{1}{Q} \right\}.$$

This is a symmetric convex body of area

$$(2Q+1)\frac{2}{Q} = 4 + \frac{2}{Q} > 4.$$

Therefore it contains some non-zero integer lattice point, (q,p) say. By symmetry, we may assume that q is positive. The definition of $\mathcal{C}$ now implies $q \le Q$ and

$$\left| \alpha - \frac{p}{q} \right| < \frac{1}{qQ}.$$

This implies the assertion of Theorem 2.1.

In some sense, we have shown above that the form $\alpha X - Y$ takes arbitrarily small values on the lattice $\mathbb{Z}^2$. Our next aim is another interpretation of Minkowski's theorem in the spirit of the application just given. For this purpose we recall some standard notion from linear algebra (which we have already used to some extent).

A **homogeneous linear form** Y_i in variables $X_1, \ldots, X_n$ with coefficients α_{ij} is of the form

$$Y_i = Y_i(X_1, \ldots, X_n) = \alpha_{i1}X_1 + \ldots + \alpha_{in}X_n.$$

Let's consider a collection of such linear forms, $Y_1, \ldots, Y_n$ in $\mathbf{X} := (X_1, \ldots, X_n)$, and assume that the coefficients α_{ij} are real. Then the points in the Y-space corresponding to integral $x_1, \ldots, x_n$ form a lattice Λ (see Section 8.2).

First, suppose that $\mathcal{C}$ is given by the conditions

$$|Y_1(\mathbf{X})| \le \lambda_1, \quad \ldots \quad , |Y_n(\mathbf{X})| \le \lambda_n,$$

where the λ_j are fixed positive real numbers. Then $\mathcal{C}$ is convex, symmetric, and of volume $2^n \lambda_1 \cdot \ldots \cdot \lambda_n$. Hence, an application of Minkowski's theorem 8.3 leads to

Corollary 8.8. *If $Y_1, \ldots, Y_n$ are homogeneous linear forms in $X_1, \ldots, X_n$ with real coefficients α_{ij}, and the numbers $\lambda_1, \ldots, \lambda_n$ are positive with*

$$\lambda_1 \cdot \ldots \cdot \lambda_n \ge \det(\Lambda),$$

then there are integers $x_1, \ldots, x_n$, not all equal to 0, such that

$$|Y_i(\mathbf{x})| = |\alpha_{i1}x_1 + \ldots + \alpha_{in}x_n| \le \lambda_i \qquad for \quad i = 1, \ldots, n.$$

In particular, we can make $|Y_i(\mathbf{x})| \le \det(\Lambda)^{\frac{1}{n}}$.

Next we shall consider quadratic forms. A **quadratic form** Q is a function in unknowns $X_1, \ldots, X_n$ of the form

$$Q = Q(X_1, \ldots, X_n) = \sum_{1 \leq i,j \leq n} \alpha_{ij} X_i X_j$$

having coefficients a_{ij} in some field (or ring); again we suppose the coefficients α_{ij} being real (however, later we shall also deal with other fields). Moreover, without loss of generality we may assume that $\alpha_{ij} = \alpha_{ji}$. A quadratic form Q is said to be **positive definite** if $Q = Q(\mathbf{x})$ takes only positive values for $\mathbf{x} \neq \mathbf{0}$. The **discriminant** of a positive quadratic form Q is defined by

$$\Delta = \det((\alpha_{ij})_{1 \leq i,j \leq n}).$$

Note that the matrix $(\alpha_{ij})_{1 \leq i,j \leq n}$ is symmetric and has positive eigenvalues, so its determinant is positive. It is a well-known fact from linear algebra (principal axis transformation) that any positive definite quadratic form Q can be expressed as

$$Q = Y_1^2 + \ldots + Y_n^2,$$

where the Y_i are homogeneous linear forms in $X_1, \ldots, X_n$ (what we already used in Section 8.1).

For real $\lambda > 0$, define $\mathcal{C}$ by the condition $Q \leq \lambda^2$. It is easily seen that $\mathcal{C}$ is convex, symmetric, and its volume is λ^n times the volume of the n-dimensional unit sphere, that is,

$$(8.8) \qquad J_n := \int \cdots \int_{y_1^2 + \ldots + y_n^2 \leq 1} dy_1 \ldots dy_n = \frac{\pi^{\frac{n}{2}}}{\Gamma\left(\frac{n}{2} + 1\right)}.$$

Applying Minkowski's theorem 8.3 after a short computation gives

Corollary 8.9. *If Q is a positive definite quadratic form in variables $X_1, \ldots, X_n$ with discriminant Δ, then there exist integers $x_1, \ldots, x_n$, not all equal to zero, such that*

$$Q(x_1, \ldots, x_n) \leq 4 J_n^{-\frac{2}{n}} \Delta^{\frac{1}{n}}.$$

Let γ_n be the least constant such that there exists a non-zero lattice point $\mathbf{x} = (x_1, \ldots, x_n) \in \mathbb{Z}^n$ with $Q(\mathbf{x}) \leq \gamma_n \Delta^{\frac{1}{n}}$. Only a few values of γ_n are explicitly known. For instance, we have

$$\gamma_2 = \frac{2}{\sqrt{3}}, \quad \gamma_3 = \sqrt[8]{2}, \quad \text{and} \quad \gamma_4 = \sqrt{2}.$$

In general, the asymptotic bound $\gamma_n < (1 + o(1)) \frac{n}{\pi e}$ is known (see [**36**] and [**85**]).

8.6. The shortest lattice vector problem

Sometimes one not only is interested in knowing that a lattice Λ has a point in a set $\mathcal{C}$ but also wants to have a number of linearly independent points in $\mathcal{C}$. For $\mathcal{C} \subset \mathbb{R}^n$ we define the numbers λ_k for $1 \leq k \leq n$ by

$$\lambda_k = \inf\{\lambda > 0 : \dim(\lambda \mathcal{C} \cap \Lambda) \geq k\}.$$

These numbers are called **successive minima** with respect to Λ and $\mathcal{C}$, which illustrates the fact that λ_k is the lower bound of all $\lambda > 0$ such that $\lambda\mathcal{C} = \{(\lambda x_1, \ldots, \lambda x_n) : (x_1, \ldots, x_n) \in \Lambda\}$ contains k linearly independent lattice points. In particular, λ_1 is the least number such that $\lambda_1\mathcal{C}$ contains a non-zero lattice vector. Clearly,

$$\lambda_1 \leq \ldots \leq \lambda_n,$$

and these successive minima exist if $\mathcal{C}$ is convex, symmetric, and bounded; otherwise it can happen that $\lambda_k = \infty$. Here we have

Theorem 8.10. *Let $\mathcal{C} \subset \mathbb{R}^n$ be convex, symmetric, and bounded with successive minima $\lambda_1, \ldots, \lambda_n$. Then*

$$\frac{2^n}{n!} \leq \lambda_1 \cdot \ldots \cdot \lambda_n \cdot \mathrm{vol}(\mathcal{C}) \leq 2^n.$$

This is Minkowski's second theorem. We leave out the proof of this rather deep theorem (which can be found in [**36**]), but note that both bounds can be attained. Now we switch to an interesting related question.

Some bases are better than others! In several applications one wants to deal with a basis having comparably small lattice vectors. However, given a lattice, algorithmically it seems rather difficult to find a shortest non-zero lattice vector. It is conjectured that there exists no *efficient* algorithm to solve this problem if the rank of the lattice is *large*.

First, recall that the Euclidean length of a vector $\mathbf{z} \in \mathbb{R}^n$ is defined by

$$|\mathbf{z}| = \sqrt{\mathbf{z}^t\mathbf{z}} = \sqrt{z_1^2 + \ldots + z_n^2},$$

where $z_1, \ldots, z_n$ are the coordinates of $\mathbf{z}$. Now let Λ be a lattice of rank r and let $\mathbf{z}_1 \in \Lambda$ be a shortest non-zero vector (of course, we cannot expect $\mathbf{z}_1$ to be unique), $\mathbf{z}_2 \in \Lambda$ be a shortest vector independent of $\mathbf{z}_1$, etc., and, finally, let $\mathbf{z}_r \in \Lambda$ be a shortest vector independent of $\mathbf{z}_1, \ldots, \mathbf{z}_{r-1}$. Then an application of Minkowski's second theorem 8.10 yields the estimate

$$\prod_{j=1}^{r} |\mathbf{z}_j| \leq \frac{2^r}{J_r} \det(\Lambda),$$

where J_r is the volume of the unit ball in $\mathbb{R}^r$ (see (8.8)). Since the unit ball of dimension r contains the hypercube whose vertices have coordinates $\frac{\pm 1}{\sqrt{r}}$, we have $J_r \geq (\frac{2}{\sqrt{r}})^r$. Thus

(8.9) $$|\mathbf{z}_1| \leq \sqrt{r}\det(\Lambda)^{\frac{1}{r}}.$$

A fundamental problem in lattice theory is, given a basis of a lattice Λ, to determine a non-zero vector in Λ with minimal length. This so-called **shortest lattice vector problem** has several important consequences (some of them we will meet in the sequel).

In dimension two there is a rather simple way to find shortest vectors in lattices due to Gauss. His algorithm shares some similarities with the Euclidean algorithm for computing the greatest common divisor. Let Λ

be a lattice in $\mathbb{R}^2$ with basis $\mathbf{z}_1, \mathbf{z}_2$. Assume that $|\mathbf{z}_1| \le |\mathbf{z}_2|$. The simple algorithm works as follows. Choose an integer k such that

$$(8.10) \qquad -\frac{1}{2}\mathbf{z}_1^{\mathrm{t}}\mathbf{z}_1 < (\mathbf{z}_2 - k\mathbf{z}_1)^{\mathrm{t}}\mathbf{z}_1 \le \frac{1}{2}\mathbf{z}_1^{\mathrm{t}}\mathbf{z}_1.$$

In the next step replace $\mathbf{z}_2$ with $\mathbf{z}_2 - k\mathbf{z}_1$. If $|\mathbf{z}_1| \le |\mathbf{z}_2|$, we are done, and otherwise interchange $\mathbf{z}_1$ and $\mathbf{z}_2$ and start again. This process is called **Gauss reduction**. It is not too difficult to prove that it indeed yields the shortest non-zero vector of the lattice. In fact, in every loop the vector $\mathbf{z}_1$ is smaller than before, and since the lattice is discrete, the algorithm terminates.

Much harder is the situation in higher dimensions. Then the best results are based on a clever lattice reduction procedure. However, before we are in the position to study the shortest lattice vector problem in higher dimensions we have to recall some facts from linear algebra.

8.7. Gram–Schmidt and consequences

Recall the **Gram–Schmidt orthogonalization** procedure.   Let $\mathbf{z}_1, \dots, \mathbf{z}_r$ be (not necessarily independent) vectors in $\mathbb{R}^n$. Then define recursively the vectors $\mathbf{z}_1^*, \dots, \mathbf{z}_r^*$ by $\mathbf{z}_1^* = \mathbf{z}_1$ and, for $2 \le i \le r$,

$$\mathbf{z}_i^* = \mathbf{z}_i - \sum_{\substack{j < i \\ \mathbf{z}_j^* \ne 0}} \frac{\mathbf{z}_i^{\mathrm{t}}\mathbf{z}_j^*}{|\mathbf{z}_j^*|^2}\mathbf{z}_j^*.$$

Among the vectors $\mathbf{z}_i^*$ there may be some zero vectors since we did not assume the vectors $\mathbf{z}_i$ to be independent, but they provide an orthogonal basis of the space spanned by the vectors $\mathbf{z}_i$, that is,

$$\langle \mathbf{z}_1, \dots, \mathbf{z}_k \rangle = \langle \mathbf{z}_1^*, \dots, \mathbf{z}_k^* \rangle$$

for all $k \le r$, and $(\mathbf{z}_i^*)^{\mathrm{t}}\mathbf{z}_j^* = 0$ for $i \ne j$. Then, by the multiplication theorem for determinants,

$$\det\left((\mathbf{z}_i^*)^{\mathrm{t}}\mathbf{z}_j^*\right)_{1 \le i,j \le r} = \det(\mathbf{z}_i^{\mathrm{t}}\mathbf{z}_j)_{1 \le i,j \le r}.$$

We define the **Gram–Schmidt coefficients** by

$$\mu_{i,j} = \mu_{i,j}(\mathbf{z}_i, \mathbf{z}_j^*) = \begin{cases} \dfrac{\mathbf{z}_i^{\mathrm{t}}\mathbf{z}_j^*}{|\mathbf{z}_j^*|^2} & \text{if } \mathbf{z}_j^* \ne \mathbf{0}, \\ 0 & \text{if } \mathbf{z}_j^* = \mathbf{0}. \end{cases}$$

A short computation shows

$$\mathbf{z}_i^{\mathrm{t}}\mathbf{z}_i = (\mathbf{z}_i^*)^{\mathrm{t}}\mathbf{z}_i^* + 2(\mathbf{z}_i^*)^{\mathrm{t}}\sum_{j<i}\mu_{i,j}\mathbf{z}_j^* + \sum_{j<i}\mu_{i,j}(\mathbf{z}_j^*)^{\mathrm{t}}\sum_{j<i}\mu_{i,j}\mathbf{z}_j^*,$$

resp.

$$|\mathbf{z}_i|^2 = |\mathbf{z}_i^*|^2 + \sum_{1 \le j < i} \mu_{i,j}^2 |\mathbf{z}_j^*|^2.$$

Thus we have $|\mathbf{z}_i^*| \le |\mathbf{z}_i|$ for $1 \le i \le r$. Now let $\mathbf{z}_1, \dots, \mathbf{z}_r$ be a basis of a lattice Λ and let $\mathbf{z}_1^*, \dots, \mathbf{z}_r^*$ be its Gram–Schmidt orthogonalization. Then, as an immediate consequence of the last estimate, we obtain Hadamard's inequality:

$$(8.11) \qquad\qquad \det(\Lambda) = \prod_{i=1}^{r} |\mathbf{z}_i^*| \le \prod_{i=1}^{r} |\mathbf{z}_i|;$$

note that equality holds if and only if the set of the $\mathbf{z}_i$ forms an orthogonal basis.

We return to the shortest non-zero lattice vector problem, but now we are interested in higher dimensions. Our first aim is to prove a lower bound for the length of a shortest non-zero lattice vector.

Lemma 8.11. *With the notation from above, for any* $0 \ne \mathbf{z} \in \Lambda$,

$$|\mathbf{z}| \ge \min\{|\mathbf{z}_1^*|, \ldots, |\mathbf{z}_r^*|\}.$$

Proof. We have

$$\mathbf{z} = \sum_{i=1}^{k} a_i \mathbf{z}_i,$$

where $k \le r$ and all a_i are non-zero integers. By the Gram–Schmidt orthogonalization we find alternatively

$$\mathbf{z} = \sum_{i=1}^{k} \ell_i \mathbf{z}_i^*,$$

where the coefficient ℓ_k is a non-zero integer (since we may assume that $\mu_{k,k} = 1$). Hence,

$$|\mathbf{z}|^2 = \sum_{i=1}^{k} \ell_i^2 \cdot |\mathbf{z}_i^*|^2 \ge \ell_k^2 \cdot |\mathbf{z}_k^*|^2 \ge |\mathbf{z}_k^*|^2,$$

which proves the lemma. •

Thus, the length of a shortest non-zero lattice vector is at least as long as the shortest vector of the Gram–Schmidt orthogonalization. However, the shortest vector in the Gram–Schmidt orthogonalization is in general not a lattice vector. Nevertheless, this observation is rather useful for our later purpose. It suggests to search for an *nearly* orthogonal basis which lies in the lattice, where *nearly* orthogonal means that we take the Gram–Schmidt orthogonalization and round the Gram–Schmidt coefficients to the nearest integer.

8.8. Lattice reduction in higher dimensions

Now we present the so-called **LLL-lattice reduction**, named after the initial letters of the inventors A.K. Lenstra, H.W. Lenstra & Lovász [**102**], which realizes the idea of searching among the lattice vectors close to the Gram–Schmidt orthogonalization in order to solve the shortest lattice vector problem. Here and in the sequel we use the notation introduced in the previous section; in particular, we denote the vectors resulting from the

Gram–Schmidt orthogonalization by an asterisk $*$ and write $\mu_{i,j}$ for the associated Gram–Schmidt coefficients.

A basis $\mathbf{z}_1, \ldots, \mathbf{z}_r$ of a lattice Λ is said to be **LLL-reduced** if $|\mu_{i,j}| \leq \frac{1}{2}$ and

$$(8.12) \qquad |\mathbf{z}_i^* + \mu_{i,i-1}\mathbf{z}_{i-1}^*|^2 \geq \frac{3}{4}|\mathbf{z}_{i-1}^*|^2$$

for all $1 < i \leq r$. Note that the vector $\mathbf{z}_i^* + \mu_{i,i-1}\mathbf{z}_{i-1}^*$ is the projection of $\mathbf{z}_i$ on the orthogonal complement of $\mathbf{z}_1, \ldots, \mathbf{z}_{i-2}$. Condition (8.12) can also be rewritten as

$$(8.13) \qquad |\mathbf{z}_i^*|^2 \geq \left(\frac{3}{4} - \mu_{i,i-1}^2 \right) |\mathbf{z}_{i-1}|^2.$$

With this definition we shall now prove some nice properties of LLL-reduced bases.

Theorem 8.12. *Let* $\mathbf{z}_1, \ldots, \mathbf{z}_r$ *be an* LLL-*reduced basis of a lattice* Λ. *Then*

$$(8.14) \qquad \det(\Lambda) \leq \prod_{i=1}^r |\mathbf{z}_i| \leq 2^{\frac{r(r-1)}{4}} \det(\Lambda),$$

and

$$(8.15) \qquad |\mathbf{z}_1| \leq 2^{\frac{r-1}{4}} \det(\Lambda)^{\frac{1}{r}}.$$

Moreover, for any set of linearly independent vectors $\mathbf{x}_1, \ldots, \mathbf{x}_t \in \Lambda$,

$$(8.16) \qquad |\mathbf{z}_j| \leq 2^{\frac{r-1}{2}} \max\{|\mathbf{x}_1|, \ldots, |\mathbf{x}_t|\}$$

for $1 \leq j \leq t$.

Proof. By definition we have

$$
\begin{aligned}
|\mathbf{z}_i|^2 &= |\mathbf{z}_i^*|^2 + \mu_{i,i-1}^2|\mathbf{z}_{i-1}^*|^2 + \ldots + \mu_{i,1}^2|\mathbf{z}_1^*|^2 \\
&\leq |\mathbf{z}_i^*|^2 + \frac{1}{4}\left(|\mathbf{z}_{i-1}^*|^2 + \ldots + |\mathbf{z}_1^*|^2 \right).
\end{aligned}
$$

Since $|\mathbf{z}_j^*|^2 \geq \frac{1}{2}|\mathbf{z}_{j-1}^*|^2$ by (8.13), we have

$$(8.17) \qquad |\mathbf{z}_j^*|^2 \leq 2^{i-j}|\mathbf{z}_i^*|^2$$

for $j \leq i$. In addition, with the next to last inequality we get for any i the estimate

$$(8.18) \qquad |\mathbf{z}_i|^2 \leq \left(1 + \frac{1}{4}(2 + \ldots + 2^{i-1}) \right) |\mathbf{z}_i^*|^2 \leq 2^{i-1}|\mathbf{z}_i^*|^2.$$

This and Hadamard's inequality (8.11) imply

$$\det(\Lambda) = \prod_{i=1}^r |\mathbf{z}_i^*| \leq \prod_{i=1}^r |\mathbf{z}_i| \leq 2^{\frac{r(r-1)}{4}} \prod_{i=1}^r |\mathbf{z}_i^*| = 2^{\frac{r(r-1)}{4}} \det(\Lambda).$$

This is inequality (8.14).

To prove the second assertion we deduce from (8.17) and (8.18) that

$$(8.19) \qquad |\mathbf{z}_j| \leq 2^{\frac{j-1}{2}}|\mathbf{z}_j^*| \leq 2^{\frac{j-1}{2}+\frac{i-j}{2}}|\mathbf{z}_i^*| = 2^{\frac{i-1}{2}}|\mathbf{z}_i^*|$$

whenever $1 \leq j < i \leq r$. Thus

$$|\mathbf{z}_1|^r \leq \prod_{i=1}^{r} 2^{\frac{i-1}{2}} |\mathbf{z}_i^*| \leq 2^{\frac{r(r-1)}{4}} \prod_{i=1}^{r} |\mathbf{z}_i^*| = 2^{\frac{r(r-1)}{4}} \det(\Lambda).$$

Taking the rth root this yields (8.15).

It remains to prove (8.16). We can choose a minimal integer k such that $\mathbf{x}_1, \ldots, \mathbf{x}_t$ lie in the span of $\mathbf{z}_1, \ldots, \mathbf{z}_k$. Now suppose that

$$\mathbf{x}_i = \sum_{j=1}^{k} a_{ij} \mathbf{z}_j = \sum_{j=1}^{k} b_{ij} \mathbf{z}_j^*$$

with some integers a_{ij} and some real numbers b_{ij}. Obviously, we can find an index i such that $a_{ik} \neq 0$. Note that $a_{ik} = b_{ik}$ (by the definition of $\mathbf{z}_j^*$). Since the $\mathbf{x}_1, \ldots, \mathbf{x}_t$ are independent, k cannot be smaller than t. Now it is easily seen that

$$|\mathbf{x}_i|^2 \geq b_{ik}^2 |\mathbf{z}_k^*|^2 = a_{ik}^2 |\mathbf{z}_k^*|^2 \geq |\mathbf{z}_k^*|^2.$$

In view of (8.19) it thus follows that

$$|\mathbf{z}_j|^2 \leq 2^{k-1} b_{ik}^2 |\mathbf{z}_k^*|^2 \leq 2^{k-1} |\mathbf{x}_i|^2 \leq 2^{r-1} \max\{|\mathbf{x}_1|^2, \ldots, |\mathbf{x}_t|^2\}$$

for $j < k$. Since $k \geq t$, this inequality holds, in particular, for $j \leq t$, which gives (8.16). The theorem is proved. •

In view of (8.16) we see that the vector $\mathbf{z}_1$ in an LLL-reduced basis is at most by a factor of $2^{\frac{r-1}{2}}$ larger than a shortest non-zero lattice vector. In fact, it is often a shortest one, and if not, then in practice it can be often used as a substitute for a shortest lattice vector.

8.9. The LLL-algorithm

The big question is how to get such a nice LLL-reduced basis. The answer is given by the following LLL-reduction algorithm:

Suppose that the vectors $\mathbf{z}_1, \ldots, \mathbf{z}_{k-1}$ are LLL-reduced (which is true if $k = 2$).

1. Replace $\mathbf{z}_k$ by $\mathbf{z}_k - \sum_{j<k} a_j \mathbf{z}_j$ with integers a_j such that the associated Gram–Schmid constants satisfy $|\mu_{k,j}| \leq \frac{1}{2}$ for all $j < k$.
2. If (8.13) holds for $i = k$, that is,

$$|\mathbf{z}_k^*|^2 \geq \left(\frac{3}{4} - \mu_{k,k-1}^2 \right) |\mathbf{z}_{k-1}|^2,$$

then $\mathbf{z}_1, \ldots, \mathbf{z}_k$ is LLL-reduced. Otherwise (if (8.13) does not hold for $i = k$), interchange $\mathbf{z}_k$ and $\mathbf{z}_{k-1}$ and apply step 1 to $\mathbf{z}_1, \ldots, \mathbf{z}_{k-1}$.

First, we have to show that this algorithm terminates. For this aim define

$$\delta_i = \det((\mathbf{z}_k^t \mathbf{z}_\ell)_{1 \leq k, \ell \leq i})$$

for $1 \leq i \leq r$. Note that $\delta_r = \det(\Lambda)^2$. Further, put

(8.20) $$\Delta = \prod_{i=1}^{r-1} \delta_i.$$

During the LLL-algorithm the value of Δ changes only when two vectors $\mathbf{z}_k$ and $\mathbf{z}_{k-1}$ are interchanged. More precisely, in this case only the value δ_{k-1} changes. It is easy to compute that the new value of δ_{k-1} is then given by

$$\delta'_{k-1} := \delta_{k-1} \frac{|\mathbf{z}_k^* + \mu_{k,k-1}\mathbf{z}_{k-1}^*|^2}{|\mathbf{z}_{k-1}^*|^2}.$$

Interchanging $\mathbf{z}_k$ and $\mathbf{z}_{k-1}$ means that condition (8.13) was not fulfilled. It thus follows from (8.12) (with $i = k$) that $\delta'_{k-1} \leq \frac{3}{4}\delta_{k-1}$ and so Δ gets reduced by a factor $\leq \frac{3}{4}$. On the contrary, Δ is bounded below by a quantity which depends only on the lattice and not on the chosen basis. This follows from an application of Minkowski's theorem 8.10 as we shall show now.

Let λ be the length of a shortest non-zero vector in Λ. It follows from Theorem 8.10, applied to the lattice generated by $\mathbf{z}_1, \ldots, \mathbf{z}_i$ in the form of the inequality (8.9), that

$$\lambda \leq \sqrt{i}\delta_i^{\frac{1}{2i}}.$$

This and (8.20) yield after a short computation

$$\Delta \geq \prod_{i=1}^{r-1} \left(\frac{\lambda^2}{i}\right)^i \geq \left(\frac{\lambda^2}{r}\right)^{\frac{r(r-1)}{2}}.$$

Hence Δ is bounded below by an absolute constant, depending only on the lattice and not on the basis, and on the other side, every loop in the LLL-reduction algorithm reduces the initial value of Δ by a factor of $\leq \frac{3}{4}$. Thus, it follows that the algorithm must terminate. There is a simpler argument if Λ is an integer lattice. In this case the numbers δ_i are integers, and so is Δ. Consequently, only finitely many loops are possible.

Given a lattice Λ with basis $\mathbf{z}_1, \ldots, \mathbf{z}_r$, an inductive application of the LLL-reduction algorithm constructs an LLL-reduced basis in polynomial time (in the input data), and, in particular, a lattice vector whose length is at most a factor of $2^{\frac{r-1}{2}}$ larger than the shortest possible length. To be more precise, the algorithm stops after at most

(8.21) $$O\left(r^2 \log \frac{M\sqrt{r}}{\lambda}\right)$$

interchanges of $\mathbf{z}_k$ and $\mathbf{z}_{k-1}$, where $M := \max\{|\mathbf{z}_1|, \ldots, |\mathbf{z}_r|\}$. This was used by A.K. Lenstra, H.W. Lenstra & Lovász [102] in order to give the first algorithm for factoring univariate polynomials over the rationals in polynomial time (a breakthrough comparable with the primality test by Agrawal et al.). In the meantime LLL has found several further applications, the most spectacular in cryptography. In the following section we shall investigate its power with respect to a diophantine question. For further details on LLL-reduction and hints for an implementation we refer to the excellent sources Lovász [104] and Beukers [20].

8.10. The small integer problem

We conclude with an application to a basic problem in the theory of dio-phantine approximations. We want to find explicitly $\mathbb{Z}$-linear combinations of a given set of real numbers which have *small* values. More precisely, given rational numbers $\alpha_1, \ldots, \alpha_m$, and positive real numbers ε and Q, find integers $q_1, \ldots, q_m$, not all equal to zero, such that

$$(8.22) \qquad\qquad |q_1\alpha_1 + \ldots + q_m\alpha_m| \leq \varepsilon,$$

and $|q_i| \leq Q$ for $1 \leq i \leq m$. This is the so-called **small integer problem**. Of course, we are interested in solutions for *large* m.

Applying Corollary 8.8 with

$$Y_1 = X_1 \,, \ \ldots \,, \ Y_m = X_m, \quad \text{and} \quad Y_{m+1} = \alpha_1 X_1 + \ldots + \alpha_m X_m,$$

and $\lambda_1 = \ldots = \lambda_m = Q, \lambda_{m+1} = \varepsilon$, we can succeed to find a solution provided that $Q \geq \varepsilon^{-\frac{1}{m}}$, even in the more general case when we allow the α_i to be arbitrary real numbers. However, no efficient algorithm is known to produce a solution for this range for Q. We shall briefly sketch how the LLL-algorithm can be used to find *explicitly* and *quickly* a solution to the small integer problem if

$$(8.23) \qquad\qquad Q \geq 2^{\frac{m}{4}} \varepsilon^{-\frac{1}{m-1}}.$$

For this purpose we relax the problem slightly by assuming that $\alpha_1 = 1$ and q_1 may take any integer value. Further we suppose that $0 < \varepsilon < 1$. We consider the matrix

$$\mathcal{M} = \begin{pmatrix} 1 & \alpha_2 & \alpha_3 & \ldots & & \ldots & \alpha_m \\ 0 & \frac{\varepsilon}{Q} & 0 & \ldots & & \ldots & 0 \\ \ldots & & & \ldots & & & \ldots \\ 0 & \ldots & & & \ldots & 0 & \frac{\varepsilon}{Q} & 0 \\ 0 & \ldots & & & \ldots & 0 & \frac{\varepsilon}{Q} \end{pmatrix}$$

and the lattice Λ generated by its column vectors. Then any lattice vector $\mathbf{z} \in \Lambda$ can be written as $\mathbf{z} = \mathcal{M}\mathbf{q}$, where $\mathbf{q} = (q_1, \ldots, q_m) \in \mathbb{Z}^m$. Now assume that $\mathbf{0} \neq |\mathbf{z}| \leq \varepsilon$. Then the entries of the vector $\mathbf{q}$ solve (8.22) and satisfy $|q_i| \leq Q$ for $2 \leq i \leq m$. In view of Theorem 8.12 and the time estimate (8.21) by the choice $\varepsilon = 2^{\frac{m+1}{4}} \det(\mathcal{M})^{\frac{1}{m}}$, which corresponds to (8.23), the LLL-algorithm can find such a vector $\mathbf{q}$ in polynomial running time. With a bit more effort one can also force q_1 to satisfy the desired estimate and also the assumption $\alpha_1 = 1$ can be removed.

Notes on sphere packings

An application we have not mentioned so far, but which is a must to be mentioned, are sphere packings. For a compact subset $\mathcal{C} \subset \mathbb{R}^n$ we define the **lattice constant** by

$$\Delta(\mathcal{C}) = \min\{\det(\Lambda) : \Lambda \cap \mathcal{C} \neq \emptyset\},$$

where the minimum is taken over all full lattices in $\mathbb{R}^n$. As shown by Mahler, this quantity exists. One can show that the number

$$2^{-n} \frac{\text{vol}(\mathcal{C})}{\Delta(\mathcal{C})}$$

is equal to the maximum **density of a lattice packing** of $\mathcal{C}$, that is, the proportion of $\mathbb{R}^n$ which can be filled by the set $\mathcal{C} + \Lambda$ for some lattice Λ without overlaps. If $\mathcal{C}$ is star-shaped, then this proportion is at least $2^{1-n}\zeta(n)$. In many applications one wants to fill as much of the space as possible. In the special case of spheres $\mathcal{C}$ we thus enter these questions: *what are the optimal sphere lattice packings* and *is the optimal sphere packing always a lattice packing?* The case of dimension two is classic mathematics. In dimension three, the last question is known as **Kepler conjecture**, and was recently answered by Hales. He showed that both the hexagonal lattice packing and the face centered cubic lattice packing have the optimal packing density, namely, $\frac{\pi}{\sqrt{18}} \approx 0.74\ldots$, among all sphere packings. Thus, the intuitive way greengrocers stack oranges to pyramids is optimal (this is mathematically not rigorous since there are only finitely many oranges in our world). Long before Hales it was known that these two lattices are optimal among the class of sphere packings which rely on lattices — in fact it was Gauss who proved this remarkable result — but Hales found a way to tackle *irregular* sphere packings, based on massive computations; his proof has not totally been checked despite the fact that 12 reviewers of a renowned journal worked on it for more than four years. We refer to [**72**] for an impression on Hales' proof, and to [**41**] for more information on this fascinating piece of mathematics in general.

Exercises

8.1. *Show that the assertion of Minkowski's convex body theorem 8.1 remains true if $\mathcal{C}$ is not bounded. Further, discuss what can be said if the strict inequality on the volume of $\mathcal{C}$ is replaced by $\text{vol}(\mathcal{C}) \geq 2^n$.*

8.2. Suppose that a given ellipse is described by the equation

$$\left(\frac{X}{\alpha}\right)^2 + \left(\frac{Y}{\beta}\right)^2 = 1,$$

where α, β are positive real numbers.
Prove that the area of the ellipse is equal to $\pi\alpha\beta$.

8.3.* *If a, b, c are integers with $a > 0$ and $\Delta := ac - b^2 > 0$, show that there exists at least one integer solution x, y, not both equal to zero, to*

$$aX^2 + 2bXY + c^2Y \leq \frac{4}{\pi}\sqrt{\Delta}.$$

8.4.* *Popeye is standing in the middle of a regular forest $\mathcal{F}$, which has the shape of a disk of diameter a centered at the origin. Trees of diameter b grow at each lattice point except for the origin where Popeye stands. For which values of $a, b > 0$ can Popeye look outside $\mathcal{F}$?*

Hint: Suppose that Popeye can look outside along some line ℓ. This defines some strip passing through the origin which contains no lattice point. What does Minkowski's theorem tell about such a strip?

8.5. *Extend Minkowski's convex body theorem as follows: if $\mathcal{C} \subset \mathbb{R}^n$ is symmetric (around the origin), convex, and bounded with $\mathrm{vol}(\mathcal{C}) > k2^n$, then $\mathcal{C}$ contains at least $2k$ different lattice points.*

8.6. *How many integral lattice points are inside a circle of a given radius? Prove that the number is approximately equal to the area of the circle. More precisely, let $\mathcal{C}$ be the circle of radius r centered at the origin; then show that*

$$\#\{(x,y) \in \mathbb{Z}^2 : x^2 + y^2 \leq r^2\} = \pi r^2 + O(r).$$

What can be said about arbitrary lattices?

The above estimate is from Gauss. The so-called **circle problem** asks for the best possible error term; it is conjectured that the true error term is of size $r^{\frac{1}{2}+\varepsilon}$. The present best upper bound is $\ll r^{131/208=0.6298...}$ due to Huxley [**87**].

8.7. *Show that $\binom{-4}{9}, \binom{2}{1}$ form a basis of $\mathbb{Z}^2$. Is $\binom{2}{12}, \binom{4}{13}$ also a basis? If so find a linear mapping that transforms one basis into the other.*

8.8.* *Given a lattice Λ with basis $\mathbf{z_1}, \ldots, \mathbf{z_n}$, show that*

$$\det(\Lambda) = \mathrm{vol}(\{\lambda_1 \mathbf{z_1} + \ldots + \lambda_n \mathbf{z_n} : 0 \leq \lambda_1, \ldots, \lambda_n \leq 1\}).$$

8.9. *Prove that integers $n \equiv 7 \bmod 8$ are not representable as sums of three integer squares.*

Hint: Apply infinite descent to integers of the form $n = 4^m(8k + 7)$.

8.10. *For an odd prime p, show that there are $\frac{p-1}{2}$ quadratic residues and as many non-residues $\bmod\, p$.*

Hint: Consider the values a^2 for $a = 1, \ldots, \frac{p-1}{2}$ and show further that $x^2 \equiv (p-x)^2 \bmod p$. (0 is not a quadratic residue!)

8.11.* *i) Prove that every prime number $p \equiv 1 \bmod 6$ can be written as $p = a^2 + 3b^2$, where a and b are positive integers.*
ii) Show that the set of numbers of the form $a^2 + 3b^2$ is closed with respect to multiplication.

8.12. Quaternions were discovered by Hamilton in the nineteenth century. In analogy to complex numbers, a **quaternion** (or **hyper complex number**) can be written as a linear combination $a + b \cdot \mathbf{i} + c \cdot \mathbf{j} + d \cdot \mathbf{k}$, where $\mathbf{i}, \mathbf{j}, \mathbf{k}$ are non-real quantities satisfying

$$\mathbf{i}^2 = \mathbf{j}^2 = \mathbf{k}^2 = \mathbf{ijk} = -1,$$

and a, b, c, d are real numbers. The set of quaternions is denoted by $\mathbb{H}$ and has quite a lot of structure (see [**51**]).

i) *Show that* $\mathbb{H}$ *is a non-commutative ring with subring* $\mathbb{C}$.

ii) *Verify the identity*

$$(a + b \cdot \mathbf{i} + c \cdot \mathbf{j} + d \cdot \mathbf{k})(A + B \cdot \mathbf{i} + C \cdot \mathbf{j} + D \cdot \mathbf{k})$$
$$= (aA - bB - cC - dD) + (aB + bA + cD - dC) \cdot \mathbf{i}$$
$$+ (aC - bD + cA + dB) \cdot \mathbf{j} + (aD + bC - cB + dA) \cdot \mathbf{k},$$

and derive formula (8.5).

iii) *Find representations of the numbers* $2004, 2005, 2006$ *as sums of four squares.*

8.13. *Prove Exercise 2.6 on simultaneous approximation by applying Corollary 8.8.*

8.14.* *In the notation from Section 8.5, prove the existence of* $0 \neq \mathbf{x} = (x_1, \ldots, x_n) \in \mathbb{Z}^n$ *such that*

$$|Y_1(\mathbf{x})| + \ldots + |Y_n(\mathbf{x})| \leq (n! \det(\Lambda))^{\frac{1}{n}},$$

and

$$|Y_1(\mathbf{x}) \cdot \ldots \cdot Y_n(\mathbf{x})| \leq n^{-n} n! \det(\Lambda).$$

Hint: In view of the arithmetic–geometric mean inequality it suffices to prove only one of the above assertions.

8.15.* *i) For the volume* $J_n(t)$ *of the* n-*dimensional sphere of radius* $\sqrt{t}$, *prove the formula*

$$J_n(t) = \beta_n t^{\frac{n}{2}},$$

where β_n *is a positive real number, not depending on* t.

Hint: Use induction on n to prove

$$
\begin{aligned}
J_{n+1}(t) &= \int \cdots \int_{\{(x_1, \ldots, x_{n+1}) : x_1^2 + \ldots + x_{n+1}^2 \leq t\}} dx_1 \ldots dx_{n+1} \\
&= \int_{-\sqrt{t}}^{\sqrt{t}} J_n(t - x^2) \, dx = 2\beta_n \int_0^{\sqrt{t}} (t - x^2)^{\frac{n}{2}} \, dx \\
&= 2\beta_n t^{\frac{n+1}{2}} \int_0^{\frac{\pi}{2}} (\cos(w))^{n+1} \, dw.
\end{aligned}
$$

ii) *Deduce from i) the recursion formula*

$$\beta_{n+1} = 2\beta_n \mathcal{I}_{n+1}, \qquad where \quad \mathcal{I}_n := \int_0^{\frac{\pi}{2}} (\cos(w))^n \, dw,$$

and conclude

$$\beta_{2k} = \frac{\pi^k}{k!} \quad and \quad \beta_{2k+1} = \frac{2^{2k+1} k! \pi^k}{(2k+1)!} \qquad for \quad k = 1, 2, \ldots.$$

iii) *Unify both formulae by means of the gamma function and deduce formula (8.8).*

Hint: Recall the functional equation of the gamma function $\Gamma(z+1) = z\Gamma(z)$, and that $\Gamma(n+1) = n!$ for $n \in \mathbb{N}$.

iv) Prove **Wallis' formula.**

$$\lim_{n\to\infty} \frac{1}{\sqrt{n}} \frac{2\cdot 4\cdot\ldots\cdot(2n)}{1\cdot 3\cdot\ldots\cdot(2n-1)} = \sqrt{\pi}.$$

8.16. *Using Stirling's formula,*

$$\Gamma(z+1) = (\sqrt{2\pi z} + o(1))\left(\frac{z}{e}\right)^z,$$

valid for $z \in \mathbb{C}$ with positive real part (and some more z), deduce from (8.8) an estimate for the volume of the n-dimensional unit sphere J_n. Show that

$$\lim_{n\to\infty} \mathrm{vol}(J_n) = 0.$$

8.17. *Give a rigorous proof that the Gauss reduction algorithm (8.10) produces a shortest non-zero lattice vector in dimension two.*

8.18. *Prove all comments on the Gram–Schmidt orthogonalization from Section 8.7. Compute the Gram–Schmidt orthogonalization of $\binom{12}{2}, \binom{13}{4}$ and make a picture.*

8.19.* *Let Λ be a lattice with basis $\mathbf{z_1}, \ldots, \mathbf{z_r}$. Prove that the length of a non-zero vector in Λ is bounded below by the smallest eigenvalue of the Gram-matrix associated with the $\mathbf{z_j}$.*

8.20. *Prove estimate (8.21) for the number of swaps in the LLL-algorithm.*

8.21. *Using the LLL-algorithm find (by hand computation) an LLL-reduced basis for the lattice generated by the vectors*

$$\mathbf{z_1} = \begin{pmatrix} 1 \\ 1 \\ 1 \end{pmatrix}, \quad \mathbf{z_2} = \begin{pmatrix} -1 \\ 0 \\ 2 \end{pmatrix}, \quad \mathbf{z_3} = \begin{pmatrix} 3 \\ 5 \\ 6 \end{pmatrix}.$$

8.22.* *Implement the LLL-algorithm and return to the last exercise.*

8.23.* *Find a uniform bound for all coefficients q_i in the small integer problem from Section 8.10 by investigating the matrix*

$$\mathcal{M}' = \begin{pmatrix} \alpha_1 & \alpha_2 & \ldots & & \alpha_m \\ \frac{\varepsilon}{Q} & 0 & \ldots & & 0 \\ 0 & \frac{\varepsilon}{Q} & 0 & & \ldots \\ & & \ldots & & \\ \ldots & & & \frac{\varepsilon}{Q} & 0 \\ 0 & & \ldots & 0 & \frac{\varepsilon}{Q} \end{pmatrix}.$$

Hint: Note that the lattice generated by $\mathcal{M}'$ is not full-dimensional.

CHAPTER 9

Transcendental numbers

Algebraic numbers are, by definition, roots of polynomials with integer coefficients. It is a priori not clear whether there exist numbers which are not algebraic. Such *hypothetical* numbers are called *transcendental* and this name dates at least back to Leibniz who wrote in 1704 *omnem rationem transcendunt* (which is Latin for *transcending everything rational*). As we shall show, transcendental numbers exist and, indeed, if we choose randomly a real number, then very likely this number is transcendental. The long-standing problem to give explicit examples of transcendental numbers was solved by Liouville. However, his examples could not prove the transcendence of such important numbers as e or π. The proofs of their transcendence, due to Hermite and Lindemann, are highlights of classical mathematics and the main theme of this chapter.

9.1. Algebraic vs. transcendental

A complex number α is said to be **algebraic** if there exists a not identically vanishing polynomial with integer coefficients and root α. For instance, all rational numbers and all quadratic irrationals are algebraic. Another example is

$$\alpha = \sqrt{2 - \sqrt{3}}$$

since $2 - \alpha^2 = \sqrt{3}$ resp. $(2 - \alpha^2)^2 = 3$. Thus the polynomial

$$X^4 - 4X^2 + 1$$

has the zero α; moreover, the latter polynomial is the minimal polynomial of α, i.e., the unique polynomial with coprime integer coefficients of least degree and root α (for this and more see Appendix A.7). The degree of an algebraic number, i.e., the degree of its minimal polynomial, gives a notion of size which will turn out to be important in our later studies.

The algebraic number α above has a representation by **radicals**, i.e., by repeated root extractions and rational operations on the coefficients of the related minimal polynomial. Not all algebraic numbers are of this simple shape. We know from school how to find the roots of a quadratic polynomial, and for polynomial equations of degree three and four we have formulae due to dal Ferrar, Tartaglia, and Cardano from the Middle Ages. All such roots obey a representation by radicals. However, Abel proved that the quintic is not solvable by repeated root extractions and rational operations on the coefficients. Thus we cannot expect that algebraic numbers are in general of such a simple form as above. By Galois theory we can find polynomials with integer coefficients of a degree greater than or equal to five having zeros (which are algebraic numbers by definition) that cannot be expressed by radicals, e.g.,

$$X^5 - 6X + 2.$$

For this amazing theory we refer to [**99**].

A complex number α is called **transcendental** if it is not algebraic, i.e., there exists no non-zero polynomial with integer coefficients and root α (of course, here we also can allow polynomials with rational coefficients). At first glance, it is not clear whether transcendental numbers exist. As we shall show now, they do, and the easiest proof of their existence relies on Cantor's notion of countability.

The **height** $\mathrm{H}(P)$ of a polynomial P with integral coefficients is defined as the maximum of all coefficients of P in absolute value. Let D, H be positive real numbers. Then it is easily seen that there exist only finitely many polynomials $P(X)$ with integral coefficients for which

$$\deg P \leq D \qquad \text{and} \qquad \mathrm{H}(P) \leq H.$$

It follows that the set of polynomials with rational coefficients is countable. In particular, the set of roots of such polynomials is also a countable set. On the other side, $\mathbb{R}$ is uncountable (as shown in Theorem 1.3). Thus

Theorem 9.1. *The set of algebraic numbers is countable, and the set of transcendental numbers is uncountable.*

This shows that *almost all* numbers are transcendental.

9.2. Liouville's theorem

So far we do not know a single transcendental number. This was one of the major problems from eighteenth/nineteenth century mathematics. In 1844, Liouville found a solution by virtue of the following:

Theorem 9.2. *For any algebraic number α of degree $d > 1$ there exists a positive constant c, depending only on α, such that*

$$\left| \alpha - \frac{p}{q} \right| > \frac{c}{q^d}$$

for all rationals $\frac{p}{q}$ with $q > 0$.

Proof. Denote by $P(X)$ the minimal polynomial of α. Then, by the mean-value theorem,

$$-P\left(\frac{p}{q}\right) = \underbrace{P(\alpha)}_{=0} - P\left(\frac{p}{q}\right) = \left(\alpha - \frac{p}{q}\right) P'(\xi)$$

for some ξ lying in between $\frac{p}{q}$ and α. Without loss of generality we may assume that the distance between α and $\frac{p}{q}$ is small, say

$$\left| \alpha - \frac{p}{q} \right| < 1.$$

Then $|\xi| < 1 + |\alpha|$ and hence $|P'(\xi)| < \frac{1}{c}$ for some positive c (since polynomials are bounded on compact sets). It follows that

$$\left| \alpha - \frac{p}{q} \right| > c \left| P\left(\frac{p}{q}\right) \right|.$$

Since $P(X)$ is irreducible of degree $d \geq 2$, $\frac{p}{q}$ cannot be a zero of $P(X)$. Hence

$$\left| q^d P\left(\frac{p}{q}\right) \right| \geq 1$$

(since $P(X)$ has integer coefficients). This proves the theorem. •

The proof even provides a value for the constant c in Liouville's theorem in terms of the height, namely,

$$(9.1) \qquad c = \frac{1}{d^2(1 + |\alpha|)^{d-1} H(\alpha)};$$

here $H(\alpha)$ is the **height** of the algebraic number α, and it is defined to be equal to the height of the minimal polynomial of α.

Liouville's theorem shows that algebraic numbers cannot be approximated *too good* by rationals. Consequently, a real number which can be *better* approximated has to be transcendental! If we assume that α is algebraic of degree d and has the continued fraction expansion $\alpha = [a_0, a_1, \ldots]$, then we have in view of equation (5.12) and Liouville's theorem the inequality

$$\frac{c}{q_n^d} < \left| \alpha - \frac{p_n}{q_n} \right| \leq \frac{1}{a_{n+1} q_n^2},$$

resp.

$$(9.2) \qquad c a_{n+1} < q_n^{d-2},$$

where the positive constant c depends only on α. If $d = 2$, the case of quadratic irrationals α, we see what we already know, namely, that the sequence of partial quotients of α is bounded. But whenever

$$\limsup_{n \to \infty} \frac{\log a_{n+1}}{\log q_n} = \infty,$$

we can find for any positive δ infinitely many convergents of α for which $a_{n+1} \geq q_n^\delta$, and in view of (9.2) such an α cannot be algebraic. For instance, consider the real number

$$(9.3) \qquad \alpha = [1, 10^{1!}, 10^{2!}, 10^{3!}, \ldots].$$

It is easily seen that $a_n = 10^{n!}$ and $q_n = 10^{n!(1 + o(1))}$. In view of the criterion above it turns out that the continued fraction (9.3) is transcendental.

9.3. Liouville numbers

A real number α is said to be a **Liouville number** if for every positive integer m there exist integers a_m and $b_m > 1$ such that

$$(9.4) \qquad\qquad 0 < \left| \alpha - \frac{a_m}{b_m} \right| < \frac{1}{b_m^m}.$$

Since the right-hand side tends to zero as $m \to \infty$, the rationals $\frac{a_m}{b_m}$ approximate α better and better. In particular, it follows that the set of the numbers b_m is unbounded. A first example for a Liouville number is given by (9.3).

A more or less immediate consequence of Liouville's theorem is

Theorem 9.3. *Every Liouville number is transcendental.*

Proof. Assume that the Liouville number α is algebraic of degree d. Combining (9.4) with the estimate in Liouville's theorem implies

$$\frac{c}{b_m^d} < \frac{1}{b_m^m},$$

where c is a positive constant depending only on α. Thus it follows that $c < b_m^{d-m}$. Since the set of the b_m is unbounded, this gives the desired contradiction. •

Liouville numbers are interesting objects. It is not too difficult to prove that the set of Liouville numbers is uncountable. On the contrary, one can prove that the set of Liouville numbers has Lebesgue measure zero. Thus it might be surprising that any real number can be written as sum, resp. as product of two Liouville numbers, as Erdös [**57**] showed. We give here only a sketchy proof of the statement about the additive representation.

Theorem 9.4. *For any real number γ, there exist Liouville numbers α, β such that $\gamma = \alpha + \beta$.*

Proof. First, we note that given a Liouville number α and a non-zero rational number λ, both their sum $\alpha + \lambda$ and their product $\lambda\alpha$ is a Liouville number (this shall be proved in Exercise 9.8). Thus we have our first representations by

$$0 = \alpha + \underbrace{(-1) \cdot \alpha}_{=\beta_0} \qquad \text{and} \qquad 1 = \alpha + \underbrace{(1 + (-1) \cdot \alpha)}_{=\beta_1},$$

where, for example, we may define α by (9.3). This easily gives the desired representation for any rational γ.

Now assume that γ is irrational. Without loss of generality we may suppose that $0 < \gamma < 1$. Every real number $\gamma \in (0, 1)$ has a dyadic representation of the form

$$(9.5) \qquad\qquad \gamma = \sum_{j=1}^{\infty} \gamma_j 2^{-j} \qquad \text{with} \quad \gamma_j \in \{0, 1\}.$$

Now define

$$\alpha_j = \begin{cases} \gamma_j & \text{if} \quad j \quad \text{is odd,} \\ 0 & \text{if} \quad j \quad \text{is even,} \end{cases}$$

and $\beta_j = \gamma_j - \alpha_j$. Putting

(9.6) $$\alpha = \sum_{j=1}^{\infty} \alpha_j 2^{-j} \quad \text{and} \quad \beta = \sum_{j=1}^{\infty} \beta_j 2^{-j},$$

we have $\gamma = \alpha + \beta$. If the series for α is infinite, we get for $m \geq 1$

$$0 < \alpha - \sum_{j=1}^{(2m)!-1} \alpha_j 2^{-j} = \sum_{j \geq (2m)!} \alpha_j 2^{-j} < 2^{1-(2m+1)!}.$$

Now let

$$b_m = 2^{(2m)!-1} \quad \text{and} \quad a_m = b_m \sum_{j=1}^{(2m)!-1} \alpha_j 2^{-j}.$$

Then a_m and b_m are integers, $b_m > 1$, and

$$0 < \alpha - \frac{a_m}{b_m} < \frac{1}{b_m^m}.$$

Hence, in this case α is a Liouville number.

In a similar manner we can deal with β (here one has only to take into account that the sequence of the β_j vanishes for odd indices, so one has to shift the index m by one). We conclude that β is also a Liouville number if its dyadic series is infinite. Thus, if neither of the series in (9.6) is finite, its values, α and β, are Liouville numbers and we are done.

If the series for α is finite, α is rational. If in addition the series for β is infinite, β is a Liouville number and thus it follows that $\gamma = \alpha + \beta$, being the sum of a rational and a Liouville number is a Liouville itself. Thus we may use the representation $\gamma = \frac{\gamma}{2} + \frac{\gamma}{2}$. The same argument applies to the case of an infinite series for α and a finite series for β.

Finally, note that the case that both series in (9.6) are finite cannot occur since then α and β are rational, and so is their sum, contradicting the assumption that $\gamma = \alpha + \beta$ is irrational. The theorem is proved. $\bullet$

9.4. The transcendence of e

The Liouville numbers from the previous paragraph look somehow artifical. It is much more difficult to decide whether a given number is transcendental or algebraic. The first and path-breaking result in this direction was proved by Hermite in 1873.

Theorem 9.5. e *is transcendental.*

Niven's proof for the irrationality of π (see Theorem 1.5) is based on Hermite's proof for the transcendence of e.

Proof. Define

$$\mathcal{I}(t) = \int_0^t \exp(t - x) f(x) \, dx$$

for $t \geq 0$, where f is a real polynomial of degree m. Integration by parts and induction show that

$$
\begin{aligned}
\mathcal{I}(t) &= -\exp(t-x)f(x)\Big|_{x=0}^{t} + \int_0^t \exp(t-x)f'(x)\,dx \\
(9.7) \qquad &= \exp(t)\sum_{j=0}^{m} f^{(j)}(0) - \sum_{j=0}^{m} f^{(j)}(t).
\end{aligned}
$$

Denote by F the polynomial obtained from f by replacing each coefficient with its absolute value. Then

$$
(9.8) \qquad |\mathcal{I}(t)| \leq \int_0^t \exp(t-x)|f(x)|\,dx \leq t\exp(t)F(t).
$$

Suppose now that e is algebraic; then there exists a polynomial P with integral coefficients a_k and leading coefficient $a_d \neq 0$ for which

$$
(9.9) \qquad \sum_{k=0}^{d} a_k\, e^k = 0.
$$

Without loss of generality we may assume that P is the minimal polynomial of e. For a *large* prime p set

$$
f(x) = x^{p-1}\prod_{\kappa=1}^{d}(x-\kappa)^p;
$$

note that the degree of f is equal to $m := (d+1)p - 1$. We shall consider the quantity

$$
\mathcal{J} := \sum_{k=0}^{d} a_k \mathcal{I}(k).
$$

In view of (9.7) and (9.9)

$$
(9.10) \quad \mathcal{J} = \sum_{k=0}^{d} a_k \left(e^k \sum_{j=0}^{m} f^{(j)}(0) - \sum_{j=0}^{m} f^{(j)}(k) \right) = -\sum_{j=0}^{m}\sum_{k=0}^{d} a_k f^{(j)}(k).
$$

Now suppose that $1 \leq k \leq d$. We define

$$
g_k(x) = \frac{f(x)}{(x-k)^p} = x^{p-1}\prod_{\substack{\kappa=1 \\ \kappa \neq k}}^{d}(x-\kappa)^p.
$$

Obviously, $g_k(x)$ is a polynomial with integer coefficients. By Leibniz' formula,

$$
f^{(j)}(x) = \sum_{i=0}^{j}\binom{j}{i}\left((x-k)^p\right)^{(i)}\left(g_k(x)\right)^{(j-i)}.
$$

It follows that $f^{(j)}(k) = 0$ for $j < p$ (since then one of the factors $x - k$ survives). Further, if $j \geq p$, then

$$
f^{(j)}(k) = \binom{j}{p} p! \cdot g_k^{(j-p)}(k).
$$

Hence, for all j, $f^{(j)}(k)$ is an integer divisible by $p!$. Similarly, $f^{(j)}(0) = 0$ for $j < p-1$, and further, for $j \geq p-1$,

$$f^{(j)}(0) = \binom{j}{p-1}(p-1)! \cdot h^{(j-p+1)}(0) , \qquad \text{where} \quad h(x) := \frac{f(x)}{x^{p-1}} ;$$

again, $h(x)$ is a polynomial and has integer coefficients. It thus follows that $h^{(j)}(0)$ is an integer divisible by p for $j > 0$, and $h(0) = (-1)^{dp}(d!)^p$. Consequently, for $j \neq p-1$, $f^{(j)}(0)$ is also an integer divisible by $p!$, and $f^{(p-1)}(0)$ is an integer divisible by $(p-1)!$ but not by p for $p > d$. Therefore, let $p > d$. By (9.10) it follows that $\mathcal{J}$ is a non-zero integer divisible by $(p-1)!$. Hence, $|\mathcal{J}| \geq (p-1)!$.

On the other hand, the trivial estimate

$$F(k) \leq k^{p-1} \prod_{\kappa=1}^{d} (k+\kappa)^p \leq (2d)^m$$

implies via (9.8)

$$|\mathcal{J}| \leq \sum_{k=0}^{d} |a_k| k \, e^k F(k) \leq \mathrm{H}(e)d(d+1)(2d)^{(d+1)p-1} \leq c^p,$$

where c is a constant independent of p. Thus,

$$(p-1)! \leq |\mathcal{J}| \leq c^p,$$

which is impossible for sufficiently large p. This contradiction shows that e cannot be algebraic. •

It should be noted that we used the divisibility properties of $\mathcal{J}$ with respect to the prime p in order to assure that $\mathcal{J} \neq 0$.

9.5. The transcendence of π

It seems that in 1761 Lambert was the first to conjecture that π is not the root of a polynomial with integer coefficients. Following Hermite's ideas, in 1882 Lindemann succeeded in showing

Theorem 9.6. π *is transcendental.*

Hermite's proof of the transcendence of e does not use any deeper arithmetic. However, rather more is needed for the proof of transcendence in case of π, namely, properties of algebraic conjugates of an algebraic number and of symmetric polynomials.

Proof. Suppose that π is algebraic. Then $\alpha := i\pi$ is also algebraic, say of degree d; however, this implication is not completely trivial (since we lazily did not prove that the set of algebraic numbers forms a field), but it follows by separating the odd from the even powers in any algebraic equation for π. Now denote the conjugates of α by $\alpha_1 = \alpha, \alpha_2, \ldots, \alpha_d$. From Euler's identity $e^{i\pi} = -1$ it follows that

$$\underbrace{(1 + e^{\alpha_1})}_{=0}(1 + e^{\alpha_2}) \cdot \ldots \cdot (1 + e^{\alpha_d}) = 0.$$

The product on the left-hand side can be written as a sum of 2^d terms e^ϱ, where

$$\varrho = \delta_1 \alpha_1 + \ldots + \delta_d \alpha_d,$$

and $\delta_j = 0$ or 1. Now we assume that precisely n of the numbers ϱ are non-zero and denote them by $\beta_1, \ldots, \beta_n$. It follows that

$$q + e^{\beta_1} + \ldots + e^{\beta_n} = 0 , \quad \text{where} \quad q := 2^d - n.$$

Now we shall proceed as in the previous proof and compare estimates for

$$\mathcal{H} := \sum_{k=1}^{n} \mathcal{I}(\beta_k),$$

where $\mathcal{I}(t)$ is defined by (9.7) with

$$f(x) := \ell^{np} x^{p-1} \prod_{k=1}^{n} (x - \beta_k)^p,$$

ℓ is the leading coefficient of the minimal polynomial of α, and p again denotes a large prime number, to be chosen later. Observe that the $\ell\beta_k$ are algebraic integers; i.e., their minimal polynomial is monic. Since an algebraic integer which is rational, is an integer (see Appendix A.7), and since

$$\prod_{\varrho} (x - \varrho) = x^q \prod_{k=1}^{n} (x - \beta_k)$$

is symmetric with respect to $\alpha_1, \ldots, \alpha_d$, it follows from the fundamental theorem of symmetric polynomials A.9 that $f(x)$ has integer coefficients.

Now put $m = (n+1)p - 1$. Then we have

$$(9.11) \qquad \mathcal{H} = -q \sum_{j=0}^{m} f^{(j)}(0) - \sum_{j=0}^{m} \sum_{k=1}^{n} f^{(j)}(\beta_k).$$

Here comes the clou of the proof. Observe that the sum over k in (9.11) is a symmetric polynomial in the algebraic integers $\ell\beta_1, \ldots, \ell\beta_n$ with integral coefficients. Thus, by the fundamental theorem on symmetric polynomials A.9, it follows that this sum is a rational number. By considering again that any rational algebraic integer is an integer, furthermore the sum over k is an integer.

Since $f^{(j)}(\beta_k) = 0$ for $j < p$, the sum in question is divisible by $p!$. Further, $f^{(j)}(0)$ is an integer divisible by $p!$ if $j \neq p - 1$, and

$$f^{(p-1)}(0) = (p-1)!(-\ell)^{np}(\beta_1 \cdot \ldots \cdot \beta_n)^p,$$

which is an integer divisible by $(p-1)!$ but not by $p!$ when p is sufficiently large. Thus, $|\mathcal{H}| \geq (p-1)!$ if $p > q$. But on the other side, (9.8) implies, as in the case of e, the upper bound

$$|\mathcal{H}| \leq \sum_{k=1}^{n} |\beta_k| F(|\beta_k|) \leq c^p$$

for some constant c independent of p. As in the previous proof, this gives the contradiction. π is not algebraic. •

In fact, Lindemann proved a stronger result than the transcendence of π, namely, that for any given non-zero complex α at least one of the numbers α and e^α is transcendent. Weierstrass extended this result to linear combinations of values of the exponential function:

Theorem of Lindemann–Weierstrass. *For any distinct algebraic numbers $\alpha_1, \ldots, \alpha_n$ and any non-zero algebraic numbers $\beta_1, \ldots, \beta_n$,*

$$\beta_1 \, e^{\alpha_1} + \ldots + \beta_n \, e^{\alpha_n} \neq 0.$$

This proves the transcendence of e, and that of π follows from Euler's identity $e^{i\pi} + 1 = 0$. Furthermore, it may be applied to prove the transcendence of values of trigonometric functions at non-zero algebraic points. Meanwhile several proofs for the theorem of Lindemann–Weierstrass are known; for a collection of them we refer to [**105**].

A set of real numbers $\alpha_1, \ldots, \alpha_n$ is said to be **algebraically independent** if there is no polynomial relation

$$P(\alpha_1, \ldots, \alpha_n) = 0$$

with coefficients in $\mathbb{Q}$. Clearly, this generalizes the concept of transcendence. Given arbitrary complex numbers $\alpha_1, \ldots, \alpha_n$, linearly independent over $\mathbb{Q}$, **Schanuel's conjecture** claims that among the $2n$ numbers

$$\alpha_1, \ldots, \alpha_n, \, e^{\alpha_1}, \ldots, \, e^{\alpha_n},$$

there are at least n which are algebraically independent over $\mathbb{Q}$. This is far more general than Lindemann's theorem, and it is said that a proof of Schanuel's conjecture would imply a hundred other open conjectures in transcendental number theory.

9.6. Squaring the circle?

In ancient Greek mathematics there are (at least) four construction problems which have become rather influential in the development of geometry and algebra, and they also might be a reason for the attention to questions concerning transcendental numbers in the nineteenth century. These classical problems are

- **doubling the cube**: constructing a cube whose volume is twice the volume of a given cube;
- **constructing the regular 7-gon**, that is, a 7-gon having edges all of the same length;
- **squaring the circle**: constructing a square whose area equals that of a given circle;
- **trisecting an angle**: constructing the third part of a given arbitrary angle.

All these constructions have to be made by use of ruler and compass only(!) — no other device is allowed. This point seems to be misunderstood by many laymen still working on these problems although a definite answer to all these problems is known. We should not mention the attempts of some doctor in solitude in 1894 to legislate a legal value of π which would have led to a *solution* of the problem of squaring the circle (but refer the reader for more information on this funny story to [**155**]).

It is a rather good idea to look at these problems *upside down*. So we ask: given two points in the Euclidean plane, having distance 1, which points can be constructed by means of ruler and compass only? In the language of algebra, all points capable of construction are defined by intersection of lines and circles. Using a ruler, we can

- draw a line through two given points, and
- find the intersection of two given lines.

Using a compass, we can

- draw a circle of a given radius with a given center, and
- find the intersection of a given circle with another given circle or line.

It is remarkable that, as it was shown by Mascheroni, all these operations can be realized by use of a compass only.

We may think of constructing new points out of the given two as mathematical operations with the numbers 0 and 1. With a little algebra it can be shown that we can only add, subtract, multiply, divide, and take square roots of numbers (coordinates of points) to get new numbers (points). Thus, a number (point) is constructable if it (its coordinates) is algebraic of degree equal to a power of two. This observation is due to young Gauss around 1798 when he constructed the regular 17-gon by ruler and compass. Furthermore he proved that the regular n-gon can be constructed with ruler and compass if and only if n is the product of a power of two and distinct prime Fermat numbers $\mathsf{F}_k = 2^{2^k} + 1$. The only known prime Fermat numbers are

$$\mathsf{F}_0 = 3, \ \mathsf{F}_1 = 5, \ \mathsf{F}_2 = 17, \ \mathsf{F}_3 = 257, \ \mathsf{F}_4 = 65\,537.$$

In particular, the 7-gon is not constructable. The second of the above listed classical problems is unsolvable.

The open question of whether squaring of the circle is possible pushed research forward on transcendental numbers in the nineteenth century. In view of Lindemann's theorem we obtain a negative answer to another problem of ancient Greek mathematics. Since π is transcendent, it is impossible to get the square root of π from rational numbers in the way described. Thus it is impossible to find the length of the side of the square having the same area as a given circle by ruler and compass. We cannot square the circle!

Finally, let us consider the problem of trisecting an arbitrary angle (and leave the remaining problem to the interested reader). Let γ be any angle. We want to trisect this angle; that means, writing $\gamma = 3\theta$, we have to construct θ. The triple angle formula in trigonometry states

$$\cos(3\theta) = 4\cos(\theta)^3 - 3\cos(\theta),$$

so we have to solve this equation with respect to $X := \cos(\theta)$. For instance, if $\gamma = 3\theta = \frac{\pi}{3}$, then $\cos(3\theta) = \frac{1}{2}$, and the above formula can be rewritten as

$$(9.12) \qquad 8X^3 - 6X - 1 = 0.$$

It is not too difficult to see that this equation has neither rational nor quadratic irrational solutions. Thus, we cannot solve the problem of trisecting an angle in general. However, certain angles can be trisected, as, for example, the right angle. If we additionally allow a parabola and a hyperbola and consider intersection points with circles and straight lines as constructed points, then much more is possible. Then we can solve any construction problem which can be reduced to the solution of a cubic equation.

Notes on transcendental numbers

In his famous talk at the 1900 International Congress of Mathematicians in Paris Hilbert asked for a proof that α^β is transcendental for algebraic $\alpha \neq 0, 1$ and $\beta \notin \mathbb{Q}$. It is said that Hilbert expected this problem, the seventh on his list, to be more difficult than a solution of the Riemann hypothesis (see Appendix A.3). However, in 1934 Gel'fond and Schneider (independently) succeeded in answering Hilbert's seventh question positively while Riemann's hypothesis is still open. Thus, for example, $2^{\sqrt{2}}$ and $e^\pi = (-1)^{-i}$ are transcendental. For the proofs of Hilbert's seventh problem due to Gel'fond and Schneider we refer to Siegel's monograph [**151**].

A far-reaching generalization of both the theorem of Lindemann–Weierstrass and the result of Gel'fond and Schneider was obtained by A. Baker for which he was awarded a Fields medal at the 1966 International Congress in Moscow. Introducing a new method he proved the transcendence of

$$e^{\beta_0} \alpha_1^{\beta_1} \cdot \ldots \cdot \alpha_n^{\beta_n}$$

and that of any non-vanishing linear form

$$\beta_1 \log \alpha_1 + \ldots + \beta_n \log \alpha_n,$$

where the α_j and β_j denote non-zero algebraic numbers. Moreover, and most importantly, Baker succeeded in first quantitative results (see Baker's monograph [**10**]). His celebrated estimates for linear forms in logarithms led to plenty of applications to diophantine equations; some of them we will briefly touch in the following chapter.

We did not mention the important results of Mahler and Shidlovski on transcendental functions (see Mahler's book [**105**]) and did not present the recent results of Nesterenko, for example, his proof of the algebraic independence of π, e^π, and $\Gamma(\frac{1}{4})$; we refer to [**122**] for an overview on these questions around transcendental numbers. As Waldschmidt said, *diophantine analysis is a very active domain of mathematical research where one finds more conjectures than results*. This holds to be true, in particular, for transcendental number theory. For a comprehensive collection of open problems we refer to [**167**].

Exercises

9.1. *Let a, b, c be odd integers. Show that then the roots of the equation*

$$aX^2 + bX + c = 0$$

are irrational.

9.2. Polynomial equations of degree ≤ 4 are solvable by radicals. The cubic equation was first solved by dal Ferro around 1515. The following solution is from Cardan.

Given the equation

$$X^3 + aX + b = 0,$$

substitute $X = U + V$ and obtain

$$U^3 + V^3 + (3UV + a)(U + V) + b = 0.$$

Show that the system

$$U^3 + V^3 = -b \qquad and \qquad 3UV = -a$$

has a solution u, v and that u^3 and v^3 are roots of the quadratic equation

$$Z^2 + bZ - \frac{a^3}{27} = 0.$$

Give an explicit formula for the solutions of the cubic. What are the solutions to $X^3 - 6X + 9 = 0$?

9.3. *i) Prove that if $\alpha \neq 0$ is algebraic (transcendental), then $\frac{1}{\alpha}$ is algebraic (transcendental) too.*
ii) Show that if α is algebraic, so is $i\alpha$ algebraic.

One can show that the set of algebraic numbers forms a field; however, this requires some knowledge from algebra (namely, the primitive element theorem).

9.4. *Prove that the set of all irrational real numbers, having a continued fraction expansion which consists only of partial quotients equal to 1 or 2, is uncountable. Show that there are uncountably many transcendental numbers which have bounded partial quotients in their continued fraction expansion.*

9.5. *Prove (9.1) for the constant c in Liouville's theorem.*

9.6. *i) Prove that the number defined by (9.3) is a Liouville number.*
ii) Show that the numbers

$$\alpha = \sum_{n=1}^{\infty} 10^{-n!} = 0.11000 \ldots \qquad and \qquad \beta = \sum_{n=0}^{\infty} 10^{-3^{n^2}} = 0.10100 \ldots$$

are transcendental.

In Exercise 2.7 we have proved that α is irrational.

9.7.* *Show that the set of Liouville numbers is uncountable.*

Hint: Use the dyadic representation (9.5) in combination with Cantor's diagonal argument.

9.8. *Let α be a Liouville number and $\lambda \neq 0$ be rational. Prove that $\lambda\alpha$ and $\alpha + \lambda$ are both Liouville numbers.*

9.9. *Given two sufficiently often differentiable functions F, G, show that*

$$(FG)(x)^{(j)} = \sum_{i=0}^{j} \binom{j}{i} F(x)^{(i)} G(x)^{(j-i)}.$$

This generalization of the product formula in differential calculus is **Leibniz' formula.**

9.10.* *Prove that the exponential function assumes transcendental values at non-zero algebraic values of the argument.*

In the meantime several other functions have been shown to share this property, e.g., the Bessel function, the elliptic modular function, and the hypergeometric function.

9.11. *Use the theorem of Lindemann–Weierstrass to prove the transcendence of the trigonometric functions $\sin\alpha, \cos\alpha$, and $\tan\alpha$ for all non-zero algebraic α.*

9.12. There are many open problems in transcendental number theory. For instance, it is yet unproved whether e and π are algebraically independent or, to be more concrete, are $e + \pi$ and $e \cdot \pi$ both transcendental?

Show that at least one of the numbers $e + \pi, e \cdot \pi$ is transcendental.

The open Schanuel conjecture implies that e and π are algebraic independent over $\mathbb{Q}$, which would show the transcendence of both numbers.

9.13. The book *Kalpa Sutra* from ancient Indian mathematics (approximately 800 B.C.) states the rule

> *Increase the measure by its third part, and this third part*
> *by its own fourth, less the thirty-fourth part of that fourth.*
> (cf. [**61**])

for doubling a given area. This rule corresponds to the approximation

$$1 + \frac{1}{3} + \frac{1}{3 \cdot 4} - \frac{1}{3 \cdot 4 \cdot 34}.$$

Prove that this number is a convergent to $\sqrt{2}$. In order to solve the problem of doubling the square construct $\sqrt{2}$ by means of ruler and compass.

9.14.* *Show that the problem of doubling the cube cannot be solved by use of ruler and compass. Find a solution with the additional device of the parabola $Y = X^2$.*

9.15. *Construct the regular 6-gon by ruler and compass.*

9.16.* *Prove that the regular 7-gon is not constructable by ruler and compass.*

Hint: Without loss of generality we may assume that the vertices of the regular 7-gon are the seventh roots of unity, i.e., the zeros of the equation

$Z^7 - 1 = 0$. One of the solutions is $z = 1$. Substituting $Y := Z + \frac{1}{Z}$ it follows that any other solution satisfies the cubic equation $Y^3 + Y^2 - 2Y - 1 = 0$.

9.17. *Prove that the roots of the equation (9.12) are irrational.*

Hint: Substitute $Y = 2X$ and assume that $y = \frac{a}{b}$ is a solution. Show that this leads to the equation $b^3 = a(a^2 - 3b^2)$. Experts in algebra may alternatively use the irreducibility criterion of Eisenstein.

9.18.* Bailey, Borwein & Plouffe [8] found a remarkable algorithm to compute the dth hexadecimal digit of π without computing any of the previous digits. This is based on a formula for π which was found by an application of LLL-reduction in order to find integer relations among irrational numbers.

i) For $\ell \in \mathbb{N}$ show that

$$\int_0^{\frac{1}{\sqrt{2}}} \frac{x^{\ell-1}}{1 - x^8}\, \mathrm{d}x = \frac{1}{\sqrt{2}^\ell} \sum_{k=0}^{\infty} \frac{1}{16^k(8k + \ell)}.$$

Deduce the identity

$$\pi \;=\; \int_0^1 \frac{16y - 16}{y^4 - 2y^3 + 4y - 4}\, \mathrm{d}y$$

$$\;=\; \sum_{k=0}^{\infty} \frac{1}{16^k}\left(\frac{4}{8k+1} - \frac{2}{8k+4} - \frac{1}{8k+5} - \frac{1}{8k+6}\right).$$

Hint: Substitute $y = \sqrt{2}x$ and apply partial fraction decomposition (which can be done by a computer algebra package).

ii) What is the ten billionth hexadecimal digit of π?

Hint: One may remember the binary scheme for fast exponentiation, where x^n is evaluated by successive squaring and multiplication.*

9.19. Let β be algebraic of degree d and α be algebraic but $\neq 0, 1$. Due to Gel'fond and Schneider the numbers $\alpha^\beta, \ldots, \alpha^{\beta^{d-1}}$ are all transcendental; furthermore Gel'fond showed that if $d \geq 3$, then at least two of these numbers are algebraically independent. **Gel'fond's conjecture** states that all those numbers are algebraically independent.

Show that Schanuel's conjecture implies Gel'fond's conjecture.

9.20.* Siegel wrote in the preface of [151] that *it would be misleading to call it a theory of transcendental numbers* (since) *our knowledge concerning transcendental numbers being narrowly restricted.* Some people believe that every statement about the transcendence of numbers, which is not obviously wrong, is true.

State some conjectures on algebraic, resp. transcendental numbers, which make sense, and try to prove them!

---

*More information on this topic can be found on P. Borwein's π-Web page http://www.cecm.sfu.ca/pi/pi.html.

CHAPTER 10

The theorem of Roth

It is natural to ask for stronger versions of Liouville's theorem. Only a slight improvement would imply the finiteness of integral solutions of certain important diophantine equations, so-called Thue equations. First improvements of Liouville's theorem were made by Thue, Siegel, and Dyson. The most far-reaching extension was found by Roth for which he was awarded a Fields medal at the 1958 International Congress of Mathematicians at Edinburgh. In this chapter we will give a proof of this deep and far-reaching highlight in the theory of diophantine approximations.

10.1. Roth's theorem

Suppose that α is an algebraic number of degree d; then we might ask for exponents $\tau(d)$ such that there exists a positive constant c, depending only on α, such that for all rationals $\frac{p}{q}$

$$(10.1) \qquad \left| \alpha - \frac{p}{q} \right| > \frac{c}{q^{\tau(d)}}.$$

The first refinement to Liouville's bound $\tau(d) = d$ from Theorem 9.2 was made by Thue who showed in 1909 that $\tau(d) > \frac{1}{2}d + 1$. In 1921, Siegel obtained $\tau(d) > 2\sqrt{d}$, and in 1947, the physicist Dyson proved that $\tau(d) > \sqrt{2d}$. Finally, Roth [138] proved in 1955 that $\tau(d)$ can be chosen independently of the degree d.

Theorem 10.1. *Let α be an algebraic number of degree $d \geq 2$ and let $\varepsilon > 0$. Then there exist only finitely many rational solutions $\frac{p}{q}$ to the inequality*

$$\left| \alpha - \frac{p}{q} \right| < \frac{1}{q^{2+\varepsilon}}.$$

This proves a conjecture of Siegel, namely, that one can take any $\tau(d) > 2$ in (10.1), which is equivalent to Theorem 10.1. Consequently, algebraic numbers are approximable of order $\kappa = 2$ but not better. Of course, Liouville's theorem 9.2 is stronger for quadratic irrational α, but here we are mainly interested in algebraic numbers of degree $d > 2$. In view of Dirichlet's approximation theorem Roth's theorem is best possible with respect to the exponent 2. It might be possible that the ε-quantity can be sharpened. Lang conjectured that, for any α of degree $d \geq 3$, the inequality

$$\left| \alpha - \frac{p}{q} \right| < \frac{1}{q^2 (\log q)^\kappa}$$

has only finitely many solutions, where $\kappa > 1$ is any constant.

Unfortunately, Roth's theorem is ineffective. The method of proof does not provide an algorithm to determine the set of solutions or the constant c in (10.1). In particular cases Baker's celebrated estimates for linear forms in logarithms yield explicit lower bounds; for example, Baker [**9**] succeeded in proving

$$(10.2) \qquad \left| \sqrt[3]{2} - \frac{p}{q} \right| \geq \frac{10^{-6}}{q^{2.9955}},$$

valid for all rationals $\frac{p}{q}$. The appearing exponent is larger than Roth's exponent $2 + \varepsilon$, but the implicit constant is absolute. We shall later see the advantage of having effective estimates for applications.

Before we give the lengthy proof of Roth's theorem we present one of its most important applications, a finiteness theorem for a special class of diophantine equations.

10.2. Thue equations

We start with a tale on cabs and cubes. During his stay in England, the ingenious Indian mathematician Ramanujan spent several weeks in the hospital (suffering from British climate and cuisine). Once his colleague Hardy came to visit and remarked that he had come in taxicab number 1729, which is *surely a rather dull number.* Ramanujan replied immediately that this is not true; 1729 is rather interesting since it is the smallest integer expressible as a sum of two cubes in two different ways:

$$(10.3) \qquad 1729 = 1^3 + 12^3 = 9^3 + 10^3.$$

This can be seen as follows. Since

$$(X + Y) \cdot (X^2 - XY + Y^2) = X^3 + Y^3 = 1729 = 7 \cdot 13 \cdot 19,$$

we have to consider all possible (but finitely many) factorizations $1729 = A \cdot B$ and solve

$$A = X + Y \qquad \text{and} \qquad B = X^2 - XY + Y^2.$$

The substitution $Y = A - X$ leads to the quadratic equation

$$X^2 - AX + \frac{1}{3}(A^2 - B) = 0,$$

so that we only have to check whether

$$\frac{1}{6}(3A \pm \sqrt{12B - 3A^2})$$

is an integer. This yields the two pairs $A = 13, B = 133$ and $A = 19, B = 91$ which correspond to the representations given above. We leave it to the reader to verify that all integers $m < 1729$ have at most one representation as a sum of two cubes.

Analyzing the equation $X^3 + Y^3 = 1729$ was rather simple since the left hand turned out to be reducible. If instead we have an equation with an irreducible polynomial, as, for example, $X^3 + 2Y^3 = m$, the situation is much more difficult. The first remarkable result for such equations was found by Thue [**160**] in 1909.

Theorem 10.2. *Let a, b, m be non-zero integers. Then the equation*

$$(10.4) \qquad\qquad aX^3 + bY^3 = m$$

has only finitely many solutions in integers.

Proof. If $x, y \in \mathbb{Z}$ is a solution to (10.4), then $X = ax, Y = y$ is a solution to $X^3 + a^2 bY^3 = a^2 m$. Thus it is enough to prove the theorem with $a = 1$. Further, since we may replace y by $-y$ and/or b by $-b$ in (10.4) if necessary, it is sufficient to consider the equation

$$(10.5) \qquad\qquad X^3 - bY^3 = m,$$

where b, m are positive integers. The factorization method which we used above worked very well. This observation motivates us to have a look at the factorization

$$(10.6) \quad X^3 - bY^3 = (X - \alpha Y) \cdot (X^2 + \alpha XY + \alpha^2 Y^2) \qquad \text{with} \quad \alpha = \sqrt[3]{b}.$$

If b is a perfect cube, α is an integer and so the latter identity is a factorization over the integers and we can factor m and proceed as above. If b is not a perfect cube, we need a different idea. Then we argue as Euler did in the case of the Pell equation (see Section 6.2). If $x, y \in \mathbb{N}$ is a large solution of (10.5), then the first factor on the right-hand side of (10.6) must have a small modulus. This follows from the estimate

$$(10.7) \quad x^2 + \alpha xy + \alpha^2 y^2 = \left(x + \frac{1}{2}\alpha y \right)^2 + \frac{3}{4}(\alpha y)^2 \geq \frac{3}{4}\alpha^2 y^2,$$

which implies via (10.6)

$$m = x^3 - by^3 = |x - \alpha y| \cdot |x^2 + \alpha xy + \alpha^2 y^2| \geq \frac{3}{4}\alpha^2 y^2 \cdot |x - \alpha y|.$$

This gives

$$(10.8) \qquad\qquad \left| \alpha - \frac{x}{y} \right| \leq \frac{4m}{3\alpha^2 y^3}.$$

In view of Roth's theorem the quantity on the left is bounded below by $cy^{-2-\varepsilon}$, where c is a positive constant depending only on α. Hence it follows that

$$y < \left(\frac{4m}{3\alpha^2 c} \right)^{\frac{1}{1-\varepsilon}},$$

and thus inequality (10.8) can only have finitely many solutions $\frac{x}{y}$. This proves the assertion of the theorem. •

It should also be noted that Thue's estimate $\tau(3) > \frac{5}{2}$ in (10.1) is sufficient (Thue did not have Roth's bound) but Liouville's $\tau(3) = 3$ is not.

Although the set of integral solutions is finite, Thue's argument does not give a complete list of solutions or determine whether the equation is solvable. This follows from the fact that Roth's theorem is ineffective: we do not know how large the constant c is. However, in particular cases —

and, of course, here we do not mean *trivial* cases when the left-hand side of
(10.4) is reducible — it is possible to obtain upper bounds for the solutions.
For instance, combining (10.2) with (10.8) shows that any integer solution
x, y to

$$X^3 - 2Y^3 = m,$$

where m is a positive integer, must satisfy the estimate

$$|y| \le 10^{1317} m^{223}.$$

This is a large bound, but at least it only grows like a power of m and a
complete analysis of such an equation is reduced to computation. Effective
estimates are a rather difficult but extremely important topic in diophantine
analysis. We refer to Bombieri [**26**] for a survey on the progress made in
the last 40 years.

Theorem 10.2 can easily be extended to a wider class of equations of
higher degree, called **Thue equations**. Using the above reasoning it is not
too difficult to prove

Theorem 10.3. *Let $P(X, Y)$ be an irreducible binary form with integer
coefficients of degree at least three and let m be any integer. Then the
equation $P(X, Y) = m$ has only finitely many solutions in integers.*

10.3. Finite vs. infinite

It makes a big difference whether we ask for *integer* or for *rational* solutions to a given diophantine equation. Obviously, there are only finitely many points on the unit circle with integer coordinates, whereas the
Pythagorean triples yield an infinitude of points with rational coordinates
(see Exercise 6.8). Here we want to sketch another more advanced example.

By Theorem 10.2 the Thue equation

$$(10.9) \qquad\qquad X^3 + Y^3 = 1729$$

has only finitely many integer solutions; one of them being $(x, y) = (1, 12)$.
The transformation

$$(10.10) \qquad X \mapsto \frac{20\,748}{X + Y} \qquad \text{and} \qquad Y \mapsto 62\,244\, \frac{X - Y}{X + Y}$$

provides a one-to-one correspondence between the solutions of the previous
equation and the solutions of

$$(10.11) \qquad\qquad Y^2 = X^3 - 1\,291\,438\,512.$$

This equation is a special example of a class of diophantine equations, named
elliptic curves (that are not ellipses but they are related to them), which
became a major field of research in the twentieth century. An appropriate
presentation is far beyond the scope of this book. Nevertheless, we take a
brief view on this interesting piece of mathematics.

We start with some common notions which we shall also use later. Given
a polynomial $P(X, Y)$ defined over a number field, the real solutions (x, y)
to the equation $P(X, Y) = 0$ form a **curve** in the Euclidean plane $\mathbb{R}^2$. A
point on this curve is called **singular** if in this point the curve crosses itself

or if the point is a cusp. All other points on the curve have a neighborhood which looks like a curved piece of the real line. Of course, one can also consider complex solutions to $P(X, Y) = 0$, or solutions in other fields. A non-singular curve over $\mathbb{C}$ looks locally like the complex plane $\mathbb{C}$, and is called a **Riemann surface**. If the polynomial P has total degree three, so $P(X, Y)$ is of the form

$$aX^3 + bX^2Y + cXY^2 + dY^3 + eX^2 + fXY + gY^2 + hX + iY + j = 0,$$

we have a **cubic curve**; and if it is a non-singular one, the Riemann surface is a torus (roughly speaking, a doughnut). In 1929, Siegel proved that any non-singular cubic curve with integral coefficients has only finitely many points with integral coordinates; in fact he proved a more general statement which also applies to affine plane curves of higher degree and which goes much beyond the statement of Theorem 10.2. On the contrary, linear or quadratic equations can have an infinitude of solutions in integers (e.g., Pell's equation).

A non-singular cubic curve, which is defined over $\mathbb{Q}$ and contains at least one rational point, is called an **elliptic curve**. We may think of the cubic equation defining an elliptic curve as given by $Y^2 = f(X)$, where $f(X)$ is a cubic polynomial. The famous theorem of Mordell states that the set of rational points on an elliptic curve forms a finitely generated abelian group. In particular, this means that we can *add* points on an elliptic curve and their sum is a further point on this curve. This surprising fact relies on the simple observation that a generic straight line has three intersection points with the cubic equation $Y^2 = f(X)$. This gives an algebraic relation between any two given points P_1 and P_2 on the elliptic curve and a third one, $\mathsf{Q} = (x, y)$ say. For some reasons we do not explain, one cannot take Q to be the sum of P_1 and P_2, but replacing Q by its conjugate with respect to the x-axis is doing the job:

$$\mathsf{P}_1 + \mathsf{P}_2 = (x, -y).$$

It can happen that an elliptic curve has only a few rational points, e.g., the only rational points on $Y^2 = X^3 - 4X$ are given by $(0,0)$ and $(\pm 2, 0)$. However, the situation is rather different in our example from above. The point $(1, 12)$ on (10.9) is mapped via (10.10) onto $\mathsf{P} = (1596, -52\,668)$ on the elliptic curve (10.11). Computing the multiple

$$(10.12) \qquad\qquad 2\mathsf{P} = \left(\frac{250\,572}{121}, \frac{-115\,950\,996}{1331} \right),$$

it follows that P has infinite order (since points with finite order must have integer coordinates by the theorem of Nagell–Lutz). Hence, there are infinitely many rational points on the elliptic curve (10.11). With regard to (10.10) it follows that there are also infinitely many rational solutions to the Thue equation (10.9).

In the last 20 years the theory of elliptic curves has found many important and unexpected applications, e.g., primality tests and factoring methods in cryptography, and, last but not least, in Frey's attack on Fermat's last

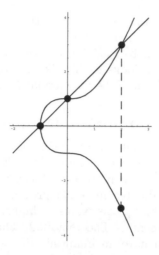

FIGURE 10.1. Adding points on an elliptic curve; here $(-1,0) + (0,1) = (2,-3)$ on the elliptic curve given by $Y^2 = X^3 + 1$.

theorem via the Shimura–Taniyama–Weil conjecture which led to Wiles' celebrated proof. The interested reader can find this and much more in the excellent monographs Silverman & Tate [152] and Washington [169].

The remainder of this chapter gives a complete proof of Roth's theorem. Our presentation follows Cassels [35], Tijdeman [161], and Roth [139]. We start with an introduction to multivariate polynomials and the fundamental notions of a differential operator and the index.

10.4. Differential operators and indices

Let m be a positive integer and denote by M the set of mappings

$$\tau : \{1,\ldots,m\} \to \mathbb{N}, \quad j \mapsto \tau(j) =: \tau_j.$$

For $\tau, \sigma \in \mathsf{M}$, we write $\tau \leq \sigma$ if $\tau_j \leq \sigma_j$ for all $1 \leq j \leq m$. Further, we define $\mathsf{M}(d)$ to be the set of all $\tau \in \mathsf{M}$ such that $\tau \leq d$, where d is the constant mapping $\tau_j = d$. Then we can write multivariate polynomials P in $\mathbf{X} = (X_1,\ldots,X_m)$ with integer coefficients in the form

$$P(\mathbf{X}) = \sum_{\tau \in \mathsf{M}(d)} c(\tau) X_1^{\tau_1} \cdot \ldots \cdot X_m^{\tau_m},$$

where $c(\tau) \in \mathbb{Z}$ and d is sufficiently large. The least d such that $c(\tau) = 0$ for all τ which do not satisfy $\tau \leq d$ is the **(maximum relative) degree** of P, denoted by $\deg P$ (which should not be mixed with the total degree, the maximum over all $\tau_1 + \ldots + \tau_m$). Hence,

$$(10.13) \qquad\qquad P(\mathbf{X}) = \sum_{\tau \leq \deg P} c(\tau) X_1^{\tau_1} \cdot \ldots \cdot X_m^{\tau_m}.$$

We introduce a norm on the ring of polynomials $\mathbb{Z}[\mathbf{X}]$ by setting

$$\mathrm{H}(P) = \max_{\tau \in \mathsf{M}} |c(\tau)|.$$

Extending the notion of height introduced in the previous chapter, $\mathrm{H}(P)$ is called the **height** of P.

For $1 \le i \le m$, we define the (formal) **partial derivative** $\frac{\partial}{\partial X_i}$ with respect to the variable X_i as a linear map on $\mathbb{Z}[X_1, \ldots, X_m]$ by the following formula:

$$\frac{\partial}{\partial X_i} cX^{\tau_i} Q = \tau_i c X_i^{\tau_i - 1} Q,$$

where c is a constant, τ_i is a non-negative integer, and Q is a polynomial not depending on X_i; in the case of $\tau_i = 0$ the right-hand side above has to be interpreted as the zero polynomial. Thus, the just described partial derivatives map $\mathbb{Z}[\mathbf{X}]$ onto itself. For $\mathbf{i} = (i_1, \ldots, i_m) \in \mathsf{M}$, we set

$$(10.14) \qquad \mathrm{D}_{\mathbf{i}} = \frac{1}{i_1!} \left(\frac{\partial}{\partial X_1} \right)^{i_1} \cdot \ldots \cdot \frac{1}{i_m!} \left(\frac{\partial}{\partial X_m} \right)^{i_m};$$

this defines a **differential operator of order** $|\mathbf{i}| := i_1 + \ldots + i_m$.

Lemma 10.4. *The differential operator* $\mathrm{D}_{\mathbf{i}}$ *maps* $\mathbb{Z}[\mathbf{X}]$ *onto itself. If* $P \in \mathbb{Z}[\mathbf{X}]$ *has degree* $\le t_j$ *in* X_j *for* $1 \le j \le m$, *then*

$$\mathrm{H}(\mathrm{D}_{\mathbf{i}}P) \le 2^{t_1 + \cdots + t_m} \mathrm{H}(P).$$

Proof. Each polynomial is a linear combination of monomials of the form

$$cX_1^{t_1} \cdot \ldots \cdot X_m^{t_m}$$

for some constant $c \in \mathbb{Z}$ and some $t \in \mathsf{M}$. By the linearity of $\mathrm{D}_{\mathbf{i}}$ it suffices to prove the lemma in the case of P being such a monomial. Then it is easy to compute

$$(10.15) \qquad \mathrm{D}_{\mathbf{i}}P(\mathbf{X}) = c \prod_{j=1}^{m} \binom{t_j}{i_j} X_j^{t_j - i_j}.$$

The right-hand side is an element of $\mathbb{Z}[\mathbf{X}]$ with regard to our convention concerning negative exponents $t_j - i_j$. By the well-known combinatorial inequality $\binom{t}{i} \le 2^t$, valid for integers $t, i \ge 0$, the product of binomial coefficients on the right-hand side of (10.15) turns out to be bounded above by $2^{t_1 + \cdots + t_m}$. This leads to the estimate in the lemma. Clearly, $\mathrm{D}_{\mathbf{i}}P \in \mathbb{Z}[\mathbf{X}]$. The lemma is proved. •

Given a non-identically vanishing polynomial P in the form (10.13), the **index** $\mathrm{ind}(P)$ of P at the point $\mathbf{x} = (x_1, \ldots, x_m)$ with respect to $(d_1, \ldots, d_m)$ is defined as the least value of κ such that there is a tuple $\mathbf{i} = (i_1, \ldots, i_m)$ with

$$\frac{i_1}{d_1} + \ldots + \frac{i_m}{d_m} = \kappa \qquad \text{and} \qquad (\mathrm{D}_{\mathbf{i}}P)(\mathbf{x}) \ne 0;$$

for short,

$$(10.16) \qquad \operatorname{ind}(P) = \inf\left\{ \sum_{j=1}^{m} \frac{i_j}{d_j} \ : \ (D_{\mathbf{i}}P)(\mathbf{x}) \neq 0 \right\},$$

where the infimum is taken over all admissible vectors $\mathbf{i}$. In view of the Taylor expansion of P at $\mathbf{x}$,

$$(10.17) \qquad P(\mathbf{X}) = \sum_{\mathbf{i} \leq \deg P} (D_{\mathbf{i}}P)(\mathbf{x})\,(X_1 - x_1)^{i_1} \cdot \ldots \cdot (X_m - x_m)^{i_m},$$

it follows that the index exists, and so the infimum is a minimum. If P does not vanish at $\mathbf{x}$, the index is equal to zero. When the order of vanishing of P at $\mathbf{x}$ is large, the index is also large. Hence, the index might be regarded as a measure for the vanishing of a multivariate polynomial. If P vanishes identically, we define the index to be $+\infty$.

Lemma 10.5. *With the notation from above, the index at some point* $\mathbf{x}$ *with respect to* $(d_1, \ldots, d_m)$ *satisfies*

$$\begin{aligned} \operatorname{ind}(D_{\mathbf{i}}P) &\geq \operatorname{ind}(P) - \sum_{j=1}^{m} \frac{i_j}{d_j}, \\ \operatorname{ind}(PQ) &= \operatorname{ind}(P) + \operatorname{ind}(Q), \\ \operatorname{ind}(P + Q) &\geq \min\{\operatorname{ind}(P), \operatorname{ind}(Q)\}. \end{aligned}$$

Proof. The first assertion is obvious. For the proof of the second claim assume that the infimum in (10.16) for the index of PQ is attained at $\mathbf{k}$. Then

$$(D_{\mathbf{i}}P)(\mathbf{x})(D_{\mathbf{j}}Q)(\mathbf{x}) \neq 0$$

for some $\mathbf{i}$ and $\mathbf{j}$ with $\mathbf{i} + \mathbf{j} = \mathbf{k}$. This implies $\operatorname{ind}(PQ) \geq \operatorname{ind}(P) + \operatorname{ind}(Q)$. Conversely, if the infimum for P and Q are taken at $\mathbf{i}$ and $\mathbf{j}$, respectively, then take among those the one with least first coefficient, among those the one for which the second coefficient is minimal, and so on. Then it follows that

$$(D_{\mathbf{i}+\mathbf{j}}PQ)(\mathbf{x}) \neq 0$$

since all terms in this sum but $(D_{\mathbf{i}}P)(\mathbf{x})(D_{\mathbf{j}}Q)(\mathbf{x})$ are equal to zero.

The third assertion is left to the reader. •

10.5. Outline of Roth's method

Roth's ingenious proof is quite technical and rather long. It might be regarded as an extension of Liouville's proof of Theorem 9.2 and Thue's refinement. Here we shall sketch the main idea.

First, we briefly review Liouville's idea in proving Theorem 9.2; that is, an irrational algebraic number α of degree d is not approximable of order d by rationals $\frac{p}{q}$ or, for short, $\tau(d) = d$ in (10.1). His proof depends mainly on

   • the construction of a polynomial $P(X)$ with integral coefficients that vanishes at α and for which $P(\frac{p}{q})$ is *small* in terms of $|\alpha - \frac{p}{q}|$,

• the fact that $P(\frac{p}{q})$ *cannot be too small* as a function of q.

We are looking for an improvement upon Liouville's theorem. Thus, instead of taking the minimal polynomial of α, we shall work with an arbitrary polynomial $P(X)$ with integer coefficients and root α. Let κ denote the multiplicity of the linear factor $X - \alpha$ in $P(X)$. Then, in terms of the index at α with respect to $\deg P$,

$$\frac{\kappa}{\deg P} = \operatorname{ind}(P).$$

We assume that there are integers p and $q > 0$ such that

(10.18)
$$\left| \alpha - \frac{p}{q} \right| < \frac{1}{q^\tau}.$$

By the Taylor expansion we have

$$P\left(\frac{p}{q} \right) = \sum_{i=0}^{\deg P} (D_i P)(\alpha) \left(\frac{p}{q} - \alpha \right)^i.$$

If $P(\frac{p}{q}) \neq 0$, then

$$\frac{1}{q^{\deg P}} \leq \left| P\left(\frac{p}{q} \right) \right| \leq C \left| \alpha - \frac{p}{q} \right|^\kappa < \frac{C}{q^{\kappa\tau}},$$

where C is some positive constant. If now (10.18) possesses infinitely many solutions p, q, we obtain

$$\tau \leq \frac{\deg P}{\kappa} = \frac{1}{\operatorname{ind}(P)}.$$

To make τ as small as possible, the index $\operatorname{ind}(P)$ has to be as large as possible. For this purpose one has to find a polynomial P which has a large index in α. This is a difficult task. Since α is a root of multiplicity κ of P, the κth power of the minimal polynomial of α divides P.

We shall briefly explain this fact from algebra since it is not trivial. If $(X - \alpha)^\kappa$ divides P, then also $(X - \alpha_j)^\kappa$ divides P, where α_j is any of the conjugates of α. This is an easy application of Gauss' lemma A.8. This holds since we assume that P has integer coefficients.

We return to our analysis of candidates for improvements to Liouville's theorem. From the divisibility of P by the κth power of the minimal polynomial of α it follows that $d\kappa \leq \deg P$; recall that d is the degree of α and of its minimal polynomial, respectively. This gives a bound for the index, namely,

$$\operatorname{ind}(P) = \frac{\kappa}{\deg P} \leq \frac{1}{d};$$

for the minimal polynomial of α equality holds. This inequality leads to $\tau = d$, which is nothing other than Liouville's theorem, and shows that polynomials in one variable are not sufficient for an improvement.

The deeper reason for the restrictions we have by using polynomials in only one variable is that any polynomial P with integer coefficients which has a certain index at α has also the same index at all conjugates α_j of

α. The index is a measure of how many linear restrictions we can attach to the coefficients of a polynomial of fixed degree; more precisely, it is the proportion of the number of restrictions and the number of coefficients. If for all conjugates the index is the same as for α, namely, $\mathrm{ind}(P)$, this proportion is bounded above by $\frac{1}{d}$. It was Thue's contribution to introduce polynomials in two variables. This provides a bit more freedom in the search for suitable polynomials and led him to his improvement. Before Roth it was already known that any progress would demand the use of polynomials in more than two variables.

We cannot directly argue as Liouville did. Assume that inequality (10.18) has infinitely many rational solutions $\frac{p_n}{q_n}$ with $\tau = 2 + \varepsilon$. If a suitable multivariate polynomial vanishes at $\underline{\alpha} := (\alpha, \dots, \alpha)$, it might also be zero at $(\frac{p_1}{q_1}, \dots, \frac{p_m}{q_m})$. But we can obtain a contradiction by means of the index. In order to prove Theorem 10.1 we shall establish the existence of a polynomial P which vanishes of *high* order at $\underline{\alpha}$ and thus has *large* index, but has *small* index at $(\frac{p_1}{q_1}, \dots, \frac{p_m}{q_m})$, where the $\frac{p_n}{q_n}$ are suitably chosen approximations to α. This will yield a contradiction which proves Roth's theorem. In some sense, the determination of the order of approximation of algebraic numbers is transformed to an approximation problem for polynomials in several variables.

10.6. Siegel's lemma

The following statement, called Siegel's lemma, provides the existence of a small integer solution of a system of homogeneous linear equations with rational coefficients and with more variables than equations.

Theorem 10.6. *Let M, N be positive integers with $N > M$ and, for $1 \leq j \leq M$, let $Y_1, \dots, Y_M$ be linear forms in $X_1, \dots, X_N$ with integer coefficients. Then there exist integers $z_1, \dots, z_N$, not all equal to zero, such that*

$$(10.19) \qquad Y_j(z_1, \dots, z_N) = 0 \qquad for \qquad 1 \leq j \leq M,$$

and

$$(10.20) \qquad \max\{|z_1|, \dots, |z_N|\} \leq (N \max\{\mathrm{H}(Y_1), \dots, \mathrm{H}(Y_M)\})^{\frac{M}{N-M}}.$$

Proof. Consider the linear transformation $\ell : \mathbb{R}^N \to \mathbb{R}^M$, given by

$$\ell(x_1, \dots x_N) = (Y_1(x_1, \dots x_N), \dots, Y_M(x_1, \dots x_N)).$$

Given a fixed positive integer X, the set

$$\mathcal{M} := \{(x_1, \dots, x_N) \in \mathbb{R}^N : 0 \leq x_1, \dots, x_N \leq X\}$$

is mapped under ℓ into

$$\ell(\mathcal{M}) = \{(y_1, \dots, y_M) \in \mathbb{R}^n : -C_j^- X \leq y_j \leq C_j^+ X, 1 \leq j \leq M\},$$

where $-C_j^-$ is the sum of the negative and C_j^+ is the sum of the positive coefficients of the linear form Y_j. Thus, $0 \leq C_j^- + C_j^+ \leq N\mathrm{H}(Y_j)$. There are $(X + 1)^N$ many points with integer coefficients in $\mathcal{M}$. However, $\ell(\mathcal{M})$ contains at most

$$(XN \max\{\mathrm{H}(Y_1), \ldots, \mathrm{H}(Y_M)\} + 1)^M$$

points with integer coefficients. So whenever

(10.21) $$(X + 1)^N > (XN \max\{\mathrm{H}(Y_1), \ldots, \mathrm{H}(Y_M)\} + 1)^M,$$

which is certainly satisfied for sufficiently large X, there exist two distinct points $\mathbf{x}_j = (x_{1,j}, \ldots, x_{N,j}) \in \mathcal{M}, j = 1, 2$, that are mapped under ℓ onto the same point in $\ell(\mathcal{M})$. Since ℓ is linear, their difference $\mathbf{z} = \mathbf{x}_1 - \mathbf{x}_2 = (x_{1,1} - x_{1,2}, \ldots, x_{N,1} - x_{N,2}) \neq \mathbf{0}$ lies in the kernel of ℓ, and

(10.22) $$\max\{|x_{1,1} - x_{1,2}|, \ldots, |x_{N,1} - x_{N,2}|\} \leq X.$$

Thus, setting $\mathbf{z} = (z_1, \ldots, z_N)$ leads to (10.19), and only proving the estimate (10.20) remains. For this aim we have to find an X such that (10.21) holds. Considering the asymptotics in (10.21) as $X \to \infty$, we are led to the choice

(10.23) $$X := \left[(N \max\{\mathrm{H}(Y_1), \ldots, \mathrm{H}(Y_M)\})^{\frac{M}{N-M}} \right].$$

Now it is not difficult to verify inequality (10.21) as follows:

$$XN \max\{\mathrm{H}(Y_1), \ldots, \mathrm{H}(Y_M)\} + 1$$
$$< (X + 1)N \max\{\mathrm{H}(Y_1), \ldots, \mathrm{H}(Y_M)\} < (X + 1)^{1 + \frac{N-M}{M}} = (X + 1)^{\frac{N}{M}}.$$

This choice of X leads via (10.22) to (10.20). Siegel's lemma is proved. $\bullet$

The argument of proof reminds us of the pigeonhole principle and Mordell's proof of Minkowski's first theorem 8.1. The considered problem has also some similarities with the small integer problem we investigated in Section 8.10.

For later convenience we reformulate Siegel's lemma as follows. Let B be an $M \times N$ matrix with integer entries, not all equal to zero, and assume $N > M$; then there exists an integer vector $\mathbf{z}$ such that

$$B\mathbf{z} = 0 \quad \text{and} \quad 0 < |\mathbf{z}| \leq (N\|B\|)^{\frac{M}{N-M}},$$

where $\|B\|$ denotes the maximum of the absolute values of the entries of B.

10.7. The index theorem

Now we shall use Siegel's lemma, Theorem 10.6, in order to construct a polynomial which has a *large* index at some given point.

Theorem 10.7. *Let α be an algebraic integer of degree $d \geq 2$, and let $\varepsilon > 0$ and m be an integer satisfying $m \geq \frac{d}{2\varepsilon^2}$. Further, let $d_1, \ldots, d_m$ be positive integers. Then there exists a polynomial P in $\mathbf{X} = (X_1, \ldots, X_m)$ with integer coefficients, not vanishing identically, of degree at most d_n in X_n with index*

(10.24) $$\mathrm{ind}(P) > \frac{1}{2}m(1 - \varepsilon)$$

at $\underline{\alpha} = (\alpha, \ldots, \alpha)$ *with respect to* $(d_1, \ldots, d_m)$, *and height*

$$\mathbf{H}(P) \leq C_1^{d_1 + \ldots + d_m}$$

for some positive constant C_1, *depending only on* α.

Proof. We may assume that P is of the form

$$P(\mathbf{X}) = \sum_{j_1=0}^{d_1} \cdots \sum_{j_m=0}^{d_m} z_{j_1, \ldots, j_m} X_1^{j_1} \cdot \ldots \cdot X_m^{j_m}.$$

P has large index satisfying (10.24) if $(D_{\mathbf{i}}P)(\underline{\alpha}) = 0$ for all tuples $\mathbf{i} = (i_1, \ldots, i_m) \in \mathbb{Z}^m$ with

$$(10.25) \qquad \frac{i_1}{d_1} + \ldots + \frac{i_m}{d_m} \leq \frac{1}{2}m(1 - \varepsilon).$$

Summing up all these expressions we arrive at an equation of the form

$$A_0 \mathbf{z} + \alpha A_1 \mathbf{z} + \ldots + \alpha^{d_1 + \ldots + d_m} A_{d_1 + \ldots + d_m} \mathbf{z} = 0,$$

where the A_j are $M \times N$ matrices with integer entries and $\|A_j\| \leq 4^{d_1 + \ldots + d_m}$, $N = (d_1 + 1) \cdot \ldots \cdot (d_m + 1)$ is the number of tuples $\mathbf{i}$ for which the inequalities

$$(10.26) \qquad 0 \leq i_n \leq d_n \quad \text{for} \quad 1 \leq n \leq m$$

hold, and M is the number of those $\mathbf{i}$ which additionally satisfy (10.25). Since α is algebraic of degree d, the equation for the matrices A_j can be rewritten as

$$B_0 \mathbf{z} + \alpha B_1 \mathbf{z} + \ldots + \alpha^{d-1} B_{d-1} \mathbf{z} = 0,$$

where the B_j are $M \times N$ matrices with integer entries and $\|B_j\| \leq C_1^{d_1 + \ldots + d_m}$ for some positive constant C_1, depending only on α. The numbers $1, \alpha, \ldots, \alpha^{d-1}$ are linearly independent over $\mathbb{Q}$, so we have

$$B_0 \mathbf{z} = \mathbf{0}, \quad B_1 \mathbf{z} = \mathbf{0}, \quad \ldots, \quad B_{d-1} \mathbf{z} = \mathbf{0}.$$

Writing all these matrices B_j in sequence leads to a $dM \times N$ matrix B such that $B\mathbf{z} = 0$ and $\|B\| \leq C_1^{d_1 + \ldots + d_m}$. Now we shall use Siegel's lemma in order to determine $\mathbf{z}$ and the coefficients $z_{j_1, \ldots, j_m}$ of P.

For this purpose we need an estimate for M. Obviously, we can replace (10.25) by

$$(10.27) \qquad \left| \frac{i_1}{d_1} + \ldots + \frac{i_m}{d_m} - \frac{m}{2} \right| \geq \varepsilon m.$$

To get an upper bound for M, the number of tuples $\mathbf{i}$ satisfying (10.26) and (10.27), we use a probabilistic argument (but on a very low level). We consider $i_1, \ldots, i_m$ as independent random variables such that any individual i_n is uniformly distributed on the set $\{0, 1, \ldots, d_n\}$. Putting

$$X = \sum_{n=1}^{m} \frac{i_n}{d_n},$$

then X is a random variable. Any individual random variable $\frac{i_n}{d_n}$ has expectation $\frac{1}{2}$ and variance

$$(10.28) \qquad \mathbf{var}\left(\frac{i_n}{d_n}\right) = \sum_{i_n=0}^{d_n}\left(\frac{i_n}{d_n}-\frac{1}{2}\right)^2\frac{1}{d_n+1} = \frac{2d_n+1}{6d_n}-\frac{1}{4} \le \frac{1}{4}.$$

Hence, X has expectation $\mathbf{E}(X) = \frac{m}{2}$ and variance

$$\mathbf{var}(X) = \sum_{n=1}^{m}\mathbf{var}\left(\frac{i_n}{d_n}\right) \le \frac{m}{4}.$$

By Chebyshev's inequality,

$$\mathbf{P}(|X-\mathbf{E}(X)| \ge \varepsilon) \le \frac{\mathbf{var}(X)}{\varepsilon^2},$$

it follows that the expectation for the number of tuples $\mathbf{i}$ satisfying (10.26) and (10.27) is equal to the number of all tuples satisfying (10.26) times the probability for (10.27), and thus

$$M \le \frac{N}{4\varepsilon m} \le \frac{N}{2d}$$

(since $4m\varepsilon^2 \ge 2d$ by assumption). This is indeed an upper bound for M since otherwise we would get a contradiction with the laws of probability. (For an alternative proof which does not use probabilistic arguments see [**35**].)

Now applying Siegel's lemma 10.6 yields the existence of an integer vector $\mathbf{z}$ satisfying

$$B\mathbf{z} = 0 \qquad \text{and} \qquad 0 < |\mathbf{z}| \le (N\|B\|)^{\frac{dM}{N-dM}} \le N\|B\| \le C_1^{d_1+\ldots+d_m},$$

by replacing C_1 by $2C_1$. This finishes the proof of the index theorem. •

As we shall see below it is the factor $\frac{1}{2}$ in the lower bound for the index of P in the index theorem which leads to the exponent 2 in Roth's theorem.

10.8. Wronskians and Roth's lemma

Given rational functions $P_1, \ldots, P_k$ in variables $X_1, \ldots, X_m$, a determinant

$$\det\left((\mathrm{D}_{\mathbf{i}_\ell}P_j)_{1\le\ell,j\le k}\right),$$

where the $\mathrm{D}_{\mathbf{i}_\ell}$ are operators of the form (10.13) and of order $\le \ell - 1$, is called a **Wronskian**. For $m = 1$ there is exactly one differential operator of order $\ell - 1$ which yields only one Wronskian

$$\det\left(\left(\frac{\partial^{\ell-1}}{\partial X_1^{\ell-1}}P_j\right)_{1\le\ell,j\le k}\right),$$

which is not vanishing identically.

Lemma 10.8. *Suppose that $P_1, \ldots, P_k$ are rational functions in $X_1, \ldots, X_m$ having real coefficients, and are linearly independent over $\mathbb{R}$. Then at least one Wronskian of $P_1, \ldots, P_k$ is not identically zero.*

Proof by induction on k. The case $k = 1$ is trivial: the only Wronskian is P_1 itself. Now assume that the assertion of the lemma is true for $k \geq 2$. Suppose that $P_1, \ldots, P_k$ are rational functions which satisfy the assumption of the lemma. Then

$$1, \frac{P_2}{P_1}, \ldots, \frac{P_k}{P_1}$$

are linearly independent over $\mathbb{R}$ too. Moreover, there exists some j with $1 \leq j \leq m$ for which

$$\frac{\partial}{\partial X_j} \frac{P_2}{P_1}, \quad \ldots, \quad \frac{\partial}{\partial X_j} \frac{P_k}{P_1}$$

are linearly independent over $\mathbb{R}$. Using the induction hypothesis, there exists a Wronskian of these $k - 1$ rational functions which is not identically zero. It is not difficult to see that this Wronskian can be written as a linear combination of Wronskians of $P_1, P_2, \ldots, P_k$ with coefficients which are rational functions involving only partial derivatives of P_1 (this is left as Exercise 10.11 to the reader). This proves the assertion of the lemma. •

The next theorem, the so-called Roth's lemma, provides a sufficient condition for a polynomial to have a *small* index at $(\frac{p_1}{q_1}, \ldots, \frac{p_m}{q_m})$, where the $\frac{p_n}{q_n}$ will later be chosen as certain rational approximations to α.

Theorem 10.9. *Let m be a positive integer and $\varepsilon > 0$. There exists a constant $C_2 := C_2(m, \varepsilon) > 1$, depending only on m and ε, with the following property: let $d_1, \ldots, d_m$ be positive integers satisfying $d_k \geq C_2 d_{k+1}$ for $1 \leq k < m$, let $(p_1, q_1), \ldots, (p_m, q_m)$ be pairs of coprime integers with*

$$q_k^{d_k} \geq q_1^{d_1} \quad \text{and} \quad q_k \geq 2^{2mC_2} \quad \text{for} \ \ 1 \leq k \leq m.$$

Further, let $P \in \mathbb{Z}[X_1, \ldots, X_m]$ be a polynomial, not identically vanishing, of degree at most d_k in X_k for $1 \leq k \leq m$, and

$$\mathrm{H}(P) \leq q_1^{\frac{d_1}{C_2}}.$$

Then the index of P at $(\frac{p_1}{q_1}, \ldots, \frac{p_m}{q_m})$ with respect to $(d_1, \ldots, d_m)$ is less than or equal to ε.

Proof by induction on m. In the case $m = 1$ we write

$$P(X) = (q_1 X - p_1)^\ell R(X),$$

where $\frac{\ell}{d_1}$ is the index of P with respect to $\frac{p_1}{q_1}$ and d_1. By Gauss' lemma A.8, the coefficients of R are integers. Since q_1^ℓ divides the leading coefficient of P, it follows from the condition on the height of P that

$$q_1^\ell \leq \mathrm{H}(P) \leq q_1^{\frac{d_1}{C_2}}.$$

Choosing $C_2 = C_2(1, \varepsilon) = \frac{1}{\varepsilon}$ the assertion follows: the index is $\frac{\ell}{d_1} \leq \varepsilon$.

Now suppose that Roth's lemma is true for $m - 1$ but that it is not true for m. In order to prove the assertion we then have to find a polynomial satisfying the hypotheses of Roth's lemma and having an index $\delta > \varepsilon$.

We may assume that $C_2 = C_2(m-1, \delta)$ is defined for any $\delta \in (0, 1)$. We shall apply the induction hypothesis to some Wronskian which we construct as follows. Suppose that there are polynomials P_j, Q_j with rational coefficients such that

$$(10.29) \qquad P(X_1, \ldots, X_m) = \sum_{j=1}^{k} P_j(X_1, \ldots, X_{m-1}) Q_j(X_m).$$

We may assume that the length k of this decomposition is minimal. Taking into account the possibility of $Q_j(X_m) = X_m^{j-1}$ for $1 \leq j \leq d_m + 1$, we note that $k \leq d_m + 1$. Clearly, by the minimality of k, both families of polynomials, $P_1, \ldots, P_k$ and $Q_1, \ldots, Q_k$ are linearly independent over $\mathbb{R}$. By Lemma 10.8 there exists a Wronskian

$$U(X_1, \ldots, X_{m-1}) := \det\left((\mathrm{D}_{\mathbf{i}_\ell} P_j)_{1 \leq \ell, j \leq k}\right) \neq 0,$$

where the differential operators $\mathrm{D}_{\mathbf{i}_\ell}$ are of the form (10.13) but, since we are dealing here with $m-1$ variables $X_1, \ldots, X_{m-1}$, the vector $\mathbf{i}$ is replaced by $\mathbf{i}_\ell = (i_1, \ldots, i_{m-1})$ with $|\mathbf{i}_\ell| = i_1 + \ldots + i_{m-1} \leq \ell - 1 \leq k - 1$ for $1 \leq \ell \leq k$. Furthermore,

$$V(X_m) := \det\left(\left(\frac{\partial^{\ell-1}}{(\ell-1)! \partial X_m^{\ell-1}} Q_j(X_m)\right)_{1 \leq \ell, j \leq k}\right) \neq 0;$$

the non-vanishing of this Wronskian follows from the fact that all other Wronskians of this type are identically zero. Let

$$W(X_1, \ldots, X_m) := \det\left(\left(\frac{\partial^{j-1}}{(j-1)! \partial X_m^{j-1}} \mathrm{D}_{\mathbf{i}_\ell} P(X_1, \ldots, X_m)\right)_{1 \leq \ell, j \leq k}\right).$$

The entries $\mathrm{D}_{(i_1, \ldots, i_{m-1}, j-1)} P$ in the determinant on the right-hand side are partial derivatives of P and so they are polynomials with integer coefficients. Hence, W is a polynomial in $X_1, \ldots, X_m$ with integer coefficients. Moreover, it follows from the decomposition (10.29) that

$$(10.30) \qquad W(X_1, \ldots, X_m) = U(X_1, \ldots, X_{m-1}) V(X_m) \neq 0.$$

Let $\vartheta = \mathrm{ind}(P)$ and $\Theta = \mathrm{ind}(W)$ denote the indices of P and W at $(\frac{p_1}{q_1}, \ldots, \frac{p_m}{q_m})$ with respect to $(d_1, \ldots, d_m)$, respectively. As we shall show now, a lower bound for ϑ gives a lower bound for Θ of W. By the induction hypothesis and Lemma 10.5, we get, for $i_1 + \ldots + i_{m-1} \leq k - 1 \leq d_m$,

$$
\begin{aligned}
\mathrm{ind}(\mathrm{D}_{(i_1, \ldots, i_{m-1}, j-1)} P) &\geq \vartheta - \frac{i_1}{d_1} - \ldots - \frac{i_{m-1}}{d_{m-1}} - \frac{j-1}{d_m} \\
&\geq \vartheta - \frac{i_1 + \ldots + i_{m-1}}{d_{m-1}} - \frac{j-1}{d_m} \\
&\geq \vartheta - \frac{d_m}{d_{m-1}} - \frac{j-1}{d_m} \geq \vartheta - \frac{1}{C_2} - \frac{j-1}{d_m}.
\end{aligned}
$$

Using once more Lemma 10.5 we get for the index Θ of W the estimate

$$\Theta \geq \sum_{j=0}^{k-1} \max\left\{ 0, \vartheta - \frac{1}{C_2} - \frac{j-1}{d_m} \right\}$$

$$\geq \sum_{0 \leq j \leq \vartheta k} \left(\vartheta - \frac{j}{d_m} \right) - \frac{k}{C_2} \geq \vartheta[\vartheta k] - \frac{1}{d_m} \sum_{j=0}^{[\vartheta k]} j - \frac{k}{C_2}.$$

Removing the Gauss brackets we obtain the lower bound

$$(10.31) \qquad\qquad \Theta \geq \frac{1}{2}\vartheta^2 k - \frac{k}{C_2}.$$

On the contrary, the index Θ cannot be too large. Let $0 < \delta < \vartheta < 1$. The factorization (10.30) yields

$$W(X_1, \ldots, X_m) = U^*(X_1, \ldots, X_{m-1})V^*(X_m),$$

where U^* and V^* have integer coefficients. By Lemma 10.4 and the induction hypothesis there exists a constant $C_2 = C_2(m-1, \delta)$ such that

$$\mathrm{H}(\mathrm{D}_{(i_1,\ldots,i_{m-1},j-1)}P) \leq 2^{d_1 + \cdots + d_m}\mathrm{H}(P) \leq 2^{d_1 + \cdots + d_m} q_1^{\frac{d_m}{C_2}}.$$

The number of summands in the determinant expansion of W is less than or equal to $k! \leq k^{k-1} \leq 2^{kd_m}$. Using the induction hypothesis once again,

$$\mathrm{H}(W) \leq 2^{kd_m} \left(2^{d_1 + \cdots + d_m} q_1^{\frac{d_m}{C_2}} \right)^k \leq \left(2^{2m} q_1^{\frac{1}{C_2}} \right)^{d_1 k} < q_1^{\frac{2d_1 k}{C_2}}.$$

Hence

$$(10.32) \qquad \mathrm{H}(U^*) \leq q_1^{\frac{2d_1 k}{C_2}} \qquad \text{and} \qquad \mathrm{H}(V^*) \leq q_1^{\frac{2d_1 k}{C_2}} \leq q_m^{\frac{2d_m k}{C_2}}.$$

Next, applying the induction hypothesis to $U^*(X_1, \ldots, X_{m-1})$ at the point $\left(\frac{p_1}{q_1}, \ldots, \frac{p_{m-1}}{q_{m-1}} \right)$ with respect to $(kd_1, \ldots, kd_{m-1})$, we find

$$\mathrm{H}(U^*) \leq q_1^{\frac{2d_1 k}{C_2(m,\varepsilon)}} \leq q_1^{\frac{d_1 k}{C_2(m-1,\delta)}}$$

by taking $C_2 = C_2(m, \varepsilon)$ such that $C_2(m, \varepsilon) \geq 2C_2(m-1, \delta)$. It follows that the index of U^* with respect to this set of values is at most δ. Analogously, we can conclude by the induction hypothesis applied to $V(X_m)$ that the index of V^* at $\frac{p_m}{q_m}$ with respect to kd_m also is at most δ. In view of the factorization $W = U^*V^*$ the index of W at $\left(\frac{p_1}{q_1}, \ldots, \frac{p_m}{q_m} \right)$ with respect to $(kd_1, \ldots, kd_m)$ is at most 2δ. Thus for the index of W at the same point but with respect to $(d_1, \ldots, d_m)$ we find the estimate

$$(10.33) \qquad\qquad \Theta \leq 2\delta k.$$

We are at the end of the proof. A suitable choice of δ leads to a contradiction with our assumption $\vartheta > \varepsilon$. Therefore we take $C_2 = C_2(m, \varepsilon)$ so large that $C_2(m, \varepsilon) > \frac{4}{\varepsilon^2}$. Then it follows from (10.31) that

$$\Theta \geq \frac{1}{2}\varepsilon^2 k - \frac{k}{C_2} \geq \frac{1}{4}\varepsilon^2 k.$$

However, choosing $\delta < \frac{\varepsilon^2}{8}$ we obtain from (10.33) $\Theta < \frac{\varepsilon^2 k}{4}$, the desired contradiction. Roth's lemma is proved. ●

10.9. Final steps in Roth's proof

Suppose that α is algebraic of degree $d \geq 2$ and, for some $\delta \in (0, \frac{1}{2})$, the inequality

$$(10.34) \qquad \left| \alpha - \frac{p}{q} \right| < \frac{1}{q^{2+\delta}}$$

has infinitely many rational solutions $\frac{p}{q}$. Using the index theorem we shall construct a polynomial P in $\mathbf{X} = (X_1, \ldots, X_m)$ with a very large index at $\underline{\alpha} = (\alpha, \ldots, \alpha)$ with respect to arbitrary $(d_1, \ldots, d_m)$. This will imply that for a suitable choice of solutions $\frac{p_1}{q_1}, \ldots, \frac{p_m}{q_m}$ of (10.34) the polynomial P has still a large index at $(\frac{p_1}{q_1}, \ldots, \frac{p_m}{q_m})$ with respect to $(d_1, \ldots, d_m)$. However, by Roth's lemma we will find $d_1, \ldots, d_m$ for which the appropriate index of P has a small index. This contradiction will prove Roth's theorem.

It is easily seen that it suffices to prove Roth's theorem for algebraic integers, i.e., for algebraic numbers α which have a monic minimal polynomial. Furthermore, we may assume that such an α has absolute value less than one. Let P be the polynomial constructed in the proof of the index theorem with respect to α with $\varepsilon = \frac{\delta}{12}, m > \frac{d}{2\varepsilon^2}$, and arbitrary $d_1, \ldots, d_m$. Then the index of P at $\underline{\alpha}$ with respect to $(d_1, \ldots, d_m)$ is greater than $\frac{1}{2}m(1-\varepsilon)$. Now choose solutions $\frac{p_1}{q_1}, \ldots, \frac{p_m}{q_m}$ of (10.34) and integers $d_1 \geq d_2 \geq \ldots \geq d_m$ as follows:

- Take $\frac{p_1}{q_1}$ such that

$$(10.35) \qquad q_1 > \max\left\{ (6C_1)^{\frac{1}{\varepsilon}}, C_1^m, 2^{2mC_2} \right\}.$$

- Choose $\frac{p_2}{q_2}, \ldots, \frac{p_m}{q_m}$ for which

$$(10.36) \qquad \log q_{k+1} > (1 + \varepsilon)C_2 \log q_k \qquad \text{for} \quad 1 \leq k \leq m - 1.$$

- Take d_1 such that

$$(10.37) \qquad \varepsilon d_1 \log q_1 \geq \log q_m.$$

- For $2 \leq k \leq m$, choose d_k such that

$$(10.38) \qquad q_1^{d_1} \leq q_k^{d_k} \leq q_1^{d_1 + \varepsilon}.$$

The last choice is possible since (10.37) implies $q_1^{\varepsilon d_1} \geq q_m \geq q_k$. Of course, the appearing constants C_1 and $C_2 = C_2(m, \varepsilon)$ are those from Theorem 10.7 and 10.9, respectively.

In view of (10.35) and (10.38) we have

$$\frac{d_k}{d_{k+1}} = \frac{d_k \log q_k}{d_{k+1} \log q_{k+1}} \cdot \frac{\log q_{k+1}}{\log q_k} \geq \frac{1}{1+\varepsilon} \cdot (1+\varepsilon)C_2 = C_2 \geq 1.$$

Furthermore, we find

$$q_k^{d_k} \geq q_1^{d_1} > 2^{2mC_2 d_1} > 2^{2mC_2 d_k}.$$

This shows that the conditions in Roth's lemma, Theorem 10.9, are fulfilled. Applying Roth's lemma we deduce that the index of P at the point $(\frac{p_1}{q_1}, \ldots, \frac{p_m}{q_m})$ with respect to $(d_1, \ldots, d_m)$ satisfies

$$(10.39) \qquad \qquad \operatorname{ind}(P) \leq \varepsilon.$$

In order to prove Roth's theorem we have to show that P cannot have an index $\leq \varepsilon$ at $(\frac{p_1}{q_1}, \ldots, \frac{p_m}{q_m})$. This contradiction will follow once more from the index theorem.

Recall that the index of P is related to the vanishing of the partial derivatives. We suppose that, for $\mathbf{i} = (i_1, \ldots, i_m)$,

$$(10.40) \qquad \qquad \frac{i_1}{d_1} + \ldots + \frac{i_m}{d_m} \leq \varepsilon.$$

In view of (10.17)

$$(\mathrm{D_i}P)(\underline{\alpha}) = \sum_{\mathbf{j} = (j_1, \ldots, j_m)} (\mathrm{D_j}P)(0) \binom{j_1}{i_1} \cdot \ldots \cdot \binom{j_m}{i_m} \alpha^{j_1 - i_1} \cdot \ldots \cdot \alpha^{j_m - i_m}$$

and $|\alpha| < 1$, we get

$$\mathrm{H}((\mathrm{D_i}P)(\underline{\alpha})) \;\leq\; \mathrm{H}(P) \max \left\{ \binom{j_1}{i_1} \cdot \ldots \cdot \binom{j_m}{i_m} : j_k \leq d_k \right\}$$

$$(10.41) \qquad \qquad \leq\; (2C_1)^{d_1 + \ldots + d_m} \leq (2C_1)^{md_m}.$$

Expanding $\mathrm{D_i}P$ into a Taylor series around $\underline{\alpha}$, we find

$$\mathrm{D_i}P(\mathbf{X}) \;=\; \sum_{\mathbf{j}} (\mathrm{D_j}P)(\underline{\alpha}) \binom{j_1}{i_1} \cdot \ldots \cdot \binom{j_m}{i_m}$$

$$\times (X_1 - \alpha)^{j_1 - i_1} \cdot \ldots \cdot (X_m - \alpha)^{j_m - i_m}.$$

According to the index theorem 10.7, we have $(\mathrm{D_j}P)(\underline{\alpha}) = 0$ if

$$\frac{j_1}{d_1} + \ldots + \frac{j_m}{d_m} \leq \frac{1}{2} m(1 - \varepsilon),$$

or, by (10.40),

$$\frac{j_1 - i_1}{d_1} + \ldots + \frac{j_m - i_m}{d_m} \leq \frac{1}{2} m(1 - 3\varepsilon).$$

Moreover, it follows from (10.41) that

$$\sum_{\mathbf{j}} |(\mathrm{D_j}P)(\underline{\alpha})| \binom{j_1}{i_1} \cdot \ldots \cdot \binom{j_m}{i_m} \leq (2C_1)^{md_1} \sum_{\mathbf{j}} 2^{j_1 + \ldots + j_m} \leq (6C_1)^{md_1}.$$

Consequently, for

$$Q(\mathbf{X}) := \mathrm{D_i}P(\mathbf{X}) = \sum_{\mathbf{j}} Q(\mathbf{j})(X_1 - \alpha)^{j_1} \cdot \ldots \cdot (X_m - \alpha)^{j_m},$$

we get

$$\sum_{\mathbf{j}} |Q(\mathbf{j})| \leq (6C_1)^{md_1},$$

and $Q(\mathbf{j}) = 0$ if

$$\frac{j_1}{d_1} + \ldots + \frac{j_m}{d_m} \le \frac{1}{2}m(1 - 3\varepsilon).$$

Denoting those $\mathbf{j}$ which do not satisfy the latter inequality by $\mathbf{j}^*$, we find

$$
\left| Q\left(\frac{p_1}{q_1}, \ldots, \frac{p_m}{q_m}\right) \right| \le (6C_1)^{md_1} \max_{\mathbf{j}^*=(j_1,\ldots,j_m)} \left| \alpha - \frac{p_1}{q_1} \right|^{j_1} \cdot \ldots \cdot \left| \alpha - \frac{p_m}{q_m} \right|^{j_m}
$$

$$
\le (6C_1)^{md_1} \max_{\mathbf{j}^*} \left((q_1^{d_1})^{\frac{j_1}{d_1}} \cdot \ldots \cdot (q_m^{d_m})^{\frac{j_m}{d_m}} \right)^{-2-\delta}
$$

$$
\le (6C_1)^{md_1} \max_{\mathbf{j}^*} (q_1^{d_1})^{(-2-\delta)\left(\frac{j_1}{d_1}+\ldots+\frac{j_m}{d_m}\right)}
$$

by (10.34) and (10.38). Using (10.35) and once again (10.38) this expression can further be bounded by

$$
\left| Q\left(\frac{p_1}{q_1}, \ldots, \frac{p_m}{q_m}\right) \right| \le q_1^{\varepsilon m d_1} (q_1^{d_1})^{(-m(1-3\varepsilon)(1+\frac{\delta}{2})}
$$

$$
\le \left(q_1^{d_1} \cdot \ldots \cdot q_m^{d_m} \right)^{\varepsilon - \frac{1-3\varepsilon}{1+\varepsilon}\left(1+\frac{\delta}{2}\right)} < (q_1 \cdot \ldots \cdot q_m)^{-1},
$$

where in the last step we used $\delta = 12\varepsilon < \frac{1}{2}$. However, the left-hand side is a rational number with denominator dividing $q_1^{d_1} \cdot \ldots \cdot q_m^{d_m}$. It thus follows that

$$
(\mathbf{D_i}P)\left(\frac{p_1}{q_1}, \ldots, \frac{p_m}{q_m}\right) = Q\left(\frac{p_1}{q_1}, \ldots, \frac{p_m}{q_m}\right) = 0
$$

whenever (10.40) holds. Consequently, P has an index $>\varepsilon$ at $(\frac{p_1}{q_1}, \ldots, \frac{p_m}{q_m})$, contradicting (10.39). Hence there do not exist infinitely many $\frac{p_n}{q_n}$ satisfying (10.34). Since $\delta = 12\varepsilon$ can be made as small as we want, this proves Roth's theorem. •

Reviewing the proof we could obtain an upper bound for the number of $\frac{p}{q}$ satisfying the inequality in Roth's theorem; however, we cannot find an upper bound for p and q. So the proof of Roth's theorem is ineffective, more precisely, the choice (10.36); and there is no way known to circumvent this step of the proof. However, the analogue of Roth's theorem for function fields (where often things are easier) is effective due to Wang [168].

Notes for further reading

An important multidimensional generalization of Roth's theorem is W.M. Schmidt's celebrated

Subspace theorem. *Let* $Y_j(X_1, \ldots, X_n) = a_{1j}X_1 + \ldots + a_{nj}X_n$ *for* $1 \le j \le n$ *be linearly independent linear forms with real or complex algebraic coefficients* a_{ij}, *and let* $\varepsilon > 0$. *Then there exist proper linear subspaces* $S_1, \ldots, S_t$ *of* $\mathbb{Q}^n$ *such that the set of solutions to the system of inequalities*

$$|Y_1(\mathbf{X}) \cdot \ldots \cdot Y_n(\mathbf{X})| < \|\mathbf{X}\|^{-\varepsilon}$$

is contained in $S_1 \cup \ldots \cup S_t$, *where* $\|\mathbf{X}\| := \max\{|X_1|, \ldots, |X_n|\}$.

Roth's theorem follows from the case $n = 2$ with $Y_1 = X_1 - \alpha X_2, Y_2 = X_2$ (see Exercise 10.14). Just as Roth's theorem, the subspace theorem is best possible which follows from the fact that Corollary 8.8 (resp. Exercise 8.14) implies the existence of linear forms $Y_1, \ldots, Y_n$ with the property that, for any finite collection of proper linear subspaces $S_1, \ldots, S_t$ of $\mathbb{Q}^n$, the set of solutions $\mathbf{x} \in \mathbb{Z}^n$ to

$$|Y_1(\mathbf{X}) \cdot \ldots \cdot Y_n(\mathbf{X})| < 1$$

is not contained in $S_1 \cup \ldots \cup S_t$. The original source for Schmidt's theorem is [142]; more information on this, its applications and improvements are given by Evertse [59] and Schlickewei [141].

There is a remarkable conjectural link to complex analysis due to Vojta that provides a rather different view of Roth's theorem and the subspace theorem. The translation of theorems in Nevanlinna's value-distribution theory according to Vojta's dictionary led to interesting statements in the theory of diophantine approximations; some of them are deep theorems while others might be viewed as challenging open problems. For instance, one of the conjectural links is the exponent **2** in Roth's theorem, which is expected to be the *same* number **2** appearing in Nevanlinna's defect relation. We shall not explain here the defect relation but state one of its consequences where the same number **2** occurs again. Picard's theorem states that any *transcendental* function takes all values from $\mathbb{C} \cup \{\infty\}$ with at most **2** exceptions, e.g., $\exp(z) \neq 0, \infty$; here the notion of **transcendental** means that it is a non-rational meromorphic function. For a more detailed account see Vojta [166].

Exercises

10.1. Roth's theorem has many important implications. For instance, following the lines of Liouville's proof of the transcendence of Liouville numbers, one can obtain a transcendence criterion for continued fractions.

Let $\varepsilon > 0$ and suppose that $\alpha = [a_0, a_1, \ldots]$ is an irrational number which has an infinity of partial quotients satisfying $a_{n+1} \geq q_n^\varepsilon$, where q_n is the denominator of the nth convergent to α. Prove that α is transcendental.

10.2.* For an integer $g \geq 2$ and an arithmetic function $h(n) : \mathbb{N} \to \mathbb{N}$ define

$$\alpha = \sum_{n=1}^{\infty} a_n g^{-h(n)}, \qquad \text{where} \quad a_n \in \{0, 1, \ldots, g - 1\}.$$

i) Show that α is transcendental if

$$\limsup_{n \to \infty} \frac{h(n+1)}{h(n)} > 2.$$

This contains the proofs of transcendence in Exercise 9.6, but goes much beyond what one can deduce from Liouville's theorem.

ii) Deduce from i) the transcendence of $\sum_{n=1}^{\infty} 2^{-3^n}$.

Note that we cannot succeed to prove transcendence if the exponent -3^n is replaced by -2^n; nevertheless, the number

$$\sum_{n=1}^{\infty} 2^{-2^n} = [0, 3, 6, 4, 4, 2, 4, 6, 4, 2, 6, 4, \ldots]$$

is transcendental, and the partial quotients a_j of its continued fraction expansion are bounded; more precisely, $a_j \in \{2, 4, 6\}$ for $j \geq 2$ satisfying a certain aperiodic pattern. For more details see Shallit [**147**].

10.3. *Find all integral solutions to the equation $X^3 + Y^3 = m$ with $m = 1729$ and $m = 403$. What can be said about the set of solutions of $X^3 - Y^3 = m$ with $m = 0$ and $m = 1729$, respectively?*

10.4. It is one thing to know that a certain diophantine equation has only finitely many solutions. However, for a complete list of all solutions one needs to have knowledge on their size.

Let m be a positive integer. Show that every solution to the equation $X^3 + Y^3 = m$ in integers x, y satisfies the inequality

$$\max\{|x|, |y|\} \leq 2\sqrt{\frac{m}{3}}.$$

10.5. *For non-zero integers a and b, prove that the equation*

$$|aX^n - bY^n| = 1$$

has at most finitely many integer solutions if $n \geq 3$.

Applying several techniques from the theory of diophantine approximations, including Baker's estimates for linear forms and lattice reduction, Bennett & de Weger [**16**] proved that this equation has at most one solution in positive integers x and y, with the possible exception of a, b, and n satisfying $b = a + 1, 2 \leq a \leq \min\{0.3n, 83\}$, and $17 \leq n \leq 347$.

10.6.* *Prove Theorem 10.3. Is the condition on the irreducibility of $P(X, Y)$ necessary?*

10.7. *Compute the inverse of the transformation (10.10). What is the point on (10.9) corresponding to the point (10.12) on the elliptic curve (10.11)? Compute more such pairs.*

10.8. *Recall all facts from Section 10.4 on multivariate polynomials, in particular, Taylor's formula (10.17) from calculus.*

10.9. *Prove the last assertion in Lemma 10.5.*

10.10. *Prove equation (10.28).*

10.11. *Given a rational function Q, show that then every Wronskian of $QP_1, \ldots, QP_k$ can be written as a linear combination of Wronskians of $P_1, \ldots, P_k$ having coefficients which are rational functions involving only partial derivatives of Q.*

Hint: Do induction on k.

10.12. *Show that it is sufficient to prove Roth's theorem for algebraic integers.*

10.13. * *In the notation of Section 10.1, show that Roth's method applied to polynomials in two variables, that is, $m = 2$, leads only to the estimate $\tau(d) < c\sqrt{d}$ with some computable constant c.*

10.14. * *Prove that Roth's theorem is equivalent to the following assertion: let $Y_1(X_1, X_2) = \alpha X_1 + \beta X_2$ and $Y_2(X_1, X_2) = \beta X_1 + \delta X_2$ be linearly independent linear forms with algebraic coefficients. Then, for any $\varepsilon > 0$, there are only finitely many integers x_1, x_2 such that*

$$0 < |Y_1(x_1, x_2) Y_2(x_1, x_2)| < \max\{|x_1|, |x_2|\}^{-\varepsilon}.$$

This formulation is very similar to the statement of Schmidt's subspace theorem.

10.15. *Deduce from the subspace theorem that, given real algebraic numbers $\alpha_1, \ldots, \alpha_n$ such that $1, \alpha_1, \ldots, \alpha_n$ are linearly independent over $\mathbb{Q}$, then, for any positive ε, the inequality*

$$\max_{1 \leq j \leq n} \left| \alpha_j - \frac{p_j}{q} \right| < \frac{1}{q^{1 + \frac{1}{n} + \varepsilon}}$$

has only finitely many solutions $p_1, \ldots, p_n, q$ in integers with $q > 0$.

10.16. * *Prove the remark from the notes on the existence of linear forms $Y_1, \ldots, Y_n$ with the property that, for any finite collection of proper linear subspaces $S_1, \ldots, S_t$ of $\mathbb{Q}^n$, the set of solutions $\mathbf{x} \in \mathbb{Z}^n$ to*

$$|Y_1(\mathbf{X}) \cdot \ldots \cdot Y_n(\mathbf{X})| < 1$$

is not contained in $S_1 \cup \ldots \cup S_t$.

10.17. * Consider the inequality

$$|(X_1 + \sqrt{2}X_2 + \sqrt{3}X_3)(X_1 + \sqrt{2}X_2 - \sqrt{3}X_3)(X_1 - \sqrt{2}X_2 - \sqrt{3}X_3)| < \|X_n\|^{-\varepsilon}.$$

Show that every triple of integers (x_1, x_2, x_3), where $x_1 > 0, x_2 < 0$ is a solution of the Pell equation $X_1^2 - 2X_2^2 = 1$ and $x_3 = 0$ satisfies this inequality. Similarly, show that this inequality also has infinitely many solutions in coprime integers in the subspaces $X_1 = 0$ and $X_2 = 0$. What can be said about integer solutions satisfying $x_1 x_2 x_3 \neq 0$? Apply the subspace theorem!

This example is from Evertse [**59**].

CHAPTER 11

The abc-conjecture

Diophantine equations are a rather difficult topic. We may ask whether there exist solutions and, if so, whether there are infinitely or only finitely many. In special examples answers to both questions were found (e.g., the Pell equation) but so far no general method for treating diophantine equations is known. With the finiteness theorem 10.3 for the solutions to Thue equations in the previous chapter we have seen a tantalizing result with respect to the second question. However, the state of knowledge for another variety of diophantine equations is totally unclear. The most far-reaching concept is the recent abc-conjecture which is the main theme of this chapter. Since its first appearance in the 1980s, many applications to several quite different problems were found — from diophantine equations via the class number problem in algebraic number theory to the prime number distribution. Besides, the abc-conjecture is a very general and rather simple statement which makes its application, saying it with the words of Granville & Tucker [**70**], *as easy as abc*.

11.1. Hilbert's tenth problem

There is no complete theory for the variety of diophantine equations known; even stronger, it seems there may exist nothing like that. At the 1900 International Congress of Mathematicians in Paris, Hilbert posed the following problem:

> *Given a diophantine equation with any number of unknown quantities and with integral numerical coefficients: to devise a process according to which it can be determined by a finite number of operations whether the equation is solvable in rational integers.*

This is problem no. 10 in Hilbert's famous list of 23 problems which shaped much of the development of mathematics in the twentieth century. In our modern language, in his tenth problem Hilbert asked if there exists a universal algorithm which can determine whether a given polynomial diophantine equation with integral coefficients has a solution in integers. The meaning of Hilbert's question is a bit unclear. What is the definition of an *algorithm*? Based on the concept of a Turing machine, we understand an *algorithm* as a definite procedure that can be implemented as a computer program, consisting of *finitely* many successive steps with mathematical legitimate manipulations, which leads to an answer *yes* or *no*.

In 1961, Davis, Putman & Robinson [45] proved, by an ingenious combination of formal logic and number theory that, on the basis of a rather technical but plausible assumption, such an algorithm does not exist. Finally, in 1970, Matiyasevich [110] gave a definite negative answer to Hilbert's tenth problem by showing that the assumption of Davis et al. is true (for the most simple approach see [106] or [89]). This means that we never can find an algorithm to decide whether a given diophantine equation has or does not have a solution! However, it does not tell us that we cannot succeed for a large class of diophantine equations or that we cannot bound the number of solutions, existing or not.

Matiyasevich's discovery is based on the notion of diophantine sets. A set $\mathcal{A}$ of m-tuples of positive integers is called **diophantine** if there exists a multivariate polynomial P in unknowns $X_1, \ldots, X_n, Y_1, \ldots, Y_m$ with integral coefficients such that the equation

$$P(X_1, \ldots, X_n, a_1, \ldots, a_m) = 0$$

has solutions $x_1, \ldots, x_n$ in positive integers if and only if $(a_1, \ldots, a_m) \in \mathcal{A}$. It is easily seen that any finite subset of $\mathbb{N}^m$ is diophantine. Much more difficult to treat are infinite sets. In 1960, Putnam showed that a set is diophantine if and only if it coincides with the set of positive values of a suitable polynomial taken at the non-negative integers. We illustrate this simple but important equivalent in the case of two unknowns, i.e., $\mathcal{A} \subset \mathbb{N}$. Define

$$Q(X, a) = (a+1)\left(1 - P(X,a)^2\right) - 1.$$

Then a simple computation shows that $a \in \mathcal{A}$ if and only if a is a positive value assumed by the polynomial $Q(X, a)$. For example, this simple criterion can be used to show that the set of Fibonacci numbers is diophantine.

Every diophantine set $\mathcal{A}$ is **listable**, i.e., there exists an algorithm which produces a list of precisely the elements of $\mathcal{A}$. Matiyasevich proved the converse implication, namely, that every listable set is diophantine. By the already known existence of a listable set which is not recursive, this led to his negative answer to Hilbert's tenth problem. However, Hilbert's tenth problem is still open if we ask for rational solutions; for more details on the state of art of this problem we refer to Poonen [128].

As a by-product of his solution to Hilbert's tenth problem, Matiyasevich proved that the set of prime numbers is diophantine. In this context one has to examine the definition of prime numbers from the viewpoint of diophantine sets. We may say that a positive integer a is prime if and only if $n > 1$ and

$$\gcd(n, (n-1)!) = 1.$$

The latter condition is equivalent to the existence of positive integers a and b such that

$$a(n-1)! - bn = 1$$

(see Section 1.9). Using this characterization, the set of prime numbers is diophantine if and only if the set of all pairs $(n, n!)$ with $n \in \mathbb{N}$ is diophantine. In view of Putnam's criterion it follows from Matiyasevich's theorem that there exists a polynomial $P(X_1, \ldots, X_n)$ whose positive values at integers $x_j \geq 0$ are primes, and every prime can be represented this way. Surprisingly, it is even possible to write such polynomials; here is an example from Jones, Sato, Wada & Wiens (cf. [131]) of degree 25 in the 26 variables $a, b, c, \ldots, z$:

$$
\begin{aligned}
(k+2) \cdot \big\{ &1 - [wz + h + j - q]^2 - [(gk + 2g + k + 1)(h + j) + h - z]^2 \\
&- [2n + p + q + z - e]^2 - [16(k+1)^3(k+2)(n-1)^2 + 1 - f^2]^2 \\
&- [e^3(e+2)(a+1)^2 + 1 - o^2]^2 - [(a^2-1)y^2 + 1 - x^2]^2 \\
&- [16r^2y^4(a^2-1) + 1 - u^2]^2 - [((a + u^2(u^2-a))^2 - 1)(n+4dy)^2 \\
&+ 1 - (x - cu)^2]^2 - [n + \ell + v - y]^2 \\
&- [(a^2-1)\ell^2 + 1 - m^2]^2 - [ai + k + 1 - \ell - i]^2 \\
&- [p + \ell(a - n - 1) + b(2an + 2a - n^2 - 2n - 2) - m]^2 \\
&- [q + y(a - p - 1) + s(2ap + 2a - p^2 - 2p - 2) - x]^2 \\
&- [z + p\ell(a - p) + t(2ap - p^2 - 1) - pm]^2 \big\}.
\end{aligned}
$$

This polynomial also takes on negative values and a prime number may appear repeatedly as a value. In general, it is not known which is the minimal possible number of variables and what the minimal degree of such a polynomial is.

With regard to Gödel's celebrated incompleteness theorem we finally note another remarkable consequence: for any given axiomization there exists a diophantine equation without integral solutions which cannot be solved in this axiomatic setting.

11.2. The ABC-theorem for polynomials

Quite often in number theory, an arithmetic problem, which is difficult to handle over number fields, is attackable over function fields. If this is the case, there is some hope for learning something about the original problem. We shall see that in many cases it is much easier to decide whether a given diophantine equation has polynomial solutions or not.

Let $\mathrm{n}(P)$ denote the number of distinct complex zeros of a polynomial P (which does not vanish identically). Recall that two polynomials are said to be coprime if they do not share any of their linear factors, resp. if they have distinct roots. Stothers [159] and then Mason [108] (in more general form) proved in the 1980s the following surprising result.

Theorem 11.1. *Let* A, B, C *be coprime polynomials over* $\mathbb{C}$, *not all constant. If*

$$A + B = C,$$

then

$$\max\{\deg A, \deg B, \deg C\} < \mathrm{n}(ABC).$$

Proof. In view of the identity $A + B = C$ the polynomials A, B, C are pairwise coprime, i.e., their zeros are pairwise distinct. By the fundamental theorem of algebra A.7 the polynomials split into linear factors. Let

$$A = a \prod_{j=1}^{n(A)} (X - \alpha_j)^{a_j}, \quad B = b \prod_{k=1}^{n(B)} (X - \beta_k)^{b_k}, \quad \text{and} \quad C = c \prod_{l=1}^{n(C)} (X - \gamma_l)^{c_l},$$

where a, b, c are complex numbers, a_j, b_k, c_l are nonnegative integers, and the $\alpha_j, \beta_k, \gamma_l$ are the distinct complex zeros of $A, B,$ and C, respectively. For later use we note the partial fraction decomposition of the logarithmic derivative

$$(\log A)' = \sum_{j=1}^{n(A)} \frac{a_j}{X - \alpha_j},$$

where the derivative is taken with respect to X; analogous formulas hold for B and C, respectively. Without loss of generality we may assume that A has maximal degree and C does not vanish identically. Setting

$$F = \frac{A}{C} \quad \text{and} \quad G = \frac{B}{C}$$

we obtain $F + G - 1 = 0$. This implies $F' + G' = 0$ and

$$\frac{A}{B} = \frac{F}{G} = -\frac{G'/G}{F'/F}.$$

We compute

$$\frac{F'}{F} = (\log F)' = (\log A)' - (\log C)' = \sum_{j=1}^{n(A)} \frac{a_j}{X - \alpha_j} - \sum_{l=1}^{n(C)} \frac{c_l}{X - \gamma_l},$$

and a similar formula holds for G'/G as well. This yields

$$\frac{A}{B} = -\frac{\displaystyle\sum_{k=1}^{n(B)} \frac{b_k}{X - \beta_k} - \sum_{l=1}^{n(C)} \frac{c_l}{X - \gamma_l}}{\displaystyle\sum_{j=1}^{n(A)} \frac{a_j}{X - \alpha_j} - \sum_{l=1}^{n(C)} \frac{c_l}{X - \gamma_l}}.$$

Multiplying both the denominator and the numerator with

$$\mathcal{N} := \prod_{j=1}^{n(A)} (X - \alpha_j) \cdot \prod_{k=1}^{n(B)} (X - \beta_k) \cdot \prod_{l=1}^{n(C)} (X - \gamma_l),$$

leads to the representation

$$\frac{\mathcal{A}}{\mathcal{B}} := \frac{A \cdot \mathcal{N}}{B \cdot \mathcal{N}} = \frac{A}{B},$$

where $\mathcal{A}$ and $\mathcal{B}$ are polynomials, both having degree $< n(A) + n(B) + n(C) = n(ABC)$ (termwise). Since A and B are coprime, and since polynomials over the complex numbers factor uniquely into irreducible (even in linear) factors, it follows that $\deg A < n(ABC)$. The theorem is proved. ●

An alternative proof relies on the Riemann–Hurwitz formula from algebraic geometry. This idea of Silverman is outlined in Granville & Tucker [70], which is also an excellent survey on the abc-conjecture.

The ABC-theorem for polynomials has several interesting applications. For instance, the speed of the deterministic polynomial primality test of Agrawal, Kayal & Saxena [2] depends on lower bounds for the size of the multiplicative semigroup generated by several polynomials modulo another polynomial F. As pointed out by Voloch, an application of the ABC-theorem shows that, under certain weak assumptions, distinct polynomials A, B, C of sufficiently small degrees cannot all be congruent $\bmod F$. For this and improvements see Bernstein [17]. In the following sections we shall consider certain diophantine applications.

11.3. Fermat's last theorem for polynomials

An immediate and interesting consequence of Theorem 11.1 is an old result of Liouville, the proof of Fermat's last theorem for polynomials. However, this approach is much simpler than Liouville's original.

Let $n > 2$ and suppose that we have polynomials $\mathcal{X}, \mathcal{Y}, \mathcal{Z}$, not all constant, with

$$(11.1) \qquad \mathcal{X}^n + \mathcal{Y}^n = \mathcal{Z}^n.$$

Without loss of generality we may assume that $\mathcal{X}, \mathcal{Y}, \mathcal{Z}$ are coprime and that $\mathcal{X}$ has maximal degree. Put $A = \mathcal{X}^n, B = \mathcal{Y}^n$, and $C = \mathcal{Z}^n$. Since the distinct zeros of $ABC = (\mathcal{X}\mathcal{Y}\mathcal{Z})^n$ coincide with the distinct zeros of $\mathcal{X}\mathcal{Y}\mathcal{Z}$, application of Theorem 11.1 yields

$$n \deg \mathcal{X} = \deg A < \mathrm{n}(ABC) = \mathrm{n}(\mathcal{X}\mathcal{Y}\mathcal{Z}) \le \deg \mathcal{X}\mathcal{Y}\mathcal{Z} \le 3 \deg \mathcal{X}.$$

This proves

Theorem 11.2. *All polynomial solutions of the Fermat equation (11.1) with exponent $n \ge 3$ are equal up to constant factors.*

The condition on the exponent is necessary as the example

$$(2X)^2 + (X^2 - 1)^2 = (X^2 + 1)^2$$

shows. It is remarkable that the solvability of the Fermat equation in polynomials coincides with the one in integers with respect to the exponents. However, this analogy breaks down if we extend the class of functions. It is not too difficult to show (by Picard's theorem) that the only solutions to the Fermat equation in entire functions are constant if the exponent is greater than or equal to three (see [32]). But if we allow meromorphic solutions, then there exist non-trivial solutions if and only if the exponent is equal to three, and all these solutions are elliptic functions of entire functions (see I.N. Baker [12]).

11.4. The polynomial Pell equation revisited

We return shortly to the polynomial Pell equation

$$(11.2) \qquad\qquad P^2 - D \cdot Q^2 = 1.$$

In Section 6.8 we were concerned with the question of the existence of non-trivial solutions P, Q when D is given. Now we shall give a simple necessary condition for its solvability due to Dubickas & Steuding [48].

For this purpose we apply the ABC-theorem for polynomials. Suppose that the polynomials $P, Q, D \in \mathbb{C}[X]$ satisfy (11.2). Then Theorem 11.1 applied with $A = P^2, B = -DQ^2$ and $C = 1$ yields

$$\deg D + 2 \deg Q = \deg DQ^2 < \mathrm{n}(P^2 DQ^2).$$

Since

$$\mathrm{n}(P^2 DQ^2) = \mathrm{n}(PDQ) \leq \deg P + \mathrm{n}(D) + \deg Q,$$

and $\deg D = 2 \deg P - 2 \deg Q$, we get the inequality

$$\mathrm{n}(D) > \deg P - \deg Q = \frac{1}{2} \deg D.$$

Thus we have shown

Theorem 11.3. *If the number* $\mathrm{n}(D)$ *of distinct zeros of* $D \in \mathbb{C}[X]$ *is less than or equal to* $\frac{1}{2} \deg D$, *then the polynomial Pell equation (11.2) has no non-trivial solution in* $\mathbb{C}[X]$.

The bound of the theorem is sharp; that means it cannot be improved for any $\deg D$. For instance, the identity

$$(X^k + 1)^2 - (X^{2k} + 2X^k) \cdot 1^2 = 1$$

shows that there is a non-trivial solution with $D(X) = X^{2k} + 2X^k$, where $\deg D = 2k$ and $\mathrm{n}(D) = k + 1$.

Further, it follows that, for $\deg D = 2$, non-trivial solutions exist only if both roots of D are distinct. This condition is not only necessary, but also sufficient. Indeed, assume that

$$D(X) = c(X - \alpha)(X - \beta)$$

with complex numbers $c \neq 0$ and $\alpha \neq \beta$. Set

$$P(X) = \frac{2X - (\alpha + \beta)}{\alpha - \beta} \qquad \text{and} \qquad Q(X) = \frac{2}{\sqrt{c}(\alpha - \beta)}.$$

A simple computation shows that this pair solves the polynomial Pell equation, and, by Corollary 6.7, this solution is minimal. We obtain

Corollary 11.4. *If* $D \in \mathbb{C}[X]$ *is a quadratic polynomial, then the Pell equation (11.2) has a non-trivial solution in* $\mathbb{C}[X]$ *if and only if* D *has distinct roots, i.e.,* $\mathrm{n}(D) = 2$.

This implies that for quartic polynomials D the condition of Theorem 11.3 is necessary but not sufficient. We give a simple example. Set

$$D(X) = (X^2 + 1)(X - \gamma)^2,$$

where γ is a positive real number. Then $n(D) = 3 > 2 = \frac{1}{2}\deg D$, but
the Pell equation has no non-trivial solutions. Indeed, assume it has. Then
there exist polynomials $P, Q \in \mathbb{C}[X]$ with $\deg P > 0$ and $Q(\gamma) = 0$, such
that

$$P(X)^2 - (X^2 + 1)Q(X)^2 = 1.$$

All solutions of this equation are, however, described by the families (6.17)
and (6.18) from Section 6.8. Every polynomial $Q(X)$ of this form satisfies
$Q(\gamma) \neq 0$, giving the desired contradiction.

11.5. The abc-conjecture

As a subsitute for a general approach toward diophantine equations one
may ask for a theory for special families of equations. The abc-conjecture,
once proved, would apply to plenty of diophantine equations. Inspired by
the easy solution of Fermat's last theorem for polynomials Oesterlé [125],
and later in refined form Masser [109], worked out a *translation* of the ABC-
theorem for polynomials to diophantine equations of the form $a + b = c$.

By the fundamental theorem of algebra A.7 each polynomial over $\mathbb{C}$ splits
into linear factors. So the degree of a polynomial is nothing but the number
of linear factors, and the distinct zeros correspond one-to-one to the distinct
linear factors. The size of a number is measured by the absolute value and
the irreducible factors of an integer are its prime factors.

$$\text{(distinct) linear factors} \quad \longleftrightarrow \quad \text{(distinct) prime divisors,}$$
$$\text{degree} \quad \longleftrightarrow \quad \text{absolute value.}$$

This dictionary is simple. It suggests replacing the degree of a polynomial by
the absolute value of a number, and the number of distinct zeros by some
quantity involving the occuring prime divisors. However, the number of
distinct prime divisors is not a good candidate as the simple example $1 + 7 =
2^3$ shows; and any other Mersenne prime larger than 7 leads to something
even worse. So we need another quantitiy that *measures* compositeness of
an integer abc corresponding to the number of distinct zeros appearing on
the right-hand side of the estimate in the ABC-theorem for polynomials.

The **radical** (resp. **squarefree kernel**) of an integer n is defined by

$$\mathsf{R}(n) := \prod_{p|n} p.$$

Now consider the *trivial* identity

$$a + b = 3^{2^n} + (-1) = 3^{2^n} - 1 = c$$

for positive integers n. By induction on n it easily turns out that 2^n divides
$3^{2^n} - 1$. Thus

$$\mathsf{R}(abc) \leq 3 \cdot 2 \cdot \frac{c}{2^n} < 6 \cdot \frac{3^{2^n}}{2^n}.$$

Obviously, this cannot be an upper bound for the number $a = 3^{2^n}$ if n
is sufficiently large. This example is from Jastrzebowski & Spielman, and
shows that the abc-conjecture cannot hold in such a simple form.

So we see that it is not so easy to state a *good* conjectural analogue; nevertheless, the given counterexamples might be regarded as first approximations. In fact, we are only an ε away from the formulation of the

abc-conjecture. *If a, b, c are coprime integers which satisfy*

$$a + b = c,$$

then, for any $\varepsilon > 0$,

$$\max\{|a|, |b|, |c|\} \leq C(\varepsilon)\mathrm{R}(abc)^{1+\varepsilon},$$

where $C(\varepsilon)$ is a constant depending only on ε (and not on a, b, c).

The form of the abc-conjecture differs slightly from the estimate in the *ABC*-theorem for polynomials by the appearance of ε and the related implicit constant. It might be possible that also the above formulation is not the final version. For instance, it would be desirable to have an explicit form of the abc-conjecture (at least with respect to the constant $C(\varepsilon)$) in order to obtain effective estimates in applications.

The abc-conjecture asserts a deep relation between the additive and the multiplicative structures of the integers; such problems are known to be very hard to tackle. In the following section we shall consider *near-*counterexamples giving some numerical and heuristic support for the truth of abc.

11.6. LLL & abc

LLL & abc is the title of a recent paper [**47**] of Dokchitser. It shows how the LLL-lattice reduction algorithm can be used to find near counterexamples to the abc-conjecture. For non-zero integers a, b, c we define

$$\mathrm{P} = \mathrm{P}(a, b, c) = \frac{\log \max\{|a|, |b|, |c|\}}{\log \mathrm{R}(abc)}.$$

This quantity is called the **power** of the triple a, b, c. We reformulate the abc-conjecture as

abc-conjecture'. *For any real $\xi > 1$, there are only finitely many triples a, b, c of coprime integers which satisfy $a + b = c$ and*

$$\mathrm{P}(a, b, c) > \xi.$$

It is easily seen that this is an equivalent formulation of the abc-formula stated in the previous section (simply by taking the logarithm of the estimate in the abc-conjecture).

Examples of integers a, b, c satisfying the above version of the abc-conjecture with $\xi = 1$ are said to be abc-**triples**. The largest known example is

$$a = 2 , \quad b = 3^{10} \cdot 109 , \quad c = 23^5 \qquad \text{with} \quad \mathrm{P} = 1.629\ldots,$$

found by Reyssat. Triples with $\mathrm{P} > 1.4$ are said to be **good** (however, in view of these near counterexamples to the truth of the abc-conjecture, one could also call them *bad*). There is some hope that we can learn something about the abc-conjecture from good abc-triples.

Dokchitser proposed applying LLL-lattice reduction in order to look for good abc-triples. His simple but nice idea is as follows. Take large prime powers

(11.3) $$a_0 = p_1^{\nu_1} \, , \quad b_0 = p_2^{\nu_2} \, , \quad \text{and} \quad c_0 = p_3^{\nu_3}$$

of *comparable* size with *small* exponents $\nu_j \in \mathbb{N}$. By the LLL-lattice reduction algorithm one can find the least integer relation between them,

(11.4) $$\alpha a_0 + \beta b_0 + \gamma c_0 = 0.$$

The coefficients α, β, γ are relatively small, and so the numbers $a = \alpha a_0, b = \beta b_0, c = \gamma c_0$ are candidates for an abc-triple with a large power; of course, one can take any number with a small radical, e.g., products of few primes. Dokchitser succeeded by this means in finding 145 from 154 known good abc-triples.

We shall illustrate this method with an example. We take $a_0 = 1$, $b_0 = 3^3$, and $c_0 = 5^3$. It is easy to solve the diophantine equation (11.4) (by means of Exercise 1.18). The set of solutions α, β, γ is given by

$$\Lambda = \begin{pmatrix} 3^3 \\ -1 \\ 0 \end{pmatrix} \mathbb{Z} + \begin{pmatrix} 5^3 \\ 0 \\ -1 \end{pmatrix} \mathbb{Z}.$$

Applying the LLL-algorithm leads to the LLL-reduced basis

$$\begin{pmatrix} -10 \\ 5 \\ -1 \end{pmatrix} \quad \text{and} \quad \begin{pmatrix} 7 \\ 9 \\ -2 \end{pmatrix}.$$

Adding both basis vectors we get the abc-equation

$$-3 + 2 \cdot 3^3 \cdot 7 - 3 \cdot 5^3 = -3 \cdot 1 + 14 \cdot 3^3 - 3 \cdot 5^3 = 0.$$

This is an abc-triple with power $\mathsf{P} = 1.10992\ldots$ This is not too bad. The reader may try to find better abc-triples.

Following Dokchitser we conclude with an empirical analysis of this approach. We start with (11.3), all three roughly of size n. In the worst case the least integer combination of the form (11.4) has coefficients α, β, γ approximately of size $\sqrt{3n}$ (see Siegel's lemma 10.6). There are about $(\sqrt{3n})^3$ combinations

$$ia_0 + jb_0 + kc_0 \quad \text{with} \quad 0 \le i, j, k \le \sqrt{3n}.$$

Since they are all of size at most $3n\sqrt{3n} = (\sqrt{3n})^3$, the pigeonhole principle yields that at least two of them are equal (as in the proof of Theorem 10.6). Now their difference gives the desired relation. In the worst case the resulting triple a, b, c has power

$$\mathsf{P}(a, b, c) \approx \frac{\log(n\sqrt{3n})}{3 \log \sqrt{3n} + \log p_1 + \log p_2 + \log p_3}.$$

For fixed primes p_j this tends to 1 as $n \to \infty$, the bound predicted by the abc-conjecture.

Now having some trust in the validity of the abc-conjecture we may consider in the following sections some of its applications.

11.7. The Erdös–Woods conjecture

Are there patterns in the prime factorization of sequences of consecutive integers? Every dth integer is divisible by d, but the local distribution of prime numbers is quite irregular, and this makes such types of questions difficult to answer. In [58], Erdös asked how many pairs of products of consecutive integers have the same prime factors.

Erdös–Woods conjecture. *There exists an integer k such that, for positive integers m and n, the conditions*

$$R(m+i) = R(n+i) \qquad for \;\; 0 \le i \le k-1$$

imply $m = n$.

This conjecture cannot hold with $k = 2$ since

$$75 = 3 \cdot 5^2, \; 1215 = 3^5 \cdot 5 \qquad and \qquad 76 = 2^2 \cdot 19, \; 1216 = 2^6 \cdot 19,$$

and in fact there are infinitely many such examples (see Exercise 11.10). However, it is believed that $k = 3$ is an admissible value.

As first noticed by Langevin [100] the abc-conjecture has an interesting impact on this problem.

Theorem 11.5. *The abc-conjecture implies the truth of the Erdös–Woods conjecture with $k = 3$, apart from possibly finitely many exceptions.*

Proof. Applying the abc-conjecture with $a = m(m+2), b = 1$, and $c = (m+1)^2$ leads to

$$(11.5) \qquad (m+1)^2 \le C(\varepsilon)R(m(m+1)(m+2))^{1+\varepsilon}.$$

We suppose that $m > n$ and show that in this case there are only finitely many m for which the statement of the Erdös–Woods conjecture with $k = 3$ holds, i.e.,

$$R(m) = R(n), \quad R(m+1) = R(n+1), \qquad and \qquad R(m+2) = R(n+2).$$

As an immediate consequence of the latter condition we find

$$m - n = (m+i) - (n+i) \equiv 0 \bmod R(m+i)$$

for $0 \le i \le 2$. Since the greatest common divisor of any two of the three numbers $R(m), R(m+1), R(m+2)$ is one or two, it follows that $R(m(m+1)(m+2))$ divides $2(m-n)$. This yields in (11.5)

$$m^2 \le C(\varepsilon)R(m(m+1)(m+2))^{1+\varepsilon} \le C(\varepsilon)(2m)^{1+\varepsilon},$$

resp. $m \le (2^{1+\varepsilon}C(\varepsilon))^{\frac{1}{1-\varepsilon}}$. Thus, m is bounded by some constant.

This proves that $k = 3$ is an admissible value in the Erdös–Woods conjecture with at most finitely many exceptions subject to the truth of the abc-conjecture. The theorem is proved. •

11.8. Fermat, Catalan & co.

In view of the rather general equation $a + b = c$ we have plenty of interesting applications of the abc-conjecture which all fit surprisingly well to the current state of knowledge on diophantine equations.

Theorem 11.6. *The abc-conjecture implies Fermat's last theorem for sufficiently large exponents; more precisely, for sufficiently large n, there are no non-trivial integer solutions to the Fermat equation*

$$X^n + Y^n = Z^n.$$

Proof. If $x^n + y^n = z^n$ with coprime integers x, y, z, then put $a = x^n, b = y^n$, and $c = z^n$ in the abc-conjecture. Without loss of generality we may assume that x, y and z are positive integers. Since each prime factor of $abc = (xyz)^n$ divides xyz, we get

$$\mathsf{R}(x^n y^n z^n) = \mathsf{R}(xyz) \le xyz.$$

So the abc-conjecture implies

$$\max\{x^n, y^n, z^n\} \le C(\varepsilon)\mathsf{R}(xyz)^{1+\varepsilon} \le C(\varepsilon)(xyz)^{1+\varepsilon}.$$

This leads to

$$(xyz)^n \le C(\varepsilon)^3 (xyz)^{3+3\varepsilon}.$$

Taking the logarithm shows

$$(n - 3 - 3\varepsilon) \log(xyz) \le 3 \log C(\varepsilon).$$

Since $xyz \ge 2$ we obtain

$$n \le 3 \frac{\log C(\varepsilon)}{\log 2} + 3 + 3\varepsilon.$$

This shows that for all exponents n larger than this upper bound the Fermat equation has no non-trivial solutions. This is the assertion. •

The proof suggests that the size of the exponents and the number of variables in diophantine equations seem to be crucial quantities when we ask for a solution in integers! For instance, if we switch from the Fermat equation with the exponent 3 to the more general equation

$$X^3 + Y^3 + Z^3 + W^3 = 0,$$

the situation changes drastically. We have already met with (10.3) a solution of this diophantine equation. It can be shown that there are infinitely many non-trivial solutions in integers; see [**76**].

Another example for the strength of the abc-conjecture is the **Catalan equation**

(11.6) $X^p - Y^q = 1.$

This equation is an example of an *exponential* diophantine equation: it has to be solved in integers X, Y, p, q all ≥ 2, so the exponents are variables too. The case $p = q = 2$ is exceptional since then the polynomial on the left splits:

$$X^2 - Y^2 = (X - Y)(X + Y).$$

It is easily seen that there are no integer solutions x, y with $xy \neq 0$ in this case. If both exponents are equal and greater than 2, then we find by an application of Thue's theorem 10.3 that there are only finitely many integer solutions. Recently, Mihăilescu [**115**] proved the **Catalan conjecture** which claims that the only solution of (11.6) in positive integers x, y, p, q with $p, q \geq 2$ is given by $3^2 - 2^3 = 1$. It should be noted that Tijdeman proved previously via Baker's bounds for linear forms that there are only finitely many solutions; however, his upper bound is much too large to attack the problem with a computer; Mihăilescu's proof follows a different line. In both cases, the fact that the right-hand side of the Catalan equation is equal to one is crucial. For an arbitrary positive integer k, Pillai conjectured that the equation

$$(11.7) \qquad\qquad X^p - Y^q = k$$

has only finitely many solutions $x, y, p,$ and q in integers, all ≥ 2. But it is yet unknown whether for any fixed $k \geq 2$ there are only finitely many solutions. Again, the *abc*-conjecture implies in a very simple way the finiteness of the set of solutions of (11.7) in integers whenever $p, q \geq 2$.

It might be natural to combine Fermat's last theorem with the Catalan problem. The **Fermat–Catalan conjecture** claims that there are only finitely many powers x^p, y^q, z^r satisfying

$$(11.8) \qquad\qquad x^p + y^q = z^r,$$

where x, y, z are positive coprime integers, and p, q, r are positive integers with

$$(11.9) \qquad\qquad \frac{1}{p} + \frac{1}{q} + \frac{1}{r} < 1;$$

the restriction on the exponents excludes certain infinite families of solutions relating to curves with small genus (we will briefly explain this notion in the following section), as, for example, the Pythagorean triples on a conic. The only known examples of solutions to the Fermat–Catalan equation so far are

$$
\begin{aligned}
1^p + 2^3 &= 3^2 \\
2^5 + 7^2 &= 3^4 \\
13^2 + 7^3 &= 2^9 \\
2^7 + 17^3 &= 71^2 \\
3^5 + 11^4 &= 122^2 \\
33^8 + 1549034^2 &= 15613^2 \\
1414^3 + 2213459^2 &= 65^7 \\
9262^3 + 15312283^2 &= 113^7 \\
17^7 + 76271^3 &= 21063928^2 \\
43^8 + 96222^3 &= 30042907^2.
\end{aligned}
$$

The last five examples were found by Beukers and Zagier (cf. [**42**]). It is conjectured that there are no solutions at all when all exponents are ≥ 3. For any fixed triple of exponents Darmon & Granville [**43**] proved that the set of solutions is finite.

What is the impact of the abc-conjecture on the Fermat–Catalan equation? First, we can replace the right-hand side < 1 of condition (11.9) by $\leq \frac{41}{42}$ (since the sum of reciprocals cannot be larger than $\frac{41}{42}$). The abc-conjecture applied to (11.8) yields

$$\max\{|x|^p, |y|^q, |z|^r\} \leq C(\varepsilon) \mathsf{R}(x^p y^q z^r)^{1+\varepsilon}.$$

We may assume that x, y, z are all positive and that z is the largest. Since

$$\mathsf{R}(x^p y^q z^r) = \mathsf{R}(xyz) \leq xyz = (x^p)^{\frac{1}{p}}(y^q)^{\frac{1}{q}}(z^r)^{\frac{1}{r}} \leq (z^r)^{\frac{1}{p}+\frac{1}{q}+\frac{1}{r}} \leq z^{\frac{41}{42}r}.$$

Hence, $z^r \leq C(\varepsilon) z^{\frac{41}{42}r(1+\varepsilon)}$. Taking $\varepsilon = \frac{1}{83}$ we get $z \leq C(\frac{1}{83})^{83}$. So we have proved

Theorem 11.7. *The abc-conjecture implies the Fermat–Catalan conjecture.*

11.9. Mordell's conjecture

Hilbert's tenth problem is unsolvable. However, we may still hope that there is a positive answer to Hilbert's question if we are limited to diophantine equations in few variables. In the special case of plane curves big progress was made in the twentieth century. We have already mentioned the work of Siegel and the progress in the theory of elliptic curves. However, the most challenging result might have been the proof of a long-standing conjecture on the set of rational points on curves of a genus greater than one.

In 1922, Mordell [**119**] stated a very general conjecture on the set of rational solutions of a huge class of diophantine equations. Let $\mathbb{K}$ be a number field and let the curve $\mathcal{C}$ be given by a homogeneous diophantine equation

$$P(X, Y) = 0,$$

where P is a polynomial with coefficients in $\mathbb{K}$. A point (x, y) on the curve $\mathcal{C}$ is singular if the gradient of P vanishes at (x, y) (this coincides with our sloppy definition of singular points in Section 10.3). A singular point (x, y) is an ordinary double point if the Hessian of P evaluated at (x, y) does not have full rank. If now the only singular points of the curve $\mathcal{C}$ over the algebraic closure of $\mathbb{K}$ are double points, then the **genus** of $\mathcal{C}$ is given by

$$g = \frac{1}{2}(n-1)(n-2) - r,$$

where $n = \deg P$ and r is the number of double points; in the case of a non-singular curve we have $r = 0$. The genus is always an integer and it contains some important geometric and topological information about the curve. For example, if $\mathcal{C}$ is defined over the complex numbers, the curve $\mathcal{C}$ can be given the structure of a compact Riemann surface, which is, topologically speaking, a torus with g holes.

Mordell conjectured that any curve defined over $\mathbb{Q}$ of a genus greater than one has only finitely many rational points. This deep conjecture was proved by Faltings [60] in 1983 by means of algebraic geometry; later Bombieri [24] gave an elementary proof (for an excellent presentation we refer the interested reader to Hindry & Silverman [81]).

Faltings' theorem. *A curve of genus at least 2 has only finitely many rational points.*

Faltings was awarded a Fields medal at the 1986 International Congress of Mathematicians in Berkeley for the proof of this far-reaching result. In order to indicate its power we shall briefly discuss its impact on Fermat's last theorem. Rational solutions to equations in two variables are equivalent to integer solutions of equations in three variables where every monomial has the same total degree (which is easily seen by multiplying with all denominators). Thus Fermat's last theorem can be reformulated as a statement on rational points on the **Fermat curve**

$$X^n + Y^n - 1 = 0.$$

It is not difficult to see that the Fermat curve with the exponent n is non-singular and so it has genus $\frac{1}{2}(n-1)(n-2)$. If $n = 2$, the genus is zero, and there are infinitely many solutions, the Pythagorean triples. For the exponent $n = 3$, the genus is equal to 1, but already Euler proved that all integer solutions are trivial. If the exponent is $n \geq 4$, the genus is greater than 2, and Faltings' theorem implies that there are only finitely many non-trivial solutions to the Fermat equation for fixed n. This was the best statement in the direction of Fermat's last theorem before Wiles' final solution.

In 1991, Elkies [53] showed that the *abc*-conjecture implies Mordell's conjecture. His proof is far beyond our scope but we shall briefly state the main idea. One can reformulate the *abc*-conjecture in terms of the rational point $\frac{a}{c}$ in $\mathbb{P}^1 \setminus \{0, 1, \infty\}$, where $\mathbb{P}^1$ stands for the one-dimensional projective line. By Belyĭ's uniformization theorem any curve defined over a number field can be realized as a covering of the projective line ramified over at most three points which we may choose without loss of generality as $0, 1, \infty$. This reduces the Mordell conjecture via the Riemann–Hurwitz formula to the *abc*-conjecture. (For a nice presentation we refer once again to [81].)

Notes on *abc*

The *abc*-conjecture has also a deep impact on diophantine approximations. Bombieri [25] proved that the *abc*-conjecture implies Roth's theorem. We shall briefly illustrate this link to Roth's theorem. Let $P \in \mathbb{Z}[X, Y]$ be a binary homogeneous form without repeated factors, which means if $P(Z, 1)$ has total degree d', then $P \in \mathbb{Z}[Z]$ is a polynomial of degree $d \geq d' - 1$ with distinct roots $\alpha_1, \ldots, \alpha_d$, and

$$P(X, Y) = Y^d P\left(\frac{X}{Y}, 1\right).$$

Now let α be any of the roots of $P(Z,1)$. Using the ideas from Section 10.2 it is not difficult to show that Roth's theorem 10.1 is equivalent to the estimate

$$|P(p,q)| = Cq^d \prod_{j=1}^{\delta} \left| \alpha_j - \frac{p}{q} \right| \gg q^{d-2-\varepsilon},$$

where C is the absolute value of the leading coefficient and the implicit constant depends only on P and $\varepsilon > 0$; here we have used the notation of Roth's theorem 10.1, although α does not appear any longer in this formulation, and $\gg$ as converse to $\ll$. On the contrary, the *abc*-conjecture can be shown to be equivalent to the estimate

(11.10) $$\mathsf{R}(P(p,q)) \gg \max\{|p|, |q|\}^{d-2-\varepsilon},$$

valid for all coprime integers p, q with an implicit constant depending on P and ε (one of the implications is simple, and the other one relies again on Belyĭ's theorem). Since $|P(p,q)| \geq \mathsf{R}(P(p,q))$ trivially, it follows that the *abc*-conjecture implies Roth's theorem.

In fact, Langevin (cf. [**167**]) showed that *abc* yields a stronger inequality than Roth's theorem, namely,

$$\left| \alpha - \frac{p}{q} \right| > \frac{c(\varepsilon)}{\mathsf{R}(pq)q^\varepsilon}.$$

Furthermore, van Frankenhuysen [**63**] unified both results, Elkies' deduction of the Mordell conjecture and Bombieri's deduction of Roth's theorem from the *abc*-conjecture. He also conjectured a refinement of Roth's statement (with respect to the appearance of ε in the exponent) implying both results.

The *abc*-conjecture implies many results which are already proved or which are conjectured by different reasoning. Observing the power of this conjecture, we might be a bit sceptical about its truth. A proof of the *abc*-conjecture seems to be out of reach with present-day methods. There is some hope to generalize Wiles' proof of Fermat's last theorem along the lines in the theory of modular forms and Galois representations. Another attempt to tackle the *abc*-conjecture are Baker's bounds for linear forms. However, both approaches are not applicable without new ideas. The best unconditional result (up to the appearing ε) is the estimate

$$\max\{|a|, |b|, |c|\} \leq \exp\left(C\mathsf{R}(abc)^{\frac{1}{3}+\varepsilon} \right),$$

where the constant C is an effectively computable positive constant, from Stewart & Yu [**158**] (using contributions by Waldschmidt and Tijdeman in advance).

We did not mention the applications of the *abc*-conjecture to the problem of existence of Siegel zeros for Dirichlet L-functions according to Granville & Stark [**69**] and Euler's problem of *numeri idonei*. We did not touch the many generalizations, e.g., to number fields due to Vojta [**166**] or the congruence analogue of Ellenberg [**54**]. More on *abc* and other open challenges is carried out in the nicely written surveys [**11**], [**66**], and [**70**]. It seems that the *abc*-conjecture holds the key to much of the future direction of diophantine analysis.*

Exercises

11.1. There is no prime producing polynomial in only one variable:

Show that any non-constant polynomial in one variable with integer coefficients represents infinitely many composite numbers.

Nevertheless, there are polynomials in one variable which represent extraordinarily many primes. For instance, Euler found the polynomial

$$X^2 + X + 41,$$

which yields prime values for each integer x in the range $-40 \le x \le 39$. This extraordinary behavior of Euler's polynomial relies on divisibility properties in the ring of integers in the number field $\mathbb{Q}(\sqrt{-163})$, an observation which is due to Rabinowitsch; for further reading we refer to [**118**].

11.2.* *Prove that a set $\mathcal{A}$ of positive integers is diophantine if and only if there exists a polynomial $Q(X_1, \ldots, X_n)$ with integer coefficients such that*

$$\mathcal{A} = \{Q(x_1, \ldots, x_n) \ge 1 : (x_1, \ldots, x_n) \in \mathbb{N}^n\}.$$

This is Putnam's theorem on diophantine sets.

11.3. This exercise shows that the set of composite numbers is diophantine.

Prove that an integer a is composite if and only if there exist integers x, y for which

$$a = (x+1)(y+1).$$

Deduce that there exists a polynomial in two variables with integer coefficients such that its values at positive integer arguments coincide with the set of positive composite numbers. Use Lagrange's theorem 8.7 to find a polynomial in eight variables whose values at all integers form the set of positive composite numbers.

11.4.* This exercise proves that the set of Fibonacci numbers is diophantine. Consider the equation

$$|(Y-X)Y - X^2| = 1.$$

For any integer solution x, y show that $y - x, x, y$ are consecutive terms in the Fibonacci sequence. Describe the set of positive values of the polynomial

$$2 - \left((Y-X)Y - X^2\right)^2.$$

Determine a polynomial $P(X, Y)$ with integer coefficients such that if x, y are integers with $P(x, y) > 0$, then $P(x, y)$ belongs to the Fibonacci sequence. Deduce that the set of Fibonacci numbers is diophantine.

---

*Nitaj's Web page www.math.unicaen.fr/~nitaj/abc.html provides a lot of information on the *abc*-conjecture and related topics.

Hint: The first part of the exercise can be shown by induction. For more hints we refer to [**13**].

11.5. Let F, G be two polynomials of positive degree with $F^3 - G^2 \neq 0$. *Show that if F and G are coprime, then*

$$\deg(F^3 - G^2) \geq \max\left\{\frac{1}{2}\deg F - 1, \frac{1}{3}\deg G - 1\right\}.$$

This estimate is from Davenport [**44**]; the application of the ABC-theorem for polynomials simplifies Davenport's original proof significantly.

What can be said for polynomials F, G which have a common factor?

11.6. *Compute the prime factorizations of $2^n \pm 1$ for $n \leq 100$. What can be said about the prime divisors and their multiplicity?*

11.7. *For $n \in \mathbb{N}$ prove that $3^{2^n} - 1$ is divisible by 2^n.*

11.8. *Show that both versions of the abc-conjecture (see Sections 11.5 and 11.6) are equivalent.*

11.9.* *Use Dokchitser's method to compute good abc-triples.*

11.10. *For $m = 2^d - 2$ and $n = m(m+2)$, where d is an integer parameter, prove that*

$$R(m) = R(n) \qquad and \qquad R(m+1) = R(n+1).$$

This shows that $k = 2$ is not admissible in the Erdős–Woods conjecture for infinitely many numbers m and n.

11.11. *Let $\varepsilon > 0$. Assuming the truth of the abc-conjecture, prove that there exists a constant $c(\varepsilon)$ such that if $n < m$ are positive integers having the same prime divisors, then*

$$m - n \geq c(\varepsilon)n^{\frac{1}{2}-\varepsilon}.$$

This application of the abc-conjecture is due to Cochrane & Dressler [**39**]. It is conjectured that between any two positive integers having the same prime divisors there is a prime number. Note that under assumption of the Riemann hypothesis any interval $[x, x + x^{\frac{1}{2}+\varepsilon}]$ contains at least one prime for sufficiently large x.

11.12.* Denote by

$$q_1 - 1 < q_2 = 2^2 < q_3 = 2^0 < q_4 = 3^2 < q_5 < \cdots$$

the sequence of positive integer powers. It is an open problem to prove that the gaps between consecutive powers become arbitrarily large.

Given $\varepsilon > 0$, show under assumption of the abc-conjecture that, for all sufficiently large n,

$$q_{n+1} - q_n > n^{\frac{1}{12}-\varepsilon}.$$

11.13.* *Show that the abc-conjecture implies the finiteness of the set of so-lutions of the Catalan equation (11.6) in integers whenever $p, q \geq 2$. Generalize the proof to the more general equation (11.7). What can be said about polynomial solutions?*

11.14.* For $k \in \mathbb{Z}$, consider the diophantine equation

$$Y^2 = X^3 - k.$$

This defines an elliptic curve which we first met in the previous chapter. For coprime solutions x, y, Hall [**74**] conjectured

$$|x|^3 \ll |k|^{6+\varepsilon} \qquad \text{and} \qquad |y|^2 \ll |k|^{6+\varepsilon}$$

for any $k \neq 0$. This can be compared with the estimate of Davenport in Exercise 11.5.

Show that the abc-conjecture implies Hall's conjecture.

11.15. *Apply estimate (11.10) to $P(X, Y) = XY(X + Y)$ and deduce the abc-conjecture.*

11.16.* *Assuming the abc-conjecture, show that the equation $n! + 1 = m^2$ has only finitely many integer solutions. Try to find some solutions!*

The quest for all solutions to this diophantine equation is called **Brocard's problem**. The application of the abc-conjecture is from Overholt [**126**].

11.17. The **generalized abc-conjecture** states that, for every $n \geq 3$ and every positive ε, there exists a constant $C_n(\varepsilon)$, depending only on n and ε, such that whenever

$$x_1 + \ldots + x_n = 0$$

in coprime integers $x_1, \ldots, x_n$, and no subsum vanishes, then

$$\max\{|x_1|, \ldots, |x_n|\} \leq C_n(\varepsilon) \mathsf{R}(x_1 \cdot \ldots \cdot x_n)^{2n-5+\varepsilon}.$$

i) Why is the condition that no subsum vanishes necessary?
ii) Apply the generalized abc-conjecture to the analogue of the Fermat–Catalan equation

$$x_1^{p_1} + \ldots + x_n^{p_n} = 0 \qquad \text{and} \qquad \frac{1}{p_1} + \ldots + \frac{1}{p_n} < \frac{1}{n-2}.$$

Shapiro & Spencer [**148**] obtained a generalization of the ABC-theorem for the sum of n polynomials.

CHAPTER 12

p-adic numbers

p-adic numbers were invented about 100 years ago by Hensel [**80**]. In order to apply the powerful analytical methods of treating power series in one complex variable to number theory he investigated their p-adic analogues. This led him to the new and rather exotic world of p-adic numbers, besides the field of real numbers. In fact, for every prime number p there exists a field $\mathbb{Q}_p$ of p-adic numbers, and these fields $\mathbb{Q}_p$ may be regarded as completions of $\mathbb{Q}$ with respect to a certain curious norm. Meanwhile the theory of p-adic numbers has plenty of applications in various mathematical fields, and even in physics and biology.

12.1. Non-Archimedean valuations

By the unique prime factorization of integers every rational number $x \neq 0$ has a unique representation

$$x = \pm \prod_p p^{\nu_p(x)} , \qquad \text{where} \quad \nu_p(x) \in \mathbb{Z},$$

and the product is taken over all prime numbers p; but in fact only finitely many of the p-exponents $\nu_p(x)$ of x are non-zero. Thus, if we fix a prime p, then there exist non-zero integers a, b for which

$$x = \frac{a}{b} \cdot p^{\nu_p(x)} \qquad \text{with} \quad ab \not\equiv 0 \bmod p.$$

We define the p-**adic absolute value** on $\mathbb{Q}$ by setting

$$|x|_p = \begin{cases} p^{-\nu_p(x)} & \text{if } x \neq 0, \\ 0 & \text{if } x = 0. \end{cases}$$

The function $x \mapsto \nu_p(x)$ is called p-**adic valuation**.

An **absolute value** on a field $\mathbb{K}$ is a function $|\,.\,| : \mathbb{K} \to \mathbb{R}$ satisfying the axioms:

- $|x| \geq 0$ for all $x \in \mathbb{K}$, and $|x| = 0$ if and only if $x = 0$;
- $|x \cdot y| = |x| \cdot |y|$ for all $x, y \in \mathbb{K}$;
- $|x + y| \leq |x| + |y|$ for all $x, y \in \mathbb{K}$.

If the triangle inequality in the last axiom can be replaced by

(12.1) $$|x + y| \leq \max\{|x|, |y|\} \qquad \text{for all} \quad x, y \in \mathbb{K},$$

the absolute value is said to be **non-Archimedean**; otherwise the absolute value is called **Archimedean** (for some reason we will explain below). Clearly, for any absolute value $|\,.\,|$ on a field $\mathbb{K}$ we trivially have $|\pm 1| = 1$

and $|-x| = |x|$ for any $x \in \mathbb{K}$. Furthermore, $|x^n| = 1$ with $n \in \mathbb{N}$ implies that $|x| = 1$.

The well-known absolute value on $\mathbb{R}$, given by

$$|x|_\infty = \begin{cases} x & \text{if } x \geq 0, \\ -x & \text{if } x < 0, \end{cases}$$

is the standard example of an Archimedean absolute value; the notation $|.|_\infty$ is tradition in the context of p-adic numbers, and in the sequel we will sometimes write $p = \infty$. A first but not important example of a non-Archimedean absolute value is the **trivial** absolute value which is constant 1 on all non-zero elements.

More interesting are p-adic absolute values. It is easy to check that $|.|_p$ is indeed an absolute value, and it is not much more difficult to prove that even (12.1) is satisfied. For this purpose we state, equivalently,

$$\nu_p(x + y) \geq \min\{\nu_p(x), \nu_p(y)\} \qquad \text{for } x, y \in \mathbb{Q}.$$

This means that the p-exponent of the sum $x + y$ cannot be less than the minimum of the p-exponents of x and of y, but this is evident.

We note that a rational number x has a *small* p-adic absolute value if and only if x is divisible by a *large* power of p; the *size* of x has no effect here. This was the underlying idea for Hensel when introducing p-adic numbers. Divisibility properties of the integers are of fundamental interest in number theory!

12.2. Ultrametric topology

An absolute value on a field $\mathbb{K}$ induces a topology. Since an absolute value $|.|$ is a norm, we may define a metric by setting

$$d(x, y) = |x - y| \qquad \text{for } x, y \in \mathbb{K}.$$

$\mathbb{K}$ together with such an absolute value is a metric space. The standard absolute value yields the usual concept of distance in Euclidean vector spaces. For instance, the triangle inequality with respect to the standard absolute value shows that the sum of any two sides of a triangle in the Euclidean plane (or $\mathbb{C}$ isomorphically) is greater than the third side. More interesting is the case of topologies coming from non-Archimedean absolute values. If $|.|$ is a non-Archimedean absolute value, then we call the corresponding metric an **ultrametric** and $\mathbb{K}$ together with this ultrametric is said to be an **ultrametric space**. In view of (12.1), a metric is ultrametric if and only if

$$d(x, y) \leq \max\{d(x, z), d(y, z)\} \qquad \text{for all } x, y, z \in \mathbb{K};$$

the latter inequality is called the **ultrametric inequality** but we will use the same name also for (12.1). What does this inequality imply for the simple geometry of triangles?

In order to answer this question we start with

Lemma 12.1. *Let $\mathbb{K}$ be a field and let $|\,.\,|$ be a non-Archimedean absolute value on $\mathbb{K}$. If $x, y \in \mathbb{K}$ and $|x| \neq |y|$, then*

$$|x + y| = \max\{|x|, |y|\}.$$

Proof. Without loss of generality we may assume that $|x| > |y|$. By (12.1),

$$|x + y| \leq \max\{|x|, |y|\} = |x|.$$

Moreover,

$$|x| = |x + \underbrace{y - y}_{=0}| \leq \max\{|x + y|, |y|\}.$$

Since $|y| < |x|$, we get $|x| \leq |x + y|$. Combining both inequalities yields $|x + y| = |x|$, what we had to prove. •

Now let x, y, z be three elements of an ultrametric space. Consider the triangle with *vertices* x, y, z and assume that there are two sides of different length, say

$$d(x, z) = |x - z| \neq |y - z| = d(y, z).$$

Then, in view of the above lemma, $d(x, y) = |x - y|$ is equal to the bigger of the two:

$$d(x, y) = \max\{d(x, z), d(y, z)\}.$$

It thus follows that in an ultrametric space all triangles are isosceles!

Next we shall consider balls in an ultrametric space. We define the **open** and the **closed ball** of radius r and center a by

$$\mathcal{B}_{<r}(a) = \{x \in \mathbb{K} : d(x, a) < r\} \quad \text{and} \quad \mathcal{B}_{\leq r}(a) = \{x \in \mathbb{K} : d(x, a) \leq r\},$$

respectively. In view of the ultrametric inequality the open and the closed ball of positive radius r and center a are both open and closed. Sets which are *closed* and *open* are said to be **clopen**. It is not difficult to prove that any two open (closed) balls are either disjoint or contained in one another.

We call two absolute values **equivalent** if they induce the same topology, i.e., every set which is open with respect to one of the absolute values is open with respect to the other one too. Our next aim is a more convenient criterion which applies to any kind of absolute value, Archimedean or non-Archimedean (however, the Archimedean case is trivial).

Lemma 12.2. *Let $|\,.\,|$ and $|\,.\,|'$ be absolute values on a field $\mathbb{K}$. These absolute values are equivalent if and only if there exists a positive real number α such that*

(12.2) $$|x|' = |x|^{\alpha} \quad \text{for all} \quad x \in \mathbb{K}.$$

Proof. First, assume that $|\,.\,|$ and $|\,.\,|'$ are equivalent. Then, any sequence which converges with respect to one of these absolute values must also converge with respect to the other one (since the underlying topologies are the same). With respect to the topology induced by an absolute value $|\,.\,|$ on $\mathbb{K}$ we have $\lim_{n \to \infty} x^n = 0$ if and only if $|x^n| < 1$, and it follows that

$$|x|' < 1 \quad \Longleftrightarrow \quad |x| < 1 \quad \text{for} \quad x \in \mathbb{K}$$

or, equivalently, $|x|' > 1$ if and only if $|x| > 1$. It is clear that the same holds true if we replace here the inequality with an equality. Now we apply this observation to $x = \beta^m \gamma^n$, where $\beta, \gamma \in \mathbb{K}$ and $m, n \in \mathbb{Z}$. Taking logarithms we find

$$m \log |\beta| + n \log |\gamma| \lesseqgtr 0$$

if and only if

$$m \log |\beta|' + n \log |\gamma|' \lesseqgtr 0.$$

We may assume that $|\gamma|' \neq 1$. Then, by comparing the sets of tuples (m, n) satisfying the inequalities from above, we find

$$\log |\beta| = \lambda \log |\beta|', \qquad \text{where} \quad \lambda := \frac{\log |\gamma|}{\log |\gamma|'} > 0.$$

Since λ depends only on γ, it follows that $|\beta|' = |\beta|^\lambda$ for all $\beta \in \mathbb{K}$. This proves the first implication.

For the converse assume (12.2). Then it follows that $|x - a|' < r$ holds if and only if $|x - a| < r^{\frac{1}{\alpha}}$. This means that any open ball with respect to $|.|'$ is an open ball with respect to $|.|$ too. Hence, the topologies defined by these two absolute values are the same. The lemma is proved. •

12.3. Ostrowski's theorem

It is natural to ask what kind of absolute values a given field can carry. In 1918, Ostrowski gave a full description of the absolute values of the field of rational numbers.

Theorem 12.3. *Every non-trivial absolute value on $\mathbb{Q}$ is equivalent to one of the absolute values $|.|_p$, where either p is a prime number or $p = \infty$.*

Proof. Let $|.|$ be a non-trivial absolute value on $\mathbb{Q}$.

First, suppose that $|n| > 1$ for some positive integer n, and denote by m the least such n. Then there exists a positive real number α for which $|m| = m^\alpha$. Now we write any positive integer n to the base m, that is,

$$n = a_0 + a_1 m + \ldots + a_\ell m^\ell,$$

where the a_k are integers satisfying $0 \leq a_k < m$ and $a_\ell \neq 0$. Then, by the triangle inequality,

$$|n| \leq |a_0| + |a_1 m| + \ldots + |a_\ell m^\ell| = |a_0| + |a_1| m^\alpha + \ldots + |a_\ell| m^{\alpha \ell}.$$

Since the a_k are all less than m, it follows that $|a_k| < 1$. This leads to the estimate

$$|n| < 1 + m^\alpha + \ldots + m^{\alpha \ell} = m^{\alpha \ell} \left(m^{-\alpha \ell} + m^{\alpha(1-\ell)} + \ldots + 1 \right)$$

$$< n^\alpha \sum_{k=0}^{\infty} (m^{-\alpha})^k$$

(since $m^\ell < n$). The appearing infinite geometric series converges since $m^\alpha = |m| > 1$. Hence,

$$|n| \leq C n^\alpha$$

for all $n \in \mathbb{N}$ with some absolute constant C. Using the above estimate with n^N in place of n, where N is a large integer, we get by the multiplicativity of the absolute value

$$|n|^N = |n^N| \leq C n^{N\alpha}, \qquad \text{resp.} \qquad |n| \leq \sqrt[N]{C} n^\alpha$$

after taking Nth roots. This implies $|n| \leq n^\alpha$ for fixed n by letting $N \to \infty$. Next we shall show that the inequality the other way holds too.

With the same notation as before, $m^\ell \leq n < m^{\ell+1}$. From

$$|m^{\ell+1}| = |m^{\ell+1} \underbrace{-n+n}_{=0}| \leq |m^{\ell+1} - n| + |n|$$

it follows that

$$\begin{aligned}
|n| &\geq |m^{\ell+1}| - |m^{\ell+1} - n| \geq m^{\alpha(\ell+1)} - (m^{\ell+1} - n)^\alpha \\
&\geq m^{\alpha(\ell+1)} - (m^{\ell+1} - m^\ell)^\alpha = m^{\alpha(\ell+1)} \left(1 - \left(1 - \frac{1}{m} \right)^\alpha \right).
\end{aligned}$$

Thus we get $|n| \geq c n^\alpha$ for some absolute constant c. With the same trick as above it follows that $|n| \geq n^\alpha$.

Putting both estimates for $|n|$ together, we get $|n| = n^\alpha$. This immediately implies $|x| = |x|_\infty^\alpha$ for all $x \in \mathbb{Q}$. Hence, $|.|$ is equivalent to the absolute value.

Now let's assume that $|n| \leq 1$ for any $n \in \mathbb{N}$. Let m be the least positive integer n such that $|n| < 1$ (this exists since $|.|$ is non-trivial). Suppose that m would be composite, say $m = ab$ with positive integers a, b both $< m$. Then $|a| = |b| = 1$ by the minimality of m, and so it would follow that

$$|m| = |a| \cdot |b| = 1,$$

contradicting $|m| < 1$. Hence, m is prime. We now write p for the prime number m.

Our next aim is to show $|q| = 1$ for any prime q not equal to p. If this is not the case, i.e., $|q| < 1$, then

$$|q^N| = |q|^N < \frac{1}{2}$$

for some sufficiently large $N \in \mathbb{N}$. The same reasoning shows $|p^M| < \frac{1}{2}$ for some $M \in \mathbb{N}$. Since p^M and q^N are coprime, by Theorem 1.6 there exist integers x, y such that

$$xp^M + yq^N = 1.$$

The triangle inequality yields

$$1 = |1| = |xp^M + yq^N| \leq |xp^M| + |yq^N| = |x| \cdot |p^M| + |y| \cdot |q^N|.$$

Since $|x|, |y| \leq 1$, it follows that

$$1 \leq |p^M| + |q^N| < \frac{1}{2} + \frac{1}{2} = 1,$$

the desired contradiction. Thus $|q| = 1$.

By the unique factorization of integers, any positive integer $n > 1$ can be written as a product of powers of its prime divisors, $n = p_1^{\nu_1} \cdot \ldots \cdot p_k^{\nu_k}$ say. This leads to

$$|n| = |p_1|^{\nu_1} \cdot \ldots \cdot |p_k|^{\nu_k}.$$

In view of our previous observations it follows that $|n| = 1$ if p does not appear in the prime factorization of n. Otherwise, if p divides n, we get

$$|n| = |p|^{\nu_p(n)} = p^{-\alpha \nu_p(n)} = |n|_p^{\alpha}$$

for some positive real number α, independent of n. It is not difficult to see that the same holds true for any non-zero rational number. The theorem is proved. ●

The proof also implies the following statement:

Corollary 12.4. *An absolute value $|.|$ on $\mathbb{Q}$ is non-Archimedean if and only if*

$$\sup\{|n| : n \in \mathbb{N}\} < \infty.$$

The name *Archimedean* absolute value has its origin in Archimedes' lemma which states that for all non-zero integers (resp. rationals) x, y there exists a positive integer m with $|mx|_\infty > |y|_\infty$. This fails for any non-Archimedean absolute value, because $|mx|_p \leq |x|_p$ for all prime p and any integer m.

12.4. Curious convergence

p-adic absolute values have curious convergence properties. We start with a nice example due to Gouvêa [68]. Consider the linear equation

$$X = 1 + pX.$$

It is a simple task to find the solution by separating all X-terms. But we shall try something completely different. The iteration

$$x_0 := 0 , \quad x_n := 1 + px_{n-1} \quad \text{for} \quad n = 1, 2, 3, \ldots$$

leads to the sequence of numbers

$$x_n = 1 + p + p^2 + \ldots + p^{n-1} = \frac{1 - p^n}{1 - p}.$$

Since

$$\left| \frac{p^n}{1 - p} \right|_p = |p^n|_p \cdot |1 - p|_p^{-1} = p^{-n}$$

tends to zero as $n \to \infty$, we see that the sequence $(x_n)_n$ is convergent with respect to the p-adic absolute value! Moreover, the limit of the x_n is a geometric series with p-adic value

$$(12.3) \qquad \sum_{k=0}^{\infty} p^k = \frac{1}{1 - p};$$

this series is p-adic convergent, but, of course, divergent with respect to the standard absolute value. Actually, the p-adic limit is the solution of the equation in question. For example,

$$-1 = \frac{1}{1-2} = 1 + 2 + 2^2 + 2^3 + \dots$$

solves the equation $X = 1+2X$. This looks a little curious since the (infinite) sum of positive integers is equal to a negative number. The feature of p-adic convergence was used by Voronoï in his theory of summation of divergent series.

Similarly, we can also consider other series of similar type with respect to the p-adic absolute value as well. Indeed, the general series

(12.4) $$\alpha = \sum_{k \geq \nu} a_k p^k \quad \text{with} \quad a_k \in \mathbb{Z}, \ 0 \leq a_k < p,$$

where $\nu \in \mathbb{Z}$ (but $\nu = -\infty$ is forbidden), is p-adic convergent. Moreover, the sequences of the partial sums $\alpha_n = \sum_{k \geq \nu}^n a_k p^k$ are all Cauchy sequences. This follows immediately from the following:

Lemma 12.5. *A sequence of rational numbers* $(\alpha_n)_n$ *is a Cauchy sequence with respect to a p-adic absolute value* $|\,.\,|_p$ *if and only if*

$$\lim_{n \to \infty} |\alpha_{n+1} - \alpha_n|_p = 0.$$

Proof. Write $m = n + r > n$. In view of the ultrametric inequality (12.1) we have

$$|\alpha_m - \alpha_n|_p = |\alpha_{n+r} \underbrace{-\alpha_{n+r-1} + \alpha_{n+r-1}}_{=0} -\alpha_{n+r-2} \pm \dots + \alpha_{n+1} - \alpha_n|_p$$

$$\leq \max\{|\alpha_{n+r} - \alpha_{n+r-1}|_p, \dots, |\alpha_{n+1} - \alpha_n|_p\}.$$

This yields the assertion of the lemma. •

These observations give a first glimpse of p-adic analysis and show how much it differs from real analysis. For instance, it follows that the series $\sum_{k=0}^{\infty} k!$ is p-adic convergent, since the difference of two consecutive partial sums

$$\alpha_{n+1} - \alpha_n = \left\{ \sum_{k=0}^{n+1} - \sum_{k=0}^{n} \right\} k! = (n+1)!$$

is divisible by increasing powers of p as $n \to \infty$.

12.5. Characterizing rationals

As an application of the observations of the previous section we shall now prove a characterization of rational numbers in terms of p-adic convergence.

As a matter of fact any rational number α has a unique representation of the type (12.4). We call this series representation the p-**adic expansion** of α and the appearing a_k are said to be the **digits** of α. Recall the characterization of rationals based on the decimal fraction expansion: a decimal fraction represents a rational number if and only if the sequence of digits is eventually periodic. We shall prove the p-adic analogue.

Theorem 12.6. *The series (12.4) has a rational limit if and only if the sequence of the digits is eventually periodic.*

Proof. Without loss of generality we may assume that $\nu = 0$ and $a_0 \neq 0$ in the p-adic expansion (12.4) of α (the verification of this step is left to the reader).

First, assume that the sequence of digits a_k is eventually periodic. Writing the digits as an infinite vector, there are integers b_j and c_ℓ with $1 \leq j < h$ and $1 \leq \ell < g$ for some non-negative integers h, g such that

$$(a_0, a_1, a_2, \ldots) = (b_0, b_1, \ldots, b_{h-1}, \overline{c_0, c_1, \ldots, c_{g-1}});$$

as in the case of continued fractions the bar denotes periodic repetition. Setting

$$b = b_0 + b_1 p + \ldots + b_{h-1} p^{h-1} \qquad \text{and} \qquad c = c_0 + c_1 p + \ldots + c_{g-1} p^{g-1},$$

an application of (12.3), with p replaced by p^g, leads to

$$(12.5) \qquad \alpha = \sum_{k \geq 0} a_k p^k = b + c p^h \left(1 + p^g + p^{2g} + \ldots\right) = b + c p^h \frac{1}{1 - p^g}.$$

This shows that α is rational.

For the converse implication assume that α is rational, say $\alpha = \frac{d}{f}$ for some coprime integers d and f. Since $\nu = 0$ and $a_0 \neq 0$, it follows that f is not divisible by p, and by Fermat's little theorem A.4 there exists a positive integer g for which

$$p^g \equiv 1 \bmod f.$$

Hence

$$\alpha = \frac{d}{f} = \frac{k}{p^g - 1}$$

for some integer k. For this k there exists a positive integer h such that either $0 \leq k < p^h$ or $-p^h \leq k < 0$ according to α being non-negative or negative. Since p^h and $p^g - 1$ are coprime, by Theorem 1.6 there exist integers b and c such that

$$k = b\left(p^g - 1\right) - c p^h,$$

where either $0 \leq c < p^g - 1$ or $0 < c \leq p^g - 1$, depending on the sign of α. This leads to

$$\alpha = \frac{k}{p^g - 1} = \frac{b\left(p^g - 1\right) - c p^h}{p^g - 1} = b + c p^h \frac{1}{1 - p^g}.$$

This is the representation (12.5) from the first part of the proof. Now expanding b and c into its finite p-adic expansions the converse implication follows. •

12.6. Completions of the rationals

Recall that a field $\mathbb{K}$ is said to be **complete** if any Cauchy sequence in $\mathbb{K}$ converges in $\mathbb{K}$. We know that $\mathbb{Q}$ is not complete with respect to the usual Archimedean absolute value (see Theorem 1.4). By Theorem 12.6 the same is true if we consider any p-adic absolute value (since any p-adic series (12.4) with an aperiodic sequence of digits has a p-adic irrational value). Thus the field of rational numbers is not complete to any of its non-trivial absolute values. As a solution to this problem we may complete $\mathbb{Q}$ with respect to a p-adic absolute value, analogously to Cantor's construction of the real numbers as completion of $\mathbb{Q}$ with respect to the standard absolute value (see [51]). We briefly sketch this efficient but in detail rather technical process (for more details we refer to [68]).

Denote by $\mathcal{C}_p$ the set of all p-adic Cauchy sequences (α_n). We define addition and multiplication by

$$(\alpha_n) + (\beta_n) = (\alpha_n + \beta_n) \qquad \text{and} \qquad (\alpha_n) \cdot (\beta_n) = (\alpha_n \cdot \beta_n).$$

This makes $\mathcal{C}_p$ to a commutative ring. The subset $\mathcal{M}_p$ consisting of all Cauchy sequences (α_n) with $\lim_{n \to \infty} \alpha_n = 0$ is an ideal (we leave these claims as an exercise to the reader). In fact, $\mathcal{M}_p$ is even a maximal ideal which can be seen as follows. (For the notion of ideals and maximality see Appendix A.1.)

Suppose we have an arbitrary Cauchy sequence $(\alpha_n) \in \mathcal{C}_p \setminus \mathcal{M}_p$, i.e., $\lim_{n \to \infty} \alpha_n \neq 0$. Then there exist a constant c and an integer N such that

$$|\alpha_n|_p \geq c > 0 \qquad \text{for} \quad n \geq N.$$

Define (β_n) by setting $\beta_n = 0$ if $n < N$, and $\beta_n = \frac{1}{\alpha_n}$ otherwise. For $n \geq N$,

$$|\beta_{n+1} - \beta_n|_p = \left| \frac{1}{\alpha_{n+1}} - \frac{1}{\alpha_n} \right|_p = \left| \frac{\alpha_{n+1} - \alpha_n}{\alpha_{n+1} \alpha_n} \right|_p \leq \frac{1}{c^2} |\alpha_{n+1} - \alpha_n|_p.$$

Hence, it follows from Lemma 12.5 that (β_n) is a Cauchy sequence since (α_n) is one. Note that $\lim_{n \to \infty} \alpha_n \beta_n = 1$, and thus

$$(1) - (\alpha_n) \cdot (\beta_n) \in \mathcal{M}_p.$$

This shows that the ideal which is generated by (α_n) and $\mathcal{M}_p$ is equal to the ideal generated by (1), which is the whole ring $\mathcal{C}_p$. This shows that $\mathcal{M}_p$ is maximal.

From algebra (Theorem A.1) we know that the quotient of a commutative ring by its maximal ideal is a field. Hence, for prime p,

$$\mathbb{Q}_p := \mathcal{C}_p / \mathcal{M}_p$$

is a field, the **field of p-adic numbers**. We can embed $\mathbb{Q}$ into $\mathbb{Q}_p$ in a natural way via the mapping $\alpha \mapsto (\alpha, \alpha, \dots)$. Next we extend the p-adic absolute value from $\mathbb{Q}$ to $\mathbb{Q}_p$ as follows. Given any sequence (α_n) in $\mathbb{Q}_p$ (resp. any representative class of sequences), the sequence $(|\alpha_n|_p)$ is a

Cauchy sequence of real numbers. Since $\mathbb{R}$ is complete, this sequence has a limit, and so we define

$$|(\alpha_n)|_p = \lim_{n \to \infty} |\alpha_n|_p.$$

It is not difficult to show that this is well-defined and defines a non-Archimedean absolute value on $\mathbb{Q}_p$ (which we still call a p-adic absolute value). Since the p-adic valuation $\nu_p(\,.\,)$ takes integer values on $\mathbb{Q}$, it follows that also $\nu_p(\alpha) \in \mathbb{Z}$ for all $\alpha \in \mathbb{Q}_p$.

Next we shall prove that $\mathbb{Q}$ is dense in $\mathbb{Q}_p$. To see this we have to show that any open ball around some $\alpha = (\alpha_n)_n \in \mathbb{Q}_p$ contains (the image of) a rational number. Since (α_n) is a Cauchy sequence, for any given $\varepsilon > 0$ there exists an integer N such that

$$|\alpha_m - \alpha_n|_p \le \frac{\varepsilon}{2} \qquad \text{for all} \quad m, n \ge N.$$

Now define a constant sequence (y) by setting $y = \alpha_N$. Then $y - \alpha$ is represented by the sequence $(y - \alpha_n)$ and we obtain, for $n \ge N$,

$$|y - \alpha|_p = \lim_{n \to \infty} |\alpha_N - \alpha_n|_p \le \frac{\varepsilon}{2} < \varepsilon.$$

Hence, (y) lies in a ball of radius ε around (α_n), which proves the claim.

Finally, we observe that $\mathbb{Q}_p$ is the completion of $\mathbb{Q}$ with respect to $|\,.\,|_p$. To see this we have to show that every Cauchy sequence in $\mathbb{Q}_p$ has a limit in $\mathbb{Q}_p$. Any Cauchy sequence $(\lambda^{(m)})$ of elements in $\mathbb{Q}_p$ is a sequence of Cauchy sequences $\lambda^{(m)} = (\alpha_n^{(m)})$ of elements in $\mathbb{Q}$. Since $\mathbb{Q}$ is dense in $\mathbb{Q}_p$, we can find a subsequence of rational numbers α_{n_m} of $(\alpha_n^{(m)})$ such that

$$\lim_{m \to \infty} |\lambda^{(m)} - \alpha_{n_m}|_p = 0.$$

Now (α_{n_m}) is a Cauchy sequence in $\mathbb{Q}$. Thus, it follows that $\lim_{m \to \infty} \lambda^{(m)}$ converges to some element of $\mathbb{Q}_p$. We leave the technical details to the reader.

We collect all observations from above in

Theorem 12.7. *For any prime p, there exists a field $\mathbb{Q}_p$ with a non-Archimedean absolute value $|\,.\,|_p$ induced by the p-adic absolute value on $\mathbb{Q}$ such that*

- *$\mathbb{Q}$ is dense in $\mathbb{Q}_p$ and*
- *$\mathbb{Q}_p$ is complete with respect to $|\,.\,|_p$.*

It is not too difficult to see that the field $\mathbb{Q}_p$ satisfying the conditions on denseness and completion is unique up to isomorphisms preserving the p-adic absolute value.

In view of Ostrowski's theorem 12.3, the non-equivalent non-trivial absolute values on $\mathbb{Q}$ lead to a family of *nice* fields lying above $\mathbb{Q}$:

$$\mathbb{Q} \quad \xrightarrow{\text{completion}} \quad \mathbb{Q}_2, \ \mathbb{Q}_3, \ \mathbb{Q}_5, \ \ldots \quad \text{and} \quad \mathbb{Q}_\infty := \mathbb{R}.$$

In the twentieth century, the p-adic perspective of considering $\mathbb{Q}$ as a subset of a spectrum of topological fields obtained by completing $\mathbb{Q}$ with respect to all possible norms became a unifying and successful theme in number theory.

There is a remarkable feature of p-adic numbers in the real world. The p-adic metric provides a notion of nearness for sequences sharing some initial segment. This idea was used by Albeverio et al. [3] in order to construct a model for memory retrieval by a p-adic dynamical system. Roughly speaking, ideas are represented by p-adic sequences and the same initial segment stands for an association between the underlying ideas.

12.7. *p*-adic numbers as power series

What do p-adic numbers look like? The depiction as representative classes of p-adic Cauchy sequences is not very convenient (as we know already from the real numbers). We shall find a representation similar to the decimal fraction expansion of real numbers. Indeed we already know some p-adic examples of this type. Recall the series representations (12.4). By Lemma 12.5 these series are the limits of Cauchy sequences of rational numbers. We may identify them with elements in $\mathbb{Q}_p$, i.e., each series (12.4) defines a p-adic number α, and any p-adic number α has a representation of the form (12.4). The first assertion is clear; the second one can be shown by an approximation argument as follows.

First, suppose that α is a p-adic number with $\nu_p(\alpha) \geq 0$ and n is a positive integer. By Theorem 12.7, $\mathbb{Q}$ is dense in $\mathbb{Q}_p$. Thus we can find a rational number $\frac{x}{y}$ such that

$$\left| \alpha - \frac{x}{y} \right|_p \leq p^{-n}.$$

In view of (12.1)

$$\left| \frac{x}{y} \right|_p = \left| \underbrace{\frac{x}{y} - \alpha + \alpha}_{=0} \right|_p \leq \max \left\{ |\alpha|_p, \left| \alpha - \frac{x}{y} \right|_p \right\} \leq 1.$$

Consequently, p does not divide y and thus y has an inverse $\bmod\, p^n$, say $y y_n \equiv 1 \bmod p^n$. Furthermore, we may assume that $0 \leq x y_n < p^n$. This implies

$$\left| \frac{x}{y} - x y_n \right|_p = \left| \frac{x(1 - y y_n)}{y} \right|_p \leq p^{-n}.$$

Put $\alpha_n = x y_n$. Then, with regard to the above estimates,

$$|\alpha - \alpha_n|_p = \left| \underbrace{\alpha - \frac{x}{y} + \frac{x}{y}}_{=0} - \alpha_n \right|_p \leq \max \left\{ \left| \alpha - \frac{x}{y} \right|_p, \left| \frac{x}{y} - \alpha_n \right|_p \right\} \leq p^{-n}.$$

By induction it follows that for any such α there exists a Cauchy sequence of integers (α_n) converging to α satisfying

$$0 \leq \alpha_n < p^n \qquad \text{and} \qquad \alpha_n \equiv \alpha_{n-1} \bmod p^{n-1}.$$

This implies the existence of a sequence of integers a_k with

(12.6) $\alpha_n = a_0 + a_1 p + \ldots + a_{n-1} p^{n-1}$ and $0 \leq a_k < p$

for $0 \leq k < n$, which leads to a representation of α of the form (12.4). It is obvious that this representation is uniquely determined. The general case $\nu_p(\alpha) \in \mathbb{Z}$ is deduced from the one above by considering $\alpha p^{-\nu_p(\alpha)}$ instead of α. Thus we have proved

Theorem 12.8. *Let p be prime and $\alpha \in \mathbb{Q}_p$. Then there exist an integer ν and a sequence of integers a_k with $0 \leq a_k < p$ such that*

$$\alpha = \sum_{n \geq \nu} a_k p^k.$$

This representation is unique.

Hensel introduced p-adic numbers in an ad hoc manner via these series representations. Hensel's idea was to transport the powerful method of power series from analysis to number theory. At each point of the complex plane, a meromorphic function in one variable has a unique Laurent expansion. Analogously, for any $p \leq \infty$ every rational number has a uniquely determined p-adic expansion; if its denominator is divisble by p we may think of this number as having a pole at p.

12.8. Error-free computing

A big problem in computing is the accumulation of rounding errors. Small rounding errors can make the final result of a complex computation absolutely useless. In the recent past several strategies for error-free computing have been developed. Here we want to sketch a p-adic method which works without any rounding errors, independent of the length or complexity of the computation.

Three out of the four fundamental mathematical operations are harmless; the only problem arises from division. Computers cannot represent a simple unit fraction as, for example,

$$\frac{1}{7} = 0.\overline{142857},$$

in the binary or in the decimal system without error (since they can only perform finitely many digits). Cutting the infinite decimal fraction expansion can lead to quite large rounding errors. The solution to this problem is simply to avoid any division! For this purpose one fixes a prime power p^r and computes the inverse of an integer n modulo p^r. For instance,

(12.7) $7^{-1} \equiv 18$ mod $5^3 (= 125)$.

This can easily and quickly be done (even for *larger* numbers) by use of the Euclidean algorithm backwards (see Section 1.9). We may compute the 5-adic expansion of the inverse,

$$18 = \mathbf{3} + \mathbf{3} \cdot 5.$$

Now we denote the Hensel code of 18 with $r = 3$ digits by

$$H(18; 5, 3) = 0.\mathbf{330}.$$

For an integer α, the **Hensel code** $H(\alpha; p, r)$ of α to the prime p (basis) with precisely r digits is given by

$$H(a_0 + a_1 p + \ldots + a_{r-1} p^{r-1} + \ldots; p, r) = 0.a_0 a_1 \ldots a_{r-1};$$

notice that the contribution of terms $a_k p^k$ with $k \geq r$ in the p-adic expansion of α are ignored. For rational α with $\nu := \nu_p(\alpha) < 0$, the Hensel code of α is given by the Hensel code of $p^\nu \alpha$ shifted ν places to the left: if $r - 1 \geq -\nu$,

$$H(a_\nu p^\nu + \ldots + a_0 + a_1 p + \ldots; p, r) = a_\nu a_{\nu-1} \ldots a_{\nu-1} . a_\nu a_{\nu+1} \ldots a_{r-1+\nu};$$

How can this be used for error-free computing? Recall that 18 was the inverse of 7 mod 5^3. Interpreting the left-hand side as $\frac{1}{7}$ in the field of rationals, we set

(12.8)
$$H\left(\frac{1}{7}; 5, 3\right) = H(18; 5, 3) = 0.330.$$

For example, we want to verify the equation

$$\frac{1}{7} + \frac{2}{7} = \frac{3}{7}.$$

For this purpose we compute

$$2 \cdot 7^{-1} \equiv 36 = 1 + 2 \cdot 5 + 1 \cdot 5^2 \qquad \text{and} \qquad 3 \cdot 7^{-1} \equiv 54 = 4 + 0 \cdot 5 + 2 \cdot 5^2$$

modulo 5^3, which yields

$$H\left(\frac{2}{7}; 5, 4\right) = 0.121 \qquad \text{and} \qquad H\left(\frac{3}{7}; 5, 3\right) = 0.402.$$

Now we verify the equation in question by

$$H\left(\frac{1}{7}; 5, 3\right) + H\left(\frac{2}{7}; 5, 4\right) = 0.330 + 0.121 = 0.402 = H\left(\frac{3}{7}; 5, 3\right);$$

here we added two Hensel codes digit by digit mod 5, with carry if the sum exceeds 5. However, by (12.8) we note that different fractions can have the same Hensel code. In order to have uniqueness one has to assure that the mantissa and the basis prime are sufficiently large. (We refer the interested reader for more details to Exercise 12.18.)

Notes on the p-adic interpolation of the zeta-function

In this chapter we have laid the fundamentals for p-adic analysis. Some problems for further steps are obvious: $\mathbb{R}$ is an ordered field, which gives a notion of *bigger than* which fits perfectly together with the standard operations, but this does not hold for $\mathbb{Q}_p$; moreover, since $\mathbb{Q}_p$ is totally disconnected, we do not have intervals, so we cannot expect an analogue of the mean-value theorem. But we should not be too pessimistic; some things are much easier to handle, for instance, Cauchy sequences and convergence. Many concepts as, for example, continuity and differentiability remain unchanged since they depend only on the metric structure. Now we shall briefly present a remarkable feature of p-adic analysis with respect to analytic number theory.

p-adic numbers also encode important arithmetic information concerning the Riemann zeta-function. The link relies on certain properties of Bernoulli numbers (see Section 4.2). In the nineteenth century von Staudt and Clausen proved that, for any even positive integer m,

$$(12.9) \qquad B_m + \sum_{\substack{p \\ m \equiv 0 \bmod (p-1)}} \frac{1}{p} \in \mathbb{Z}.$$

For instance,

$$B_{30} = \frac{8\,615\,841\,276\,005}{14\,322} \qquad \text{and} \qquad B_{30} + \frac{1}{2} + \frac{1}{3} + \frac{1}{7} + \frac{1}{11} + \frac{1}{31} = 601\,580\,875.$$

A p-adic proof relies on the identity

$$(12.10) \qquad S_m(n) := \sum_{j=0}^{n-1} j^m = \sum_{j=0}^{m} \binom{m}{j} B_j \frac{n^{m+1-j}}{m+1-j}.$$

From (12.9) it follows that

$$pB_m \equiv -1 \bmod p$$

for m divisible by $p - 1$. A further family of congruences for Bernoulli numbers are the celebrated **Kummer congruences**

$$(12.11) \qquad \frac{B_m}{m} \equiv \frac{B_n}{n} \bmod p \qquad \text{if} \quad m \equiv n \not\equiv 0 \bmod (p-1).$$

These remarkable congruences were regarded as something *mysterious* before it turned out that they naturally appear in the context of p-adic numbers. In fact, in the 1960s Kubota & Leopoldt [95] used (4.5) and the Kummer congruences to prove the existence of a p-adic analytic function $\zeta_p(s)$ which interpolates the values of the Riemann zeta-function $\zeta(s)$ at the negative integers:

$$\zeta_p(1 - m) = \zeta(1 - m)(1 - p^{m-1}) \qquad \text{if} \quad m \equiv 0 \bmod p - 1;$$

here the factor $1 - p^{m-1}$ on the right-hand side is exactly the Euler factor of $\zeta(1 - m)$ at p. Generalizations to Dirichlet L-functions are important with respect to the p-adic analogue of the class number formula, and elliptic analogues of the p-adic zeta-function are a major ingredient in Wiles' solution of Fermat's last theorem.

Exercises

12.1. *Compute the following p-adic absolute values:*

$$|75|_p \quad \text{for} \quad p = 2, 3, 5 \qquad \text{and} \qquad \left| \frac{28}{75} \right|_p \quad \text{for} \quad p = 2, 3, 5, 7.$$

12.2 *Let p be prime. Give a rigorous proof that the p-adic absolute value is a non-Archimedean absolute value.*

12.3. For any polynomial $P \in \mathbb{Q}[X]$ define

$$\nu(P) = \begin{cases} 0 & \text{if } P \equiv 0, \\ -\deg P & \text{otherwise.} \end{cases}$$

Extend this definition to the field of rational functions in X and show that

$$|P| = \exp(-\nu(P)) \quad \text{for } P \in \mathbb{Q}(X)$$

defines a non-Archimedean absolute value on $\mathbb{Q}(X)$.

12.4. Let $\mathbb{K}$ be a field. A function $\varphi : \mathbb{K} \to \mathbb{R}$ is called a **valuation** on $\mathbb{K}$ if, for all $x, y \in \mathbb{K}$,

$$\begin{aligned} \varphi(xy) &= \varphi(x) + \varphi(y), \\ \varphi(x+y) &\geq \min\{\varphi(x), \varphi(y)\}. \end{aligned}$$

A valuation is said to be **discrete** if the image of $\mathbb{K}$ under φ is discrete.

i) Show that the p-adic valuation is a discrete valuation.
ii) Using Lemma 10.5 show that the index is a valuation on the ring of polynomials. Is it discrete?

12.5 *Prove that the only absolute value on a finite field is the trivial absolute value.*

12.6. *i) Show that any open (closed) ball is clopen.*
ii) Prove that any two closed open (or closed) balls in an ultrametric space are either disjoint or contained in one another.
iii) In the case $\mathbb{K} = \mathbb{Q}$ and the p-adic valuation show that the closed unit ball with center 0 can be written as a finite disjoint union of open balls.

12.7.* *Given pairwise non-equivalent absolute (not necessarily p-adic) values $|.|^{(1)}, \ldots, |.|^{(n)}$ of a field $\mathbb{K}$ and arbitrary elements $\beta_1, \ldots, \beta_n$, prove that for any given $\varepsilon > 0$ there exists $\alpha \in \mathbb{K}$ such that*

$$|\alpha - \beta_k|^{(k)} < \varepsilon \quad \text{for } 1 \leq k \leq n.$$

This so-called **weak approximation theorem** might be regarded as an analogue of the Chinese remainder theorem A.5. In fact, if the absolute values are p-adic absolute values and $\mathbb{K} = \mathbb{Q}$, then the above assertion is equivalent to Theorem A.5.

12.8. There are 12 coins and a balance scale. One of the coins is counterfeit, but all look the same. It is not known whether the counterfeit coin is heavier or lighter than the others.

Try to find the counterfeit coin in at most three weighings. Find a strategy for n coins and k weighings. Given n, what is the minimal value of needed weighings?

12.9. *Let p be prime, d a divisor of $p-1$, and n an integer with p-adic expansion $n = \sum_{k=0}^{m} a_k p^k$. Prove that*

$$n \equiv 0 \bmod d \quad \Longleftrightarrow \quad \sum_{k=0}^{m} a_k \equiv 0 \bmod d.$$

Note that for each d such a prime p exists by Dirichlet's prime number theorem for arithmetic progressions. In particular, we obtain a well-known result from school: an integer n is divisible by 9 if and only if its sum of (decimal) digits is divisible by 9.

12.10.* *Assume that $n \in \mathbb{N}$ has the p-adic expansion $n = \sum_{k=0}^{m} a_k p^k$. Show that*

$$n - \sum_{k=0}^{m} a_k = (p-1)\nu_p(n!).$$

This result is due to Legendre and improves the assertion of the previous exercise.

12.11. *Consider the sequence of 5-adic numbers $x_n = 3 + 3 \cdot 5 + \ldots + 3 \cdot 5^n$. Show that $3a_n \equiv -1 \bmod 5^n$ and deduce that x_n tends to $-\frac{1}{3}$ as $n \to \infty$.*

12.12. *i) Calculate the 3-adic expansions of*

$$\pm\frac{5}{9}, \quad \frac{9}{5}, \quad \frac{5}{9} + \frac{9}{5}, \qquad and \qquad \frac{5}{9} \cdot \frac{9}{5}.$$

ii) What is the p-adic value of

$$\sum_{k=0}^{\infty} (p-1)p^k \ ?$$

12.13. *Fill all gaps we have left in Section 12.6, i.e., show that $\mathcal{C}_p$ is a ring and $\mathcal{M}_p$ is an ideal, and give rigorous proofs that the p-adic absolute value extends to $\mathbb{Q}_p$ and that $\mathbb{Q}_p$ is complete.*

12.14. *Compute the 7-adic expansion of the sum, the product, and the quotient of*

$$5 + 4 \cdot 7 + 3 \cdot 7^2 + 2 \cdot 7^3 + \ldots \qquad and \qquad 3 + 5 \cdot 7 + 2 \cdot 7^2 + 4 \cdot 7^3 + \ldots$$

to four digits.

12.15. *Prove that a p-adic series $\sum_{k=0}^{\infty} c_k$ converges if and only if $|c_k|_p$ tends to zero as k to infinity, in which case*

$$\left| \sum_{k=0}^{\infty} c_k \right|_p \leq \max_k |c_k|_p.$$

12.16. *Prove that the series*

$$\sum_{k=0}^{\infty} p^{k^2} \qquad and \qquad \sum_{k=0}^{\infty} p^{k!}$$

are both p-adic convergent and p-adic irrational.

12.17. *Prove that the series*

$$\beta_m := \sum_{k=0}^{\infty} k^m \cdot k!$$

is p-adic convergent for any non-negative integer m. Furthermore, show that $\beta_1 = -1$.

Hint: For the latter assertion, use induction in order to show that

$$\sum_{k=0}^{n-1} k \cdot k! = n! - 1.$$

Remarkably, it is still unknown whether β_0 is p-adic irrational or not. The **Kurepa conjecture** on the **left factorial function**

$$!n := \sum_{k=0}^{n-1} k!$$

states that $!p \not\equiv 0 \bmod p$ for any odd prime p or, equivalently in terms of p-adic language, $\nu_p(!p) = 0$ for any prime $p > 2$. This conjecture was recently solved by Barsky & Benzaghou [15].

12.18. Let p be prime, r be a positive integer, and let N be the largest integer satisfying $N \le \sqrt{\frac{1}{2}(p^r - 1)}$.

Prove that every Farey fraction $\frac{a}{b} \in \mathcal{F}_N$ can be represented uniquely by its Hensel code $\mathsf{H}(\frac{a}{b}; p, r)$. For $N = 17$ and $r = 4$ compute

$$\frac{2}{3} + \frac{3}{4}, \quad \frac{3}{13} + \frac{1}{12}, \quad \frac{5}{2} + \frac{5}{7}, \quad \text{and} \quad \frac{2}{3} \cdot \frac{3}{2}$$

by use of their Hensel codes. Verify that the results are unique and correct provided that all terms lie in $\mathcal{F}_{17}$.

12.19.* *i) In the notation of the notes of this chapter, prove identity (12.10) and deduce that with respect to p-adic convergence*

$$B_m = \lim_{n \to 0} \frac{1}{n} S_m(n).$$

Hint: Extend both sides of the identity

$$\sum_{j=0}^{n-1} \exp(jx) = \frac{\exp(nx) - 1}{x} \frac{x}{\exp(x) - 1}$$

into their power series expansions in order to find an expression for $\sum_{m=0}^{\infty} S_m(n) \frac{x^m}{m!}$ in terms of the Bernoulli numbers; for this purpose recall (4.4).

ii) Apply division with remainder to the j's in $S_m(n) = \sum_{j=0}^{n-1} j^m$ to prove that

$$S_m\left(p^{k+1}\right) \equiv p S_m\left(p^k\right) \bmod p^{k+1}.$$

Deduce from this in combination with i) that

$$\left| B_m - p^{-1} S_m(p) \right|_p \le 1.$$

iii) Use Fermat's little theorem A.4 to show that $S_m(p) \equiv -1 \bmod p$ if m is divisible by $p - 1$, and congruent 0 otherwise. Deduce from this and ii) identity (12.9).

12.20.* *What is wrong in the following "proof" of the irrationality of* π?
Suppose that $\pi = \frac{a}{b}$ with integers a, b and let p be an odd prime not dividing a. Then

$$0 = \sin(pb\pi) = \sin(pa) = \sum_{n=0}^{\infty} (-1)^n \frac{(pa)^{2n+1}}{(2n+1)!} \equiv pa \bmod p^2,$$

a contradiction.

This exercise is from Koblitz [91]; the interested reader can also find there a fallacious proof of the transcendence of e (due to Hensel). Notice that Bézivin & Robba [22] gave a *p*-adic proof of the theorem of Lindemann–Weierstrass, including the transcendence of π and e. In a subsequent article Beukers, Bézivin & Robba [21] removed the difficult *p*-adic part by an elementary argument which led to a new and rather different proof.

CHAPTER 13

Hensel's lemma and applications

The main importance of p-adic numbers lies within the theory of diophantine equations. So far we have treated p-adic numbers only by means of analysis. In this chapter we will investigate p-adic numbers from the view of algebra and introduce the set of p-adic integers $\mathbb{Z}_p$. The central theme of this chapter is Hensel's *lemma*. This theorem is very likely the most important algebraic property of the p-adic numbers. By Hensel's lemma, in many circumstances one can decide very easily whether a polynomial has roots in $\mathbb{Z}_p$. It has many applications, from characterizing roots of unity to factoring polynomials.

13.1. p-adic integers

Integers are an important feature whenever we deal with rational numbers. As we shall see below it makes sense to introduce a notion of p-adic integers in $\mathbb{Q}_p$ for $p < \infty$, generalizing rational integers. For this purpose we shall briefly give a second, purely algebraic construction of $\mathbb{Q}_p$.

The result that we want to prove, Theorem 13.1 below, can be obtained by elementary means (see Exercise 13.1); however, we present here a simple and short argument which makes use of *projective limits*; readers who are not familiar with this notion are referred to Robert [**137**] (or, alternatively, to solve Exercise 13.1).

For $n \in \mathbb{N}$, consider the tower of rings of residue classes $\bmod\, p^n$ with respect to the decreasing sequence

$$\ldots \to \mathbb{Z}/p^{n+1}\mathbb{Z} \to \mathbb{Z}/p^n\mathbb{Z} \to \ldots \to \mathbb{Z}/p^2\mathbb{Z} \to \mathbb{Z}/p\mathbb{Z},$$

where each map is the natural projection

$$\pi_n : \mathbb{Z}/p^{n+1}\mathbb{Z} \to \mathbb{Z}/p^n\,\mathbb{Z} , \quad x \mapsto x \bmod\, p^n.$$

Then we can define the **projective limit**

$$\varprojlim \mathbb{Z}/p^n\mathbb{Z} = \left\{ (x_n) \in \prod_{n \geq 1} \mathbb{Z}/p^n\mathbb{Z} : \pi(x_{n+1}) = x_n \right\}.$$

Being a formal product of rings the projective limit inherits the ring structure of its factors. As a matter of fact we see that each $(x_n) \in \varprojlim \mathbb{Z}/p^n\mathbb{Z}$ gives raise to a Cauchy sequence of integers α_n (with respect to the p-adic absolute value) such that

$$x_n \equiv \alpha_n \bmod\, p^n,$$

resp. a sequence of integers a_k satisfying (12.6). It follows that

$$(13.1) \qquad \varprojlim \mathbb{Z}/p^n\mathbb{Z} \cong \left\{ \sum_{k=0}^{\infty} a_k p^k : 0 \leq a_k < p \right\}.$$

We denote the right-hand side above by $\mathbb{Z}_p$. From the properties of the projective limit we get

Theorem 13.1. $\mathbb{Z}_p$ *is a commutative ring.*

Now we can define $\mathbb{Q}_p$ as the field of fractions of $\mathbb{Z}_p$. According to this construction every element α of $\mathbb{Z}_p$ (i.e., $|\alpha|_p \leq 1$, resp. $\nu_p(\alpha) \geq 0$) is said to be a p-**adic integer**, and $\mathbb{Z}_p$ is called the **ring of p-adic integers**. This marks an important difference between the non-Archimedean p-adic fields $\mathbb{Q}_p$ and the Archimedean field $\mathbb{Q}_\infty = \mathbb{R}$, where we do not have any meaningful extension of $\mathbb{Z}$.

For $n \in \mathbb{Z}$ the sets

$$p^n \mathbb{Z}_p := \{ \alpha \in \mathbb{Z}_p : |\alpha|_p \leq p^{-n} \}$$

form, topologically speaking, a system of neighborhoods of zero in $\mathbb{Q}_p$ which covers all of $\mathbb{Q}_p$. Algebraically speaking, they form a chain of ideals which covers all ideals of $\mathbb{Q}_p$. This observation ties topology with algebra.

13.2. Solving equations in p-adic numbers

The p-adic way of thinking gives a new strategy to attack diophantine equations. Consider the equation

$$(13.2) \qquad P(X_1, X_2, \ldots, X_r) = 0,$$

where P is a polynomial in several variables with integral coefficients. We are interested in the solvability in integers. We can weaken this difficult problem by replacing (13.2) by the system of congruences

$$P(X_1, X_2, \ldots, X_r) \equiv 0 \bmod m,$$

where m runs through all positive integers, or equivalently, by the Chinese remainder theorem A.5,

$$(13.3) \qquad P(X_1, X_2, \ldots, X_r) \equiv 0 \bmod p^n \qquad \text{with} \quad n = 1, 2, \ldots,$$

where p runs through all prime numbers. This approach yields sometimes certain information about the original equation. We shall illustrate this by an example.

We ask for the integral solutions of

$$Y^2 = X^3 + 7.$$

We may rewrite this as

$$Y^2 + 1 = (X + 2)((X - 1)^2 + 3).$$

If $x, y \in \mathbb{Z}$ is a solution, then it is not difficult to see that x has to be odd. Thus $(x - 1)^2 + 3 \equiv 3 \bmod 4$. Hence there is a prime number $p \equiv 3 \bmod 4$

dividing $N := (x - 1)^2 + 3$. Reducing the equation in question modulo p shows that

$$y^2 = N(x + 2) - 1 \equiv -1 \mod p.$$

Consequently, -1 is a square mod p, contradicting the first supplementary law in the theory of quadratic residues (see Theorem A.6). Thus, there are no integer solutions to the equation in question.

So far out of one equation we have made infinitely many congruences. But we can go further. It is an astonishing and useful fact that all these unpleasantly many congruences (13.3) for a fixed prime can be summed up to one p-adic equation. Indeed, solving an equation in $\mathbb{Z}_p$ is equivalent to solving conguences mod p.

Theorem 13.2. *With the conditions on P from above, the system of congruences (13.3) is solvable if and only if the equation (13.2) has a solution in p-adic integers.*

Proof. Via the projective limit (13.1) we have

$$\mathbb{Z}_p \cong \lim_{\leftarrow} \mathbb{Z}/p^n\mathbb{Z} \subset \prod_{n=1}^{\infty} \mathbb{Z}/p^n\mathbb{Z}.$$

The equation (13.2) splits over the ring on the right-hand side into the system of congruences (13.3). Thus, each p-adic solution of (13.2) leads also to a solution of (13.3).

Conversely, for any positive integer n, let $\left(x_1^{(n)}, x_2^{(n)}, \ldots, x_r^{(n)}\right)$ be a solution of

$$P(X_1, X_2, \ldots, X_r) \equiv 0 \mod p^n.$$

If all elements

$$\left(x_j^{(n)}\right)_n \in \prod_{n=1}^{\infty} \mathbb{Z}/p^n\mathbb{Z}$$

lie in $\lim_{\leftarrow} \mathbb{Z}/p^n\mathbb{Z} \cong \mathbb{Z}_p$ for $1 \leq j \leq r$, we are done. Otherwise, we construct a subsequence with this property as follows. We only consider the case $r = 1$ and write x instead of x_1; the general case can be treated similarly.

Since $\mathbb{Z}/p\mathbb{Z}$ is finite, there are infinitely many terms of $x^{(n)}$ which are congruent to some fixed $y_1 \in \mathbb{Z}/p\mathbb{Z}$. Thus we can find a subsequence $\left(x_1^{(n)}\right)$ of $\left(x^{(n)}\right)$ for which

$$x_1^{(n)} \equiv y_1 \quad \text{and} \quad P\left(x_1^{(n)}\right) \equiv 0 \mod p.$$

Obviously, we can continue and obtain, for any $k \geq 2$, a subsequence $\left(x_k^{(n)}\right)$ of $\left(x_{k-1}^{(n)}\right)$ such that

$$x_k^{(n)} \equiv y_k \quad \text{and} \quad P\left(x_k^{(n)}\right) \equiv 0 \mod p^k,$$

where the $y_k \in \mathbb{Z}/p^k\mathbb{Z}$ satisfy

$$y_k \equiv y_{k-1} \mod p^{k-1}.$$

The y_k define a p-adic integer $y = (y_k)_k \in \lim_{\leftarrow} \mathbb{Z}/p^k\mathbb{Z}$ satisfying

$$P(y_k) \equiv 0 \bmod p^k \qquad \text{for every} \quad k \in \mathbb{N}.$$

By continuity this implies $P(y) = 0$. The theorem is proved. ●

If the polynomial P is homogeneous, the equation (13.2) has always the trivial solution $x_1 = \ldots = x_r = 0$. In this context it is more natural to ask for non-trivial solutions, i.e., solutions where not all x_j equal zero. In this case a little variation of the proof above gives the corresponding statement for a non-trivial p-adic solution.

We give an example for Theorem 13.2. Consider the system of congruences

(13.4) $$X^2 \equiv 2 \bmod 7^n \qquad (n = 1, 2, \ldots).$$

Since 2 is a quadratic residue mod 7, the congruence is solvable for $n = 1$. Indeed, we find the solutions $\pm 3 \bmod 7$. We start with $x_1 \equiv +3 \bmod 7$. Now let $n = 2$. Any solution x_2 of (13.4) with $n = 2$ also satisfies (13.4) with $n = 1$. The ansatz $x_2 \equiv 3 + 7z$ leads via (13.4) to

$$2 \equiv (3 + 7z)^2 = 9 + 6z \cdot 7 + 7^2 z^2 \equiv 2 + (1 + 6z) \cdot 7 \bmod 7^2,$$

resp.

$$0 \equiv 1 + 6z \bmod 7.$$

Thus, $z \equiv 1 \bmod 7$ and we obtain $x_2 \equiv 3 + 1 \cdot 7 \bmod 7^2$. Since here and in any further step we only have to solve linear congruences mod 7, this process continues ad infinitum and leads to a 7-adic solution

$$x = 3 + 1 \cdot 7 + 2 \cdot 7^2 + \ldots$$

to the system of congruences (13.4) and to the equation $X^2 = 2$, respectively; we may write $x = \sqrt{2}$ but notice that this is not the square root of 2 in the field of real numbers. The other solution can be found the same way starting with $x_1 \equiv -3 \bmod 7$.

13.3. Hensel's lemma

Recall the construction of the 7-adic square root of two from the last section. In some sense, solving the linear congruences step by step can be understood as applying the well-known Newton iteration method from real analysis (see also Exercise 13.9) to the polynomial $P(X) = X^2 - 2$. Starting with $x_1 = 3$ the iteration

$$x_n = x_{n-1} - \frac{P(x_{n-1})}{P'(x_{n-1})} = x_{n-1} - p \frac{P(x_n)}{p} \left(P'(x_{n-1})\right)^{-1},$$

yields

$$x_2 = 3 - 7 \frac{3^2 - 2}{7} 6^{-1} = 3 + 1 \cdot 7,$$

where $-6^{-1} = 1$ has to be understood as an identity in $\mathbb{Z}/7\mathbb{Z}$. Continuing this iteration process gives the same 7-adic value as we computed above.

This observation leads to a very important theorem due to Hensel, called Hensel's lemma. It plays a crucial role in the algebraic theory of p-adic numbers.

Theorem 13.3. *Let $P(X)$ be a polynomial with coefficients in $\mathbb{Z}_p$ and suppose that there is a p-adic integer x_1 such that*

$$P(x_1) \equiv 0 \bmod p\mathbb{Z}_p \qquad and \qquad P'(x_1) \not\equiv 0 \bmod p\mathbb{Z}_p.$$

Then there exists a p-adic integer x with

$$x \equiv x_1 \bmod p\mathbb{Z}_p \qquad and \qquad P(x) = 0.$$

Here and in the sequel the notation $b \equiv a \bmod p^n\mathbb{Z}_p$ means that $b - a \in p^n\mathbb{Z}_p$; this is a simple generalization of the notion of modular arithmetic over the integers.

In many circumstances one can very easily decide by Hensel's lemma whether a polynomial has roots in $\mathbb{Z}_p$. The test involves finding an *approximation* x_1 of a root x and then verifying a condition on the (formal) derivative of the polynomial in question. This is certainly a highlight in combining diophantine approximations and diophantine equations! Note that Newton's iteration method can fail (if the starting point of the iteration is not carefully chosen or the underlying function is not *nice*); however, Hensel's lemma always succeeds if the conditions are fulfilled.

Proof. The existence of the root x will follow from a construction of an appropriate Cauchy sequence converging to x. More precisely, we have to show that there are integers x_n satisfying

$$x_n \equiv x_{n+1} \bmod p^n \qquad and \qquad P(x_n) \equiv 0 \bmod p^n\mathbb{Z}_p.$$

The existence of x_1 follows from the assumption of the theorem. For x_2 we again do the ansatz $x_2 = x_1 + zp$ (as in the example of the previous section). In view of the formal identity (resp. the Taylor expansion)

$$P(X + h) = P(X) + hP'(X) + \frac{h^2}{2!}P''(X) + \dots,$$

we get

$$P(x_2) = P(x_1 + zp) \equiv P(x_1) + zpP'(x_1) \bmod p^2\mathbb{Z}_p.$$

Since $P(x_1) \equiv 0 \bmod p\mathbb{Z}_p$, there is some y for which $P(x_1) = yp$. Thus we have to solve the linear congruence

$$p\,(y + zP'(x_1)) \equiv 0 \bmod p^2\mathbb{Z}_p,$$

resp.

$$y + zP'(x_1) \equiv 0 \bmod p\mathbb{Z}_p.$$

Since $P'(x_1)$ is not divisble by p, it is invertible in $\mathbb{Z}_p$. Thus we can find some z satisfying $0 \leq z < p$ and

$$z \equiv -y(P'(x_1))^{-1} \bmod p\mathbb{Z}_p.$$

This defines x_2 and by the same procedure we obtain the desired sequence by induction. It is clear that this sequence is Cauchy and, by continuity, the limit x satisfies $P(x) = 0$. Hensel's lemma is proved. •

Theorem 13.2 gave a criterion for the solvability of a diophantine equation in p-adic numbers by solving the corresponding congruences modulo all powers of p. Hensel's lemma yields a solution from just one of those congruences provided that this solution is sufficiently *non-singular*.

13.4. Units and squares

As a first and easy application of Hensel's lemma we shall characterize the units in the ring of p-adic integers. For this purpose we consider the linear polynomial $P(X) = \alpha X - 1$, where $\alpha \neq 0$ is a p-adic integer. If there is a root of the equation $P(X) = 0$, then α is invertible and the root is the inverse of α. Obviously, any $\alpha \in p\mathbb{Z}_p$ cannot be invertible in $\mathbb{Z}_p$, so we may assume that $\alpha \notin p\mathbb{Z}_p$. In this case

$$P'(X) = \alpha \not\equiv 0 \bmod p\mathbb{Z}_p,$$

and furthermore we can solve the congruence

$$P(X) = \alpha X - 1 \equiv 0 \bmod p\mathbb{Z}_p.$$

Thus Hensel's lemma 13.3 implies the existence of a p-adic integer α^{-1} such that $P(\alpha^{-1}) = 0$, resp. $\alpha\alpha^{-1} = 1$. This proves

Theorem 13.4. *A p-adic integer α is invertible in $\mathbb{Z}_p$ if and only if $\alpha \in \mathbb{Z}_p \setminus p\mathbb{Z}_p$. The **ring of p-adic units** in $\mathbb{Z}_p$ is given by*

$$\mathbb{Z}_p^* := \mathbb{Z}_p \setminus p\mathbb{Z}_p = \{\alpha \in \mathbb{Z}_p : |\alpha|_p = 1\}.$$

Next we shall investigate the squares in $\mathbb{Q}_p$. In Section 13.2 we have seen that a positive integer α is a square in $\mathbb{Z}_7$ if and only if α is a quadratic residue mod 7. Now suppose more generally that p is an odd prime and

$$\alpha = a_0 + a_1 p + a_2 p^2 + \ldots \in \mathbb{Z}_p^*$$

is a p-adic unit. Consider the polynomial $P(X) = X^2 - \alpha$. Then it follows that the congruence

$$X^2 - a_0 \equiv P(X) \equiv 0 \bmod p\mathbb{Z}_p$$

can be solved if and only if a_0 is a quadratic residue mod p. In this case the solution, say x_1, is coprime with p (since a_0 is), and thus

$$P'(x_1) = 2x_1 \not\equiv 0 \bmod p\mathbb{Z}_p.$$

Hence, Hensel's lemma yields

Theorem 13.5. *Let p be an odd prime. A p-adic unit $\alpha \in \mathbb{Z}_p^*$ is a square of an element in $\mathbb{Z}_p$ if and only if $\alpha \bmod p\mathbb{Z}_p$ is a quadratic residue mod p.*

The above argument does not apply to $p = 2$ (since then the derivative P' vanishes identically), and even the analogue is not true (see Exercise 13.10). It is a bit *odd* that the case of the one and only even prime $p = 2$ is slightly different. Actually, this prime behaves in the theory of p-adic numbers quite often irregularly, and so it is justified to call 2 the *oddest* prime.

13.5. Roots of unity

A further nice application of Hensel's lemma is to determine the roots of unity in $\mathbb{Q}_p$. Recall that ζ is called an nth **root of unity** if $\zeta^n = 1$. For an nth root of unity ζ we have $|\zeta^n|_p = 1$, and therefore $\zeta \in \mathbb{Z}_p^*$. It is easily seen that the roots of unity in $\mathbb{Q}_2$ are given by $\zeta = \pm 1$. The case of odd primes is a bit more difficult. Here we have to study $P(X) = X^n - 1$. Suppose that $p \geq 3$ and let $x \in \mathbb{Z}_p$. Obviously, $P'(x) = nx^{n-1}$ is congruent zero modulo $p\mathbb{Z}_p$ if either p divides x, in which case x will not be a root of P anyway, or p divides n.

First, suppose that p and n are coprime. Then, by Hensel's lemma 13.3, for every $x_1 \not\equiv 0 \bmod p\mathbb{Z}_p$ we can find an $x \in \mathbb{Z}_p$ for which $P(x) = 0$ and $x \equiv x_1 \bmod p\mathbb{Z}_p$. It follows that the equation $P(X) = 0$ has as many distinct solutions in $\mathbb{Z}_p^*$ as in the group of prime residue classes. In view of Fermat's little theorem A.4 $(\mathbb{Z}/p\mathbb{Z})^*$ consists exactly out of $p - 1$ distinct $(p - 1)$th roots of unity. Consequently, each one corresponds to a different $(p - 1)$th root of unity in $\mathbb{Q}_p$.

Now assume that n and p are not coprime. Without loss of generality we may suppose that $n = p$ (since each pth root of unity is also a (kp)th root of unity for any $k \in \mathbb{N}$). We shall show that we cannot find any pth root of unity $\neq 1$ in $\mathbb{Q}_p$. Let $x = x_1 + zp$, where $0 \leq x_1 < p$ and $z \in \mathbb{Z}_p$. Then, by the binomial theorem,

$$x^p - x_1^p = \sum_{k=1}^{p} \binom{p}{k} x_1^{p-k} (zp)^k,$$

which lies in $p\mathbb{Z}_p$. If $x^p = 1$, then it follows from Fermat's little theorem A.4 that $x_1 = 1$. This leads to

$$1 = x^p = (1 + zp)^p = 1 + zp^2 + \sum_{k=2}^{p-1} \binom{p}{k} (zp)^k + z^p p^p.$$

Suppose that $z \neq 0$, then

$$-zp^2 = \sum_{k=2}^{p-1} \binom{p}{k} (zp)^k + z^p p^p,$$

and, by the ultrametric inequality (12.1),

$$|z|_p - 2 = |-zp^2|_p \leq \max_{2 \leq k \leq p-1} \left\{ \left| \binom{p}{k} z^k p^k \right|_p, |z^p p^p|_p \right\}.$$

Since $p > 2$, this is impossible. So we have $z = 0$ which implies $x = 1$. Thus we have proved

Theorem 13.6. *Let p be an odd prime. $\mathbb{Q}_p$ contains exactly the $(p-1)$th roots of unity.*

In particular, we see that $\mathbb{Q}_p$ is not algebraically closed; of course, we could have seen this before! (Where?) This leads to new adventures (as in the real case) but this is another story for which we refer to Gouvêa [68].

13.6. Hensel's lemma revisited

Now we shall even prove a slightly stronger version of Hensel's lemma. Recall Theorem 1.6 which gave an equivalent statement for two integers being coprime in terms of the solvability of a linear diophantine equation. Analogously, we say that two polynomials $Q, R \in \mathbb{Z}_p[X]$ are **coprime** mod p if there exist polynomials $A, B \in \mathbb{Z}_p[X]$ such that

$$A(X)Q(X) + B(X)R(X) \equiv 1 \bmod p;$$

this or other polynomial congruences in the sequel have to be understood coefficient-by-coefficient, i.e., the corresponding coefficients on both sides are congruent mod p. This definition is equivalent to Q and R having relatively prime factors.

Theorem 13.7. *Assume that there exist polynomials P, Q_1, and R_1 in $\mathbb{Z}_p[X]$ such that $Q_1(X)$ is monic, $Q_1(X)$ and $R_1(X)$ are coprime mod p, and*

$$P(X) \equiv Q_1(X)R_1(X) \bmod p.$$

Then, for $n \in \mathbb{N}$, there exist coprime polynomials Q_{n+1}, R_{n+1} in $\mathbb{Z}_p[X]$ such that $Q_{n+1}(X)$ is monic,

$$(13.5) \quad Q_{n+1}(X) \equiv Q_n(X) \bmod p^n \quad and \quad R_{n+1}(X) \equiv R_n(X) \bmod p^n,$$

and

$$(13.6) \qquad\qquad P(X) \equiv Q_{n+1}(X)R_{n+1}(X) \bmod p^{n+1}.$$

Furthermore, there exist polynomials Q, R in $\mathbb{Z}_p[X]$ such that $Q(X)$ is monic,

$$Q(X) \equiv Q_1(X) \bmod p \qquad and \qquad R(X) \equiv R_1(X) \bmod p,$$

and

$$P(X) = Q(X)R(X).$$

The strategy of proof is rather similar to those of Theorem 13.2 and the previous Hensel lemma.

Proof. First, we notice, by the assumptions of the theorem, that $\deg Q_1 = \deg Q$. It follows that $\deg R_1 \leq \deg P - \deg Q_1$. We shall construct sequences of polynomials Q_n and R_n such that each Q_n is monic of degree $\deg Q$, and the congruences (13.5) and (13.6) hold. It is clear that such sequences, once constructed, yield the desired polynomials Q and R by taking the limit; namely, the coefficients of Q and R are defined by taking the limits of the corresponding coefficients of the Q_n and R_n as $n \to \infty$ (of course, for this purpose we have to assure that the degrees of the Q_n are all the same).

Given Q_1 and R_1 we construct Q_2 and R_2 as follows. Clearly, if they exist, they have to be of the form

$$Q_2(X) = Q_1(X) + pF_1(X) \qquad and \qquad R_2(X) = R_1(X) + pG_1(X),$$

where $F_1, G_1 \in \mathbb{Z}_p[X]$. In order to show the existence of Q_2 and R_2 we have to solve the polynomial congruence (13.6) with $n = 1$, resp.,

$$
\begin{aligned}
P(X) &\equiv (Q_1(X) + pF_1(X))(R_1(X) + pG_1(X)) \\
&\equiv Q_1(X)R_1(X) + p(F_1(X)R_1(X) + G_1(X)Q_1(X)) \bmod p^2.
\end{aligned}
$$

Since $P(X) \equiv Q_1(X)R_1(X) \bmod p$ it follows that

$$
P(X) - Q_1(X)R_1(X) = pH(X)
$$

for some $H \in \mathbb{Z}_p[X]$. Hence, we have to solve

(13.7) $H(X) \equiv F_1(X)R_1(X) + G_1(X)Q_1(X) \bmod p.$

By assumption, Q_1 and R_1 are coprime modulo p; thus there exist poylnomials $A, B \in \mathbb{Z}_p[X]$ such that

$$
A(X)Q_1(X) + B(X)R_1(X) \equiv 1 \bmod p.
$$

Now define

$$
F(x) = B(X)H(X) \qquad \text{and} \qquad G(x) = A(X)H(X).
$$

Then it is easily seen that all congruences above hold to be true, so F and G are good candidates for our unknowns F_1 and G_1, respectively. But what about the condition on Q_2 being monic? We have no control over the degree of F and so we cannot guarantee that $Q_2(X) = Q_1(X) + pF(X)$ is monic.

A tricky argument solves this problem: division with remainder for polynomials (see Appendix A.6). Applied to F and Q_1 it yields the existence of polynomials $C, F_1 \in \mathbb{Z}_p[X]$ such that

$$
F(X) = C(X)Q_1(X) + F_1(X),
$$

where $\deg F_1 < \deg Q_1$. Setting

$$
G_1(X) = G(X) + C(X)R_1(X),
$$

a short computation yields

$$
\begin{aligned}
F_1(X)R_1(X) + G_1(X)Q_1(X) &\equiv F(X)R_1(X) + G(X)Q_1(X) \\
&\equiv H(X) \bmod p.
\end{aligned}
$$

This solves the congruence (13.7), and since F_1 has sufficiently small degree, $\deg F_1 < \deg Q_1$, it follows that $Q_2(X) = Q_1(X) + pF_1(X)$ is monic (since Q_1 is monic). This proves the existence of Q_2 and R_2.

Now the assertion of the theorem follows by induction on n. •

13.7. Hensel lifting: factoring polynomials

Factoring polynomials over $\mathbb{Z}$ or $\mathbb{Q}$ is an interesting problem, especially if the degree is large. By Gauss' lemma A.8 factoring polynomials over $\mathbb{Q}$ is essentially the same as factoring over $\mathbb{Z}$ (see the remark below Theorem A.8). Thus we may be restricted to polynomials with integer coefficients.

Given a polynomial $P \in \mathbb{Z}[X]$, the first approach toward a factorization of P into irreducible polynomials might be reduction $\bmod p$ for some prime p. Any factorization over $\mathbb{Z}$ yields also a factorization in the field $\mathbb{Z}/p\mathbb{Z}$ but, of course, the converse does not hold in general. Nevertheless, once found a factorization of P mod p one may hope to *lift* the factorization to $\mathbb{Z}$. In practice, it is useful to combine information obtained by reduction mod p for several p (see Exercise 13.14).

At first sight factoring mod p appears to be an easy problem since it is finite; nevertheless, simply searching for a factorization by trial and error takes much too long. Here we do not want to discuss algorithms for this problem which would carry us too far from our main interest; for an excellent account on factoring over finite fields, including the celebrated algorithm of Berlekamp, we refer to [**40**]. Our main aim is to show how Hensel's lemma can be used in practice for lifting a factorization mod p to a factorization over $\mathbb{Z}$. We shall illustrate this by two examples.

First, we consider the polynomial

$$(13.8) \qquad P(X) = X^4 - X^3 - 8X^2 + X + 1.$$

In general, a linear factor $X - \alpha$ of a polynomial $P \in \mathbb{Z}[X]$ corresponds to a zero $\alpha \in \mathbb{Z}$ of P, and hence α must be a divisor of the constant term $P(0)$. Here we have $P(0) = 1$ and since $P(\pm 1) \neq 0$, it thus follows that P is either irreducible or splits into two factors of degree two, both having a constant term equal to $+1$ or equal to -1.

To go further it is essential to have some upper bound on the absolute value of the coefficients which can occur in a factor of P. Let

$$Q(X) = \sum_{j=0}^{\deg Q} b_j X^j \qquad \text{and} \qquad P(X) = \sum_{j=0}^{\deg P} a_j X^j$$

be polynomials, both having integer coefficients, and assume that Q is a factor of P. By a theorem of Mignotte [**114**] (see also [**40**]),

$$(13.9) \qquad |b_j| \leq \binom{\deg Q - 1}{j} \left(\sum_{j=0}^{d} |a_j|^2 \right)^{\frac{1}{2}} + \binom{\deg Q - 1}{j - 1} |a_{\deg P}|$$

for all $j = 0, \ldots, \deg Q$; of course, $|.|$ denotes here the standard absolute value. Mignotte's bound reduces the set of polynomials which may possibly divide P significantly. In fact, if we denote by B the maximum over all bounds (13.9) for $j \leq \deg Q$ and consider all (finitely many) possible factorizations of P mod $2B$ — with the convention that all coefficients are taken in the interval $[-B, B]$ — , then one of them must give the factorization of P, provided P is reducible!

In our example (13.8) it follows that the coefficients of a hypothetical polynomial factor of P have absolute value less than or equal to 8. Now we choose a prime p which is larger than twice this bound, $p = 17$ say. We find (by use of a computer algebra package)

$$P(X) \equiv (X + 3)(X + 7)(X - 5)(X - 6) \bmod 17;$$

note that here and in the sequel we use the residues mod p of least absolute value. Now we can try all pairs of the appearing linear factors to construct a polynomial of degree two which divides P. Observing that $3 \cdot (-6)$ and $7 \cdot (-5) \equiv -1 \bmod 17$, we are led to

$$(X+3)(X-6) = X^2 - 3X - 1 \qquad \text{and} \qquad (X+7)(X-5) = X^2 + 2X - 1$$

in $\mathbb{Z}/17\mathbb{Z}$. Indeed, these quadratic polynomials give the factorization of P given by (13.8) over $\mathbb{Z}$.

That was rather simple. Note that we have chosen the prime p larger than twice the bound of Mignotte's estimate (13.9) in order to get coefficients of the right size. However, in practice this would often lead to *large* primes, reducing the speed of our algorithm. For instance, we may consider

$$(13.10) \qquad P(X) = X^5 + 5X^4 - 3X^3 - 18X^2 + 25X + 14.$$

In this case Mignotte's bound (13.9) turns out to be quite *large* so that we would have to work with a *large* prime p. However, by Hensel's lemma 13.7 a factorization mod p yields also to a factorization mod p^2 and, by iteration, to moduli p^n with n as large as we want. Hence we even can take $p = 2$ provided that we lift a factorization mod 2 to a moduli 2^n with sufficiently large n. The step from $\mathbb{Z}/p\mathbb{Z}$ to $\mathbb{Z}/p^n\mathbb{Z}$ is called **Hensel lifting**; this idea is due to Zassenhaus [**175**]. In our example (13.10) we find

$$P(X) \equiv X^5 + X^4 + X^3 + X \bmod 2.$$

We easily find the factorization of P into the product $Q_1 R_1$ modulo 2, where

$$Q_1(X) = X(X+1) = X^2 + X \quad \text{and} \quad R_1(X) = X^3 + X + 1$$

are coprime. We observe that their product does not give (13.10). This might indicate that a factor of P has a coefficient larger than 2. Anyway, since Q_1 and R_1 are coprime, by the proof of Hensel's lemma 13.7 we can construct polynomials Q_2, R_2 such that $P \equiv Q_2 R_2 \bmod 2^2$. For this aim we compute by the extended greatest common divisor for polynomials

$$(X+1) \cdot Q_1(X) + 1 \cdot R_1(X) \equiv 1 \bmod 2.$$

Following the construction of the proof we obtain

$$Q_2(X) = X^2 + X + 2 \qquad \text{and} \qquad R_2(X) = X^3 - X - 1$$

in $\mathbb{Z}/2^2\mathbb{Z}$. However, this does not lead to a factorization of P. Doing further iterations we get

$$Q_3(X) = X^2 - 3X + 2 \qquad \text{and} \qquad R_3(X) = X^3 + 3X - 1 \qquad \bmod 8,$$
$$Q_4(X) = X^2 + 5X + 2 \qquad \text{and} \qquad R_4(X) = X^3 - 5X + 7 \qquad \bmod 16.$$

In fact, the last step gives the desired factorization of (13.10):

$$X^5 + 5X^4 - 3X^3 - 18X^2 + 25X + 14 = (X^2 + 5X + 2) \cdot (X^3 - 5X + 7).$$

It should be noted that A.K. Lenstra, H.W. Lenstra & Lovász [**102**] discovered a first deterministic polynomial time algorithm for factoring polynomials over $\mathbb{Z}$. Their approach relies on the LLL-lattice reduction algorithm presented in Section 8.9. Most of their approach coincides with former factorization techniques. First, the polynomial P in question is reduced modulo p for a small prime p. Then this reduced polynomial is factored and the result is lifted to a factorization over $\mathbb{Z}/p^n\mathbb{Z}$ for a sufficiently large n, like we did above. The final step is to use this p-adic approximation to obtain the factorization of the initial polynomial. Only the last step uses exponential (and not polynomial) time in previous algorithms. However, using LLL-reduction one can proceed as follows. Let Q be an irreducible factor of P mod p^n. The polynomials of degree one which reduce modulo p^n to a multiple of Q form a lattice Λ which contains a vector of *relatively short* length if and only if it contains a multiple of the irreducible factor of P corresponding to Q. For obtaining the irreducible factors of P it thus suffices to apply LLL in order to find a vector of short length in Λ. This factorization method was improved by van Hoeij [84]. His refinement uses the LLL-algorithm to reduce the time needed for choosing the right subsets of factors mod p^n.

Notes on p-adics: what we leave out

Almost everything we have done in the first ten chapters can be extended to p-adic numbers. This is almost clear for topics as diophantine approximation or continued fractions. So let us have a brief look at what is obtained in more advanced questions. For instance, Ridout obtained a p-adic version of Roth's theorem, and Schlickewei proved a p-adic subspace theorem; the most general formulation of both results can be found in [81]. Remarkably, there are still some open ties. For example, the p-adic analogue of the theorem of Lindemann–Weierstrass is still unproved.

Nevertheless, p-adic numbers have a great impact on modern number theory. They became an important tool in class field theory, for instance, the p-adic proof of the celebrated Kronecker–Weber theorem which claims that every finite abelian extension of $\mathbb{Q}$ is contained in a cyclotomic extension over $\mathbb{Q}$. For this and more we refer to [37] and [123].

Exercises

13.1. *Give a direct proof of Theorem 13.1 which makes only use of the p-adic expansion from Theorem 12.8.*

13.2. *Prove that $\mathbb{Z}$ is dense in $\mathbb{Z}_p$.*

Hint: Recall the proof of Theorem 12.7 for the corresponding result for $\mathbb{Q} \subset \mathbb{Q}_p$.

13.3. *Show that $\mathbb{Q}_p = \mathbb{Z}_p[p^{-1}] = \mathbb{Z}_p + \mathbb{Z}\left[p^{-1}\right]$. Furthermore, prove that $\mathbb{Q}_p$ is uncountable.*

13.4. *i) Prove that for $n \in \mathbb{N}$ the set $p^n\mathbb{Z}_p$ is an ideal in $\mathbb{Z}_p$. Further prove that $\mathrm{m} := p\mathbb{Z}_p$ is a maximal ideal of $\mathbb{Z}_p$ and that any ideal of $\mathbb{Z}_p$ is a power of m.*

Commutative rings R with only one maximal ideal m are called **local rings**. The quotient ring R/m is a field, the so-called **residue field**. Thus, $\mathbb{Z}_p$ is a local ring and $\mathbb{Z}_p/p\mathbb{Z}_p$ is its residue field.

ii) For $n \in \mathbb{N}$, prove that $\mathbb{Z}_p / p^n \mathbb{Z}_p \cong \mathbb{Z}/p^n\mathbb{Z}$.

13.5. This and the following exercise explore the topology of the world of p-adic numbers:

Prove that for $\gamma \in \mathbb{Q}$ and $n \in \mathbb{N}$ the balls

$$\gamma + p^n \mathbb{Z}_p := \{\alpha \in \mathbb{Q}_p : |\alpha - \gamma|_p \leq p^{-n}\}$$

are clopen and cover $\mathbb{Q}_p$. Show further that the connected component of any element in $\mathbb{Q}_p$ contains only the point itself. Finally, prove that for any two distinct points one can find two non-intersectiong balls around them which separate these points.

Thus, $\mathbb{Q}_p$ is a **totally disconnected Hausdorff topological field**.

13.6. *Prove that $\mathbb{Z}_p$ is compact and $\mathbb{Q}_p$ is locally compact, i.e., any collection of open sets which covers $\mathbb{Q}_p$ has a finite subcollection which also covers $\mathbb{Q}_p$.*

Hint: Since $\mathbb{Z}_p$ is a neighborhood of zero, it suffices to prove the first assertion.

13.7.* The p-adic fields have some properties in common with the Cantor set. The **Cantor set** $\mathcal{C}$ is given by taking the unit interval $[0,1]$, removing the middle third $(\frac{1}{3}, \frac{2}{3})$, removing the middle third of each of the remaining intervals, and continuing this procedure ad infinitum.

Show that the Cantor set $\mathcal{C}$ consists of those points in the unit interval whose ternary expansions do not contain digits equal to 1. Prove that $\mathcal{C}$ is an uncountable, totally disconnected compact metric space which is nowhere dense and has Lebesgue measure zero.

For some help see, for example, [**145**].

13.8. *Extend Theorem 13.2 in order to show that the system of congruences (13.3) has a non-trivial solution (i.e., not all coordinates are equal to zero) if and only if the equation (13.2) has a non-zero solution in p-adic integers.*

13.9. This exercise has to be solved in the real world $\mathbb{R}$. Assume that P is a differentiable function in some neighborhood of $\xi \in \mathbb{R}$ with $P(\xi) = 0$ and $P'(\xi) \neq 0$. Define a sequence (x_n) recursively by

$$x_n = x_{n-1} - \frac{P(x_{n-1})}{P'(x_{n-1})} \qquad \text{for} \quad n = 1, 2, \ldots$$

for some initial value x_0, sufficiently close to ξ. If P is a *nice* function, then the sequence of x_n tends to the zero ξ of P. This is the important and self-correcting **Newton iteration method**.

i) If P is a polynomial, show that $\lim_{n\to\infty} x_n = \xi$, provided that x_0 is chosen sufficiently close to ξ.
ii) Find good rational approximations to $\sqrt{2}$ by Netwon's iteration method.
iii) Let a, b be positive integers. Apply Newton's method to the polynomial $P(X) = X^2 + (b - 2a)X + a^2 - ab - 1$ with the initial value $x_0 = a$. Show that the x_n are the convergents to the continued fraction expansion of the positive root $[a, \overline{b}]$ of P.

13.10.* *Prove that a 2-adic integer $\alpha \in \mathbb{Z}_2^*$ is a square if and only if $\alpha \in 1 + 8\mathbb{Z}_2$. Give also a characterization of the squares in $\mathbb{Q}_2$.*

13.11. *Prove that the roots of unity in $\mathbb{Q}_2$ are ± 1.*

13.12. *Show that there exists a cubic root of 2 in $\mathbb{Q}_5$. What about other p-adic fields?*

13.13. *Deduce Hensel's lemma in the form of Theorem 13.3 from its generalization Theorem 13.7.*

13.14.* *Show that the polynomial $X^4 + 1$ factors into four linear factors modulo primes $p \equiv 1 \bmod 8$ or $p = 2$, and into two irreducible polynomials of degree two if $p \not\equiv 1 \bmod 8$. Deduce that $X^4 + 1$ is irreducible over $\mathbb{Z}$.*

Hint: Use the theory of quadratic reciprocity and identities like

$$X^4 + 1 = (X^2 + \sqrt{-1})(X^2 - \sqrt{-1}).$$

13.15. *Use the techniques from Section 13.7 to factor the polynomial $X^6 - 6X^4 - 2X^3 - 7X^2 + 6X + 1$ over $\mathbb{Z}$.*

13.16.* *Instead of (13.9) prove under the same assumptions in addition with the restriction $2 \deg Q \leq \deg P$ the upper bound*

$$|b_j| \leq \max_{0 \leq j \leq m} \binom{\deg Q}{j} (\mathsf{H}(P) + 1)^{\deg Q - j},$$

where $\mathsf{H}(P)$ is the height of P. Compare this bound with Mignotte's bound (13.9).

CHAPTER 14

The local–global principle

A diophantine equation cannot be solvable over the field of rationals if it has no p-adic solutions (since $\mathbb{Q} \subset \mathbb{Q}_p$). In this chapter we will prove that *quite often* the converse implication holds to be true as well. This means that the information on solvability of certain equations over all $\mathbb{Q}_p$, $p \leq \infty$, gives information on the solvability over $\mathbb{Q}$. We will prove a special case of this principle, the celebrated theorem of Hasse–Minkowski, for ternary quadratic forms.

14.1. One for all and all for one

Now we come to a simple but remarkable property of p-adic numbers. Recall that on behalf of Ostrowski's theorem the p-adic absolute values with $p \leq \infty$ give all non-trivial absolute values on $\mathbb{Q}$ up to equivalence. Next we shall see that all p-adic values are related by means of the following **product formula**.

Theorem 14.1. *For any rational $\alpha \neq 0$,*

$$\prod_{p \leq \infty} |\alpha|_p = 1.$$

The Archimedean standard absolute value $|\,.\,|_\infty$ on $\mathbb{Q}$ holds information about the sign and the non-Archimedean absolute values $|\,.\,|_p$ with $p < \infty$ information on the divisibility by primes. As the product formula shows all this information is linked. If we know the absolute values $|\alpha|_p$ for *all* but *one* $p \leq \infty$, we can compute the remaining one from all the others by taking the reciprocal of their product.

Proof. By the unique prime factorization of any rational $\alpha \neq 0$,

$$\alpha = \text{sign}\,(\alpha) \prod_{p < \infty} p^{\nu_p(\alpha)},$$

where $\text{sign}\,(\alpha)$ is the sign of α according to $\alpha = \text{sign}\,(\alpha)|\alpha|_\infty$. In view of the definition of the p-adic absolute value we may rewrite this equation as

$$|\alpha|_\infty = \prod_{p < \infty} |\alpha|_p^{-1}.$$

The theorem is proved. ●

The primes p are also called the **finite places** and $p = \infty$ stands for the **infinite place**. The product formula combines the information with respect to all places. Actually, we can go a bit further by the following criterion for rational squares in terms of squares in $\mathbb{Q}_p$.

Theorem 14.2. *A rational number α is a square if and only if it is a square in all $\mathbb{Q}_p, p \leq \infty$.*

Proof. One implication follows easily from the embedding $\mathbb{Q} \subset \mathbb{Q}_p$. For the other one we have a look on the prime factorization of α:

$$\alpha = \pm \prod_{p<\infty} p^{\nu_p(\alpha)}.$$

If α is a square in the field of real numbers, α is non-negative. Furthermore, if α is a square in $\mathbb{Q}_p$, the exponent $\nu_p(\alpha)$ is even. This proves the other implication. •

Theorem 14.2 has interesting consequences for the structure of p-adic fields. Comparing the squares in different fields $\mathbb{Q}_p$ for $p \leq \infty$, one can show that

- two distinct p-adic fields are non-isomorphic,
- $\mathbb{R} = \mathbb{Q}_\infty$ and any $\mathbb{Q}_p$ with $p < \infty$ are non-isomorphic.

For instance, 2 is a quadratic residue mod 7 and, by Theorem 13.5, a square in $\mathbb{Q}_7$, but it is a non-residue mod 13 and thus not a square in $\mathbb{Q}_{13}$. Since $2 \in \mathbb{Q}$, there cannot exist an isomorphism between $\mathbb{Q}_7$ and $\mathbb{Q}_{13}$ fixing $\mathbb{Q}$.

14.2. The theorem of Hasse–Minkowski

Number fields and function fields of curves over finite fields are called **global** fields (but here we shall only consider $\mathbb{Q}$ itself), and the completions of a global field with discrete valuation and finite residue field are said to be **local**. The local fields contain deep information of the underlying global field:

> The **local–global** *principle is the idea of putting together information from all local fields $\mathbb{Q}_p$ and additionally $\mathbb{Q}_\infty = \mathbb{R}$, to get information in the global field $\mathbb{Q}$.*

This principle is extraordinarily successful and seems to date back to Hensel but was first clearly stated by Hasse. Theorem 14.2 is a first example in this direction: a rational number is a square if and only if it is a square locally, i.e., at all places $p \leq \infty$. This is only the very beginning but gives a first glimpse of its power. With this concept Hasse [**77**] was able to give in the 1920s an important characterization of quadratic forms.

Let $\mathbb{K}$ be a field. Then we say that a quadratic form $P \in \mathbb{K}[X_1,\ldots,X_r]$ is **isotropic** over $\mathbb{K}$ if there exist $x_1,\ldots,x_r \in \mathbb{K}$, not all equal to zero, such that

$$P(x_1,\ldots,x_r) = 0.$$

It can be shown that isotropic quadratic forms take all values; i.e., for any $\beta \in \mathbb{K}$, the equation

$$P(X_1,\ldots,X_r) = \beta$$

has a solution in $\mathbb{K}$ (see Exercise 14.2).

A first classification of isotropic quadratic forms was found by Minkowski [**116**] in 1890; he proved: if $P(X_1, \ldots, X_r)$ is a quadratic form with integer coefficients for which (13.3) with any prime p has a non-trivial solution, and if $P(X_1, \ldots, X_r) = 0$ has a non-trivial real solution, then P is isotropic over $\mathbb{Q}$. In view of Theorem 13.2 this can be translated into the celebrated

Theorem of Hasse–Minkowski. *A quadratic form is isotropic over $\mathbb{Q}$ if and only if it is isotropic over all $\mathbb{Q}_p$, $p \leq \infty$.*

This theorem gives a solution of Hilbert's eleventh problem on representations of integers in algebraic number fields by quadratic forms with rational coefficients; the case of quadratic forms with integer coefficients is still an open problem.

There are some important consequences of this deep theorem. For instance, one can deduce Lagrange's theorem 8.7 that every positive integer can be written as a sum of at most four squares. Another application is Gauss' theorem that every positive integer has a representation as a sum of at most three triangular numbers which, by the way, is the first entry (dated July 10, 1796) in his mathematical diary:

$$** \quad \text{EYPHKA} \quad \quad \text{num} = \Delta + \Delta + \Delta.$$

For proofs of these corollaries we refer to Serre [**146**]. However, the main impact of the theorem of Hasse–Minkowski lies in algebraic number theory.

Hasse's p-adic proof is much easier and more natural than Minkowski's approach, but even Hasse's proof is far beyond the scope of these notes; it uses deeper knowledge on the theory of quadratic forms, the multiplicative structure of $\mathbb{Q}_p$ and its subgroup of squares, and even Dirichlet's prime number theorem for arithmetic progressions. We refer the interested reader to [**146**]. However, in the following section we shall prove the theorem of Hasse–Minkowski for ternary quadratic forms.

14.3. Ternary quadratics

Given a ternary quadratic form $P \in \mathbb{Q}[X, Y, Z]$, it is clear that any non-trivial solution to $P = 0$ over $\mathbb{Q}$ yields also a non-trivial solution in $\mathbb{Q}_p$ for any $p \leq \infty$ (since $\mathbb{Q}$ is embedded in $\mathbb{Q}_p$). So for a proof of the local–global principle for ternary quadratic forms it remains to show

Theorem 14.3. *Let $P(X, Y, Z)$ be a quadratic form with rational coefficients. Suppose that in every $\mathbb{Q}_p$, $p \leq \infty$, there is a non-trivial solution of $P = 0$. Then there is a non-trivial solution in $\mathbb{Q}$.*

After a linear transformation of the variables, we may assume that the ternary quadratic form P has the shape

(14.1) $$P(X, Y, Z) = aX^2 + bY^2 + cZ^2.$$

From algebraic geometry we know that every curve of genus 0 defined over $\mathbb{Q}$ is birationally equivalent over $\mathbb{Q}$ to a curve

$$aX^2 + bY^2 + cZ^2 = 0,$$

where $a, b, c \in \mathbb{Q}$ (a proof can be found in Cassels [**37**]). So the local–global principle holds for curves of genus 0. This fact was, however, in a different language, already known to Legendre (see Exercise 14.11).

If all coefficients in (14.1) are zero, there is nothing to show. If any two of the three coefficients are equal to zero, all solutions of the equation $P = 0$ are trivial. Thus we may assume that at most one of the coefficients is zero.

First, suppose that exactly one of the coefficients a, b, c is zero; then we are in the case of a binary quadratic form. Without loss of generality we may assume that $c = 0$. Then we can rewrite the equation $P = 0$ as

$$(14.2) \qquad aX^2 = -bY^2.$$

For every $p < \infty$, the assumption of the theorem yields the existence of a non-trivial solution x, y in $\mathbb{Q}_p$ such that

$$\left(\frac{x}{y}\right)^2 = -\frac{b}{a}.$$

It follows from Theorem 14.2 that the number on the right-hand side is a non-zero rational square. This proves the existence of a non-trivial rational solution of (14.2).

Now assume that the coefficients a, b, c in (14.1) are all non-zero. On replacing P by dP and the variables X, Y, Z by eX, fY, gZ for suitable integers d, e, f, g, respectively, we may suppose that the coefficients a, b, c are all squarefree integers. Before we continue with the proof of Theorem 14.3 we have to reformulate the existence of non-trivial solutions to the equation $P = 0$ over the local fields.

Lemma 14.4. *Suppose that a, b, c and P satisfy the conditions from above.*

(i) *If $p \neq 2$ divides c, then there is an integer r such that*

$$ar^2 + b \equiv 0 \bmod p.$$

(ii) *If $p = 2$ does not divide abc, then*

$$a + b, \quad a + c, \quad or \quad b + c \equiv 0 \bmod 4.$$

(iii) *If $p = 2$ divides c, then*

$$a + b + cs^2 \equiv 0 \bmod 8$$

holds with $s = 0$ or $s = 1$.

Proof. First, we consider the case that c is divisble by a prime $p \neq 2$. Since P is isotropic over $\mathbb{Q}_p$, there exist integers $x_0, y_0, z_0 \in \mathbb{Q}_p$ such that

$$ax_0^2 + by_0^2 + cz_0^2 = 0;$$

multiplying with an appropriate power of p we may suppose that x_0, y_0, z_0 lie in $\mathbb{Z}_p$. If at least two of the numbers x_0, y_0, z_0 lie in $p\mathbb{Z}_p$, then also the third and we can divide through p. Thus we may assume that at most one of them is not divisible by p. Since p divides c, we have $cz_0^2 \in p\mathbb{Z}_p$. $\mathbb{Z}$ is dense in $\mathbb{Z}_p$ (see Exercise 13.2). Thus there exist integers x and y for which

$$ax^2 + by^2 \equiv 0 \bmod p,$$

where, without loss of generality, y is assumed not to be divisible by p. Therefore, we can multiply this congruence with y^{-2} which leads by setting $r = xy^{-1} \bmod p$ to the first assertion of the lemma.

As usual, the *oddest* prime $p = 2$ causes the biggest trouble. Since P is isotropic over $\mathbb{Q}_2$, there are $x_0, y_0, z_0 \in \mathbb{Z}_2$ such that $P(x_0, y_0, z_0) = 0$; as before we may assume that they are all 2-adic integers which are not all in $2\mathbb{Z}_2$. By reducing modulo $2\mathbb{Z}_2$ exactly two of x_0, y_0, z_0 are 2-adic units (since all coefficients a, b, c are odd), and the other, say $z_0 \in 2\mathbb{Z}_2$. The square of a 2-adic unit belongs to $1 + 4\mathbb{Z}_2$ (by Exercise 13.10). This gives

$$ax_0^2 + by_0^2 + cz_0^2 \equiv a + b \bmod 4\mathbb{Z}_2.$$

This proves the second assertion of the lemma. The third assertion is left to the reader. •

Proof of Theorem 14.3. Define a subset Λ of triples $(x, y, z) \in \mathbb{Z}^3$ by the following congruence conditions corresponding to the primes p dividing $2abc$:

(i) If $p \neq 2$ and p divides c, then

$$x \equiv ry \bmod p,$$

and similarly if p divides b or a.

(ii) If $p = 2$ does not divide abc and $a + b \equiv 0 \bmod 4$, then

$$z \equiv 0 \bmod 2,$$

and similarly if $a + b \not\equiv 0$ but $a + c$ or $b + c \equiv 0 \bmod 4$.

(iii) If $p = 2$ divides c, then

$$x \equiv y \bmod 4 \quad \text{and} \quad z \equiv sy \bmod 2,$$

and similarly if p divides b or a.

Here r and s are the quantities from Lemma 14.4. If, for example, $p \neq 2$ divides a but not c, then in the first condition x and y have to be replaced by y and z; this explains what we mean by *similar*. Of course, by symmetry, the analogue of Lemma 14.4 (i) holds in this case too.

In view of Lemma 8.5 the set Λ is a lattice in $\mathbb{Z}^3$. Indeed, every congruence condition defines a sublattice of $\mathbb{Z}^3$. The corresponding matrix representations are

$$\begin{pmatrix} p & r & 0 \\ 0 & 1 & 0 \\ 0 & 0 & 1 \end{pmatrix}, \quad \begin{pmatrix} 1 & 0 & 0 \\ 0 & 1 & 0 \\ 0 & 0 & 2 \end{pmatrix}, \quad \text{and} \quad \begin{pmatrix} 4 & 1 & 0 \\ 0 & 1 & 0 \\ 0 & s & 2 \end{pmatrix},$$

having determinants $p, 2$, and 8, respectively. Hence, Λ has determinant $\det(\Lambda) \leq 4|abc|$. By Lemma 14.4, for any $(x, y, z) \in \Lambda$,

(14.3) $$P(x, y, z) \equiv 0 \bmod 4|abc|.$$

For instance, if p is an odd prime dividing c, we deduce from the first assertion of Lemma 14.4 that there exist integers r, y, z such that

$$P(ry, y, z) = a(ry)^2 + by^2 + cz^2 \equiv y^2(ar^2 + b^2) \equiv 0 \bmod p.$$

The other cases follow in a similar manner and their verification is left to the reader.

Recall the use of such congruence conditions in the proof of Theorem 8.7 in order to obtain the existence of integral solutions of diophantine equations. Here we consider the set $\mathcal{C}$ of triples $(x, y, z) \in \mathbb{R}^3$ given by the inequality

$$(14.4) \qquad\qquad |a|x^2 + |b|y^2 + |c|z^2 < 4|abc|.$$

$\mathcal{C}$ is an ellipsoid and so it is a convex symmetric body. Its volume is

$$(14.5) \qquad\qquad \mathrm{vol}(\mathcal{C}) = \frac{32\pi}{3}|abc|$$

(which should be checked by the reader). Since this quantity is greater than $32|abc|$, by Minkowski's convex body theorem 8.3 there exists a non-zero lattice point $\mathbf{x}$ in $\Lambda \cap \mathcal{C}$. It follows from (14.3) that $P(\mathbf{x}) = 0$. This proves Theorem 14.3. •

Reviewing the proof, we note that we have only used that the quadratic form P is isotropic over $\mathbb{Q}_p$ for the primes p dividing $2abc$. As we shall see in the following section, for other non-Archimedean places p the solvability of $P = 0$ is guaranteed by a theorem of Chevalley. Moreover, we did not require a non-trivial solution over $\mathbb{R} = \mathbb{Q}_\infty$ in the proof, and so Theorem 14.3 goes beyond the local–global principle.

14.4. The theorems of Chevalley and Warning

In this section we study polynomial equations over finite fields. Any finite field is isomorphic to a Galois field $\mathbb{F}_q$ having $q = p^f$ elements, where p is prime and f is a positive integer; $\mathbb{F}_q$ is a field extension of $\mathbb{Z}/p\mathbb{Z}$ of degree f. The reader who is not familiar with finite fields may only consider the case $q = p$, resp. $\mathbb{F}_q = \mathbb{Z}/p\mathbb{Z}$, which is sufficient for our purpose. In 1935, Warning proved

Theorem 14.5. *Let $P \in \mathbb{F}_q[X_1, \ldots, X_n]$ have (total) degree $d < n$. Then the number N of solutions $(x_1, \ldots, x_n)$ to the equation $P(X_1, \ldots, X_n) = 0$ is divisible by p.*

Proof. It suffices to show that the number M of $(x_1, \ldots, x_n)$ for which $P(x_1, \ldots, x_n)$ does not vanish is divisible by p. This follows simply from the fact that $M + N = q^n \equiv 0 \bmod p$. Now recall that the multiplicative group $\mathbb{F}_q^*$ is cyclic of order $q - 1$. Thus, by Lagrange's theorem (A.1),

$$x^{q-1} = \begin{cases} 0 & \text{if } x = 0, \\ 1 & \text{otherwise.} \end{cases}$$

Hence M is modulo p congruent to

$$\sum_{x_1, \ldots, x_n \in \mathbb{F}_q} P(x_1, \ldots, x_n)^{q-1}.$$

On expanding $P(x_1, \ldots, x_n)$, we are led to evaluate sums of the type

(14.6)
$$\sum_{x_1, \ldots, x_n \in \mathbb{F}_q} x_1^{\nu_1} \cdot \ldots \cdot x_n^{\nu_n},$$

where

$$\nu_1 + \ldots + \nu_n \leq d(q-1) < n(q-1).$$

Consequently, there is an index j such that $\nu_j < q - 1$. Hence it suffices to prove that

(14.7)
$$S(\nu) := \sum_{x \in \mathbb{F}_q} x^\nu = 0$$

for $0 \leq \nu < q - 1$ (since then any sum of the form (14.6) vanishes mod p).

If $\nu = 0$, all summands are equal to one, and so $S(0) = q = 0$ (since the characteristic p divides q). If $1 \leq \nu < q - 1$, there exists an element $0 \neq y \in \mathbb{F}_q$ such that $y^\nu \neq 1$ (since the multiplicative group $\mathbb{F}_q^*$ is cyclic of order $q - 1$). Thus, in this case,

$$S(\nu) = \sum_{\alpha \in \mathbb{F}_q^*} x^\nu = \sum_{\alpha \in \mathbb{F}_q^*} (xy)^\nu = y^\nu S(\nu);$$

here we have used the fact that with x also xy runs through $\mathbb{F}_q^*$. This implies $S(\nu) = 0$ and completes the proof of (14.7). •

As an immediate consequence we obtain Chevalley's theorem:

Corollary 14.6. *Suppose that $P \in \mathbb{F}_q[X_1, \ldots, X_n]$ is a homogeneous polynomial of (total) degree $d < n$. Then there is a non-trivial solution to $P = 0$.*

Proof. Since P is homogeneous, $P(\mathbf{0}) = 0$, and thus by Warning's theorem there are at least $p - 1$ vectors $\mathbf{x} = (x_1, \ldots, x_n) \neq \mathbf{0}$ for which $P(\mathbf{x}) = 0$. •

In the particular case of ternary quadratic forms the corollary asserts the existence of non-trivial solutions over any $\mathbb{F}_q$. Given

$$P(X, Y, Z) = aX^2 + bY^2 + cZ^2,$$

where a, b, c are rationals, there exist $x, y, z \in \mathbb{Z}/p\mathbb{Z}$ such that $P(x, y, z) = 0$. Now assume $abc \neq 0$ and that at least one of the partial derivatives of P with respect to X, Y, Z does not vanish at (x, y, z); then Hensel's lemma, Theorem 13.3, gives a zero of the polynomial in $\mathbb{Q}_p$. This proves that P is isotropic over all $\mathbb{Q}_p$ for which $2abc$ is not divisible by p. We note this as

Corollary 14.7. *If p is an odd prime, then any quadratic form in at least three variables with at least three coefficients in $\mathbb{Z}_p^*$ is isotropic over $\mathbb{Q}_p$.*

14.5. Applications and limitations

There is another aspect in the proof of Theorem 14.3 which is remarkable — it is constructive. We illustrate this by an example. Consider

$$P(X, Y, Z) = 3X^2 - 5Y^2 - 7Z^2.$$

In view of Theorem 14.3 and Corollary 14.7 it suffices to ask for non-trivial solutions in $\mathbb{Q}_p$ for $p = 2, 3, 5, 7$. We discuss here only the case $p = 3$; the other cases can be treated similarly. Since

$$-\frac{7}{5} \in 1 + 3\mathbb{Z}_3,$$

the number on the left is a square in $\mathbb{Q}_3$. Thus there exists a 3-adic number y such that

$$y^2 = -\frac{7}{5}.$$

This leads to

$$P(0, y, 1) = 3 \cdot 0^2 - 5y^2 - 7 \cdot 1^2 = 0.$$

The other cases can be treated similarly. It follows that P is isotropic over all $\mathbb{Q}_p$ for $p = 2, 3, 5, 7$, and so, by the proof of Theorem 14.3, P is isotropic over $\mathbb{Q}$ too. If we are interested in an explicit solution, the proof of Theorem 14.3 shows us that there is an integer solution inside the ellipsoid $\mathcal{C}$ given by (14.4). Since $\mathcal{C}$ is finite, there are only finitely many candidates of integers inside $\mathcal{C}$ satisfying the conditions, and so we can check each of them until we find a solution. See Mordell [**120**] for the magnitude of integer solutions to homogenous quadratic equations in three variables.

At first glance it seems that the Hasse–Minkowski theorem reformulates the solvability of a quadratic equation over $\mathbb{Q}$ into a more difficult question. However, as we just have seen, it is easier to decide whether a given quadratic form is isotropic over p-adic fields than over the field of rationals, and the proof of Theorem 14.3 gives us means for finding rational solutions explicitly.

It is known that Hilbert's tenth problem is solvable over $\mathbb{Q}_p$ for $p \leq \infty$ (again we refer to [**128**] for more information). Since Hilbert's tenth problem is unsolvable over $\mathbb{Z}$ (and probably also over $\mathbb{Q}$), we cannot expect that the detour via p-adic will always work. Indeed, the local–global principle has limitations also. For example, by the theory of quadratic residues it is not too difficult to see that the equation

$$(14.8) \qquad (X^2 - 2)(X^2 - 17)(X^2 - 34) = 0$$

has solutions in all local fields $\mathbb{Q}_p$, $p \leq \infty$, but obviously it has no solution in $\mathbb{Q}$. One might argue that the above example is somehow special. Selmer gave

$$(14.9) \qquad 3X^3 + 4Y^3 - 5Z^3 = 0$$

as another example of an equation which is solvable locally but not globally. We will give another example where this principle fails in the following section.

The quest for an appropriate substitute for cubic or higher forms for the local–global principle is still an open problem. There are some remarkable results if the number of variables is sufficiently large. We illustrate this interesting fact by **Waring's problem**, posed by Waring in 1770: what is the least positive integer n such that any positive integer N can be written as a sum of n kth powers of non-negative integers? We denote the least such n by $g(k)$. It was first proved by Hilbert that $g(k)$ indeed exists. In view of Lagrange's theorem 8.7 we see that $g(2) \leq 4$. Since any integer $n \equiv 3 \bmod 4$ cannot be represented by three squares, it follows that $g(2) = 4$. It is a classic result from Dickson, Niven, and Pillai that $g(3) = 9$, and some more values of $g(k)$ are known for small k; however, in general the determination of $g(k)$ is an open problem.

Denote by $R(N, k, n)$ the number of solutions $(x_1, \ldots, x_n)$ to the equation

$$N = \sum_{j=1}^{n} X_j^k$$

with $x_j \geq 0$. In 1920, Hardy and Littlewood invented their famous circle method to attack Waring's problem. For $n > 2^k$, they obtained an asymptotic formula

(14.10) $$R(N, k, n) \sim N^{\frac{n}{k}-1} \frac{\Gamma(1 + \frac{1}{k})^n}{\Gamma(\frac{n}{k})} \mathfrak{S}(N);$$

here the so-called **singular series** $\mathfrak{S}(N)$ is given by the infinite Euler product

$$\mathfrak{S}(N) = \prod_p \delta_p, \quad \text{where} \quad \delta_p := 1 + \sum_{m=1}^{\infty} \left(\omega(p^m) - \omega(p^{m-1}) \right)$$

and $\omega(d)$ equals d^{1-n} times the number of solutions to the congruence

$$X_1^k + \ldots + X_n^k \equiv 0 \bmod d.$$

Since

$$\delta_p = \lim_{m \to \infty} \omega(p^m),$$

the quantity δ_p is the density for the p-adic solutions of the equation $X_1^k + \ldots + X_n^k = 0$; the gamma-factor in (14.10) can be regarded as the contribution of the infinite place $p = \infty$. Using a variant of the circle method, Birch proved that the local–global principle holds for quadratic forms of a given degree if the number of variables is sufficiently large. For details and more information we refer to [88] and [164].

There is a famous open conjecture due to Birch and Swinnerton-Dyer (which is one of the seven millennium problems) on elliptic curves which claims, roughly speaking, that the number of global solutions can be determined from local information. More precisley, it is expected that the order of vanishing of the L-function $L(s, E)$ associated with an elliptic curve E (a relative of the zeta-function) at the central point $s = 1$ is equal to the rank of the group of rational points on E. Ignoring convergence, this value

$L(1, E)$ can be regarded as an Euler product similar to the one in Waring's problem.[*]

14.6. The local Fermat problem

We conclude with the same topic with which we started, namely, Fermat's last theorem. In 1770, Euler proved Fermat's last theorem for the exponent $n = 3$; however, there is a gap in his argument concerning divisibility properties of integers of the form $a^2 + 3b^2$. This observation led to considering the Fermat equation over the field $\mathbb{Q}[\sqrt{-3}]$ (since $N(a + b\sqrt{-3}) = a^2 + 3b^2$). Gauss succeeded in proving the cubic case of Fermat's last theorem rigorously over the field $\mathbb{Q}(\sqrt{-3})$ and, consequently, there are no other solutions in the subfield $\mathbb{Q}$ besides the trivial ones.

It is a natural question to ask whether the Fermat equation has non-trivial solutions in rings other than $\mathbb{Z}$ or number fields. Now we shall show that indeed the p-adic case is rather different.

Theorem 14.8. *For every prime p and every prime $q \neq p$, the equation*

$$X^q + Y^q = Z^q$$

has non-trivial solutions in $\mathbb{Z}_p$.

Proof. We know that the Fermat equation has non-trivial integer solutions if $q = 2$, namely, the Pythagorean triples from Theorem 1.1. Since $\mathbb{Z} \subset \mathbb{Z}_p$, we may regard these solutions also as p-adic integers. So we are done if $q = 2$. In the sequel we thus may suppose q to be odd.

We note that

$$X^q + p^q - 1 \equiv X^q - 1 = (X - 1)(X^{q-1} + \ldots + X + 1) \bmod p.$$

Obviously, 1 is not a root of the polynomial $X^{q-1} + \ldots + X + 1$ modulo p (since $p \neq q$). Thus we can apply Hensel's lemma, Theorem 13.3, and obtain the existence of a p-adic integer α such that $\alpha \equiv 1 \bmod p$ and

$$\alpha^q + p^q + (-1)^q = 0.$$

Since q is odd, we deduce

$$\alpha^q + p^q = 1^q.$$

This proves the assertion. •

The statement of the theorem is also true for $q = p$. However, in this case the proof requires a bit more algebra; for this case we refer to Ribenboim [**132**]. Hence, the Fermat equation has non-trivial solutions for any local field $\mathbb{Q}_p$ with $p \leq \infty$ but, as Wiles proved, only trivial solutions over the global field $\mathbb{Q}$. Thus the famous Fermat equation is an interesting example for the failure of a *higher* local–global principle.

---

[*]For more information see http://www.claymath.org/millennium/.

Exercises

14.1. *Prove that any two distinct p-adic fields are non-isomorphic. Further, prove that $\mathbb{R}$ and any $\mathbb{Q}_p$ are non-isomorphic.*

14.2.* *Using the theorem of Hasse–Minkowski,*
i) prove that a quadratic form P over $\mathbb{Q}$ represents a non-zero rational number α, i.e., the diophantine equation $P(X) = \alpha$ is solvable, if and only if it does in each of the $\mathbb{Q}_p$;
ii) show that an isotropic quadratic form P over $\mathbb{Q}$ represents any rational number.

Hint: Assume that $P(x_1, \ldots, x_r) = 0$ with $x_1 \neq 0$; then consider the transformation $y_1 := x_1(1 + t)$ and $y_j := x_j(1 - t)$ for $2 \leq j \leq r$ and arbitrary $t \in \mathbb{Q}$.

14.3. *Assuming the theorem of Hasse–Minkowski, prove Lagrange's theorem 8.7.*

14.4. *i) Prove formula (14.5) on the volume of an ellipsoid.*

This requires some knowledge of calculus; the method of proof is a generalization of the computation of the volume of a three-dimensional sphere (see Exercises 8.2 and 8.15).

ii) Fill all further gaps in the proof of Theorem 14.3, e.g., the complete verification of (14.3).

14.5. *Prove that the equation*

$$3X^2 - 5Y^2 - 7Z^2 = 0$$

has a non-trivial solution in rational numbers x, y, z, while it fails to have a solution different from the trivial one when we change the sign of $5Y^2$.

14.6. *Show that (14.8) is solvable over all $\mathbb{Q}_p$, $p \leq \infty$, but that it is unsolvable over $\mathbb{Q}$.*

Hint: See Appendix A.5, especially the remark at the end.

14.7. *Show that the local–global principle fails for forms of total degree six independent of the number of variables.*

Hint: Use Selmer's result on the failure of the local–global principle for (14.9).

14.8. *Prove the identity*

$$60\left(x_1^2 + x_2^2 + x_3^2 + x_4^2\right)^3 = 36 \sum_{1 \leq i \leq 4} x_i^6 + 2 \sum_{1 \leq i_1 < i_2 \leq 4} (x_{i_1} \pm x_{i_2})^6$$

$$+ \sum_{1 \leq i_1 < i_2 < i_3 \leq 4} (x_{i_1} \pm x_{i_2} \pm x_{i_3})^6.$$

Using the fact that every positive integer can be written as a sum of nine cubes, deduce that every positive integer has a representation as a sum of at most 1715 sixths-powers, i.e., $g(6) \leq 1715$.

14.9.* *Prove that any positive integer can be written as a sum of at most* 53 *biquadratic powers of integers.*

Hint: Use Lagrange's theorem 8.7 in combination with an appropriate identity (similar to the one from the previous exercise). The estimate $g(4) \leq 53$ is due to Liouville.

14.10.* Let p, q be distinct odd primes such that $q \equiv 3 \bmod 4$ and $-p$ is a quadratic residue modulo q.

Show that
$$P(X, Y, Z) = X^2 + pY^2 + qZ^2 = 0$$
is isotropic over $\mathbb{Q}_2$ *and* $\mathbb{Q}_q$ *but not over* $\mathbb{R} = \mathbb{Q}_\infty$. *Deduce that there is no non-trivial solution over* $\mathbb{Q}_p$, *and that* $-q$ *is a quadratic nonresidue* $\bmod\, p$.

14.11. Legendre proved the following variant of Theorem 14.3.

Let a, b, c *be squarefree, pairwise coprime non-zero integers. Show that then*
$$aX^2 + bY^2 + cZ^2 = 0$$
has a non-trivial solution (i.e., $(x, y, z) \neq \mathbf{0}$*) if and only if* $a, b,$ *and* c *do not all have the same sign, and* $-ab, -bc,$ *and* $-ca$ *are quadratic residues modulo* $a, b,$ *and* c, *respectively.*

Hint: Use Theorem 14.3.

14.12. *Use the previous exercise in order to find those equations which do have a non-trivial solution:*
$$X^2 + 3Y^2 + 2Z^2 = 0 \qquad \text{and} \qquad -X^2 - 3Y^2 + Z^2 = 0.$$

14.13.* *Let* $d \equiv 3 \bmod 4$. *Show that the Pell minus equation* $X^2 - dY^2 = -1$ *has no rational solutions.*

Hint: Consider the quadratic form $X^2 - dY^2 + Z^2$ in $\mathbb{Q}_p$ for a prime divisor p of d. Use Exercise 8.19.

14.14.* *For prime* p *show that the Fermat equation with exponent* p *has a non-trivial* p-adic solution.

APPENDIX A

Algebra and number theory

In this appendix we list some basic facts from algebra and number theory. Furthermore, we give some more details on topics we have already touched but have not discussed deeply.

A.1. Groups, rings, and fields

A **composition** $*$ on a set G is a map $* : \mathsf{G} \times \mathsf{G} \to \mathsf{G}$. A pair $(\mathsf{G}, *)$ of a set G and a composition $*$ is said to be a **group** if the composition is associative, if there exists a **neutral element** $e \in \mathsf{G}$, i.e.,

$$e * g = g = g * e \qquad \text{for all} \quad g \in \mathsf{G},$$

and if for any $g \in \mathsf{G}$, there exists an **inverse element** $g^{-1} \in \mathsf{G}$, i.e.,

$$g * g^{-1} = e = g^{-1} * g.$$

Notice that the neutral element of a group and the inverse of any given element of a group are uniquely determined. We often write G in place of $(\mathsf{G}, *)$ if it is clear which the underlying composition $*$ is.

A group G is called **abelian** if $g_1 * g_2 = g_2 * g_1$ for any $g_1, g_2 \in \mathsf{G}$. For instance, $\mathbb{Z}$ is a group with respect to addition, more precisely, an abelian group, but it is not a group with respect to multiplication. A group G is called **cyclic** if there exists an element $g \in \mathsf{G}$ which generates the whole group G.

The number $\sharp\mathsf{G}$ of elements of a group G is called the **order** of G. If $\sharp\mathsf{G}$ is a finite number, G is said to be **finite**; otherwise, G is an **infinite** group. The important theorem of Lagrange states that if G is a finite group, then

$$(\text{A.1}) \qquad (\sharp\mathsf{G})g := \underbrace{g * g * \ldots * g}_{\sharp\mathsf{G} \text{ times}} = e \qquad \text{for any} \quad g \in \mathsf{G}.$$

The proof simply follows from the equality of the products taken over h and gh, respectively, where h runs through G.

A **ring** $\mathcal{R}$ is an abelian group $(\mathcal{R}, +)$ with an additional associative composition "$\cdot$", called multiplication, such that there exists an element $1 \in \mathcal{R}$ for which

$$1 \cdot r = r = r \cdot 1 \qquad \text{for all} \quad r \in \mathcal{R},$$

and

$$r \cdot (s + t) = r \cdot s + r \cdot t, \quad (s + t) \cdot r = s \cdot r + t \cdot r \qquad \text{for all} \quad r, s, t \in \mathcal{R}.$$

For abbreviation, we often leave out the multiplication symbol. The neutral element of $\mathcal{R}$ with respect to addition is usually denoted by 0. Note that in general $\mathcal{R}$ is not a group with respect to multiplication.

If $xy = yx$ holds for all elements x, y in some ring $\mathcal{R}$, then $\mathcal{R}$ is called **commutative**. For example, $\mathbb{Z}$ is a commutative ring with respect to standard addition and multiplication, whereas the ring of $N \times N$ matrices with real entries is for $N \geq 2$ a non-commutative ring with matrix addition and multiplication.

An element r of a ring $\mathcal{R}$ is said to be a **unit** if there exists an inverse of r with respect to multiplication; that is, the existence of $r^{-1} \in \mathcal{R}$ with $r \cdot r^{-1} = 1$. The set of units of a ring is denoted by $\mathcal{R}^*$, and forms clearly a multiplicative group. A ring $\mathcal{R}$ with $\mathcal{R}^* = \mathcal{R} \setminus \{0\}$, i.e., a ring where all non-zero elements have a multiplicative inverse, is called a **field**. For instance, the set of rational numbers $\mathbb{Q}$ is a field. An element $r \in \mathcal{R} \setminus \mathcal{R}^*$ is called **irreducible** if any factorization $r = st$ with $s, t \in \mathcal{R}$ implies that either s or t is a unit; otherwise r is said to be **reducible**. A non-zero element $p \in \mathcal{R} \setminus \mathcal{R}^*$ is called **prime** if whenever p divides st with $s, t \in \mathcal{R}$, then p divides s or p divides t. Examples for these notions appear in Sections A.2 and A.6.

Assume that $\mathcal{R}$ is a commutative ring. An **ideal** I in $\mathcal{R}$ is a subgroup of $(\mathcal{R}, +)$ for which $ri \in I$ for any $r \in \mathcal{R}$ and $i \in I$. Note that $\mathcal{R}$ is trivially an ideal and that an ideal I in $\mathcal{R}$ is equal to $\mathcal{R}$ if and only if $1 \in I$. An ideal m is said to be **maximal** if the ideal generated by m and any element in $\mathcal{R} \setminus$ m is the whole ring $\mathcal{R}$.

Theorem A.1. *An ideal* m *is a maximal ideal of a ring* $\mathcal{R}$ *if and only if the quotient ring* $\mathcal{R}/\mathrm{m} := \{r + \mathrm{m} \ : \ r \in \mathcal{R}\}$ *is a field.*

An ideal $P \subsetneq \mathcal{R}$ is said to be a **prime ideal** if $rs \in P$ implies that $r \in P$ or $s \in P$. Any maximal ideal is a prime ideal. For example, the set of multiples $m\mathbb{Z} := \{mz \ : \ z \in \mathbb{Z}\}$ of a given integer m is an ideal of $\mathbb{Z}$. It is a maximal ideal (and so also a prime ideal) if and only if $m = p$ is prime.

Now we have a short look at substructures. A **subgroup** H of a group G is a non-empty subset of G such that the composition of G makes H to a group. Similarly, a subset $\mathcal{S}$ of a ring $\mathcal{R}$ is called a **subring** if $\mathcal{S}$ is a subgroup of $(\mathcal{R}, +)$, $\mathcal{S}$ contains the element 1, and $\mathcal{S}$ is multiplicatively closed. If $\mathbb{K} \subset \mathbb{L}$ are fields, then we say that $\mathbb{K}$ is a **subfield** of $\mathbb{L}$ if $\mathbb{K}$ is a subring of $\mathbb{L}$; in this case $\mathbb{L}$ is said to be an **extension field** of $\mathbb{K}$ and the field extension is denoted by $\mathbb{L}/\mathbb{K}$.

If $f : \mathsf{G} \to \mathsf{H}$ is a map between groups $(\mathsf{G}, \oplus)$ and $(\mathsf{H}, \otimes)$, then f is called a **group homomorphism** if

$$f(x \oplus y) = f(x) \otimes f(y) \qquad \text{for all} \quad x, y \in \mathsf{G}.$$

A map $f : \mathcal{R} \to \mathcal{S}$ between rings $\mathcal{R}$ and $\mathcal{S}$ is said to be a **ring homomorphism** if it is a group homomorphism with respect to the additive groups $(\mathcal{R}, +)$ and $(\mathcal{S}, +)$, and it satisfies

$$f(xy) = f(x)f(y) \qquad \text{for all} \quad x, y \in \mathcal{R},$$

and $f(1) = 1$. If such a ring homomorphism is bijective, it is called an **isomorphism**, and $\mathcal{R}$ and $\mathcal{S}$ are said to be **isomorphic**, denoted by $\mathcal{R} \cong \mathcal{S}$.

The **kernel** $\ker f := \{r \in \mathcal{R} : f(r) = 0\}$ of a ring homomorphism f is an ideal of $\mathcal{R}$ and the **image** $f(\mathcal{R}) = \{s \in \mathcal{S} : s = f(r)\}$ is a subring of $\mathcal{S}$. The **isomorphism theorem** states that the map

$$\tilde{f} : \mathcal{R}/\ker f \to f(\mathcal{R})$$

defined by $\tilde{f}(r + \ker f) = f(r)$ is a ring isomorphism.

Let $\mathbb{K}$ be a field. There exists a ring homomorphism $\chi : \mathbb{Z} \to \mathbb{K}$ which sends $1 \in \mathbb{Z}$ to $1 \in \mathbb{K}$. This simply means that we can embed $\mathbb{Z}$ into any field $\mathbb{K}$ by adding ones in $\mathbb{K}$ according to:

$$\mathbb{N} \ni n = \underbrace{1 + \ldots + 1}_{n \text{ times}} \in \mathbb{K}.$$

If the homomorphism χ is injective, i.e., $\chi^{-1}(0) = \{0\}$, then $\mathbb{K}$ is said to have **characteristic zero**. If χ is not injective, there exists a smallest positive integer p such that $\chi(p) = 0$. In this case we say that $\mathbb{K}$ has **positive characteristic** p and it is easily seen that the characteristic has to be prime. For instance, $\mathbb{C}$ and $\mathbb{Q}_p$ are fields of characteristic zero whereas finite fields have prime characteristic.

A.2. Prime numbers

An integer n is said to be **prime** if it has no other positive divisors than one and itself. The prime numbers are the multiplicative atoms of the integers. The **fundamental theorem of arithmetic** states that factorization into primes is unique up to order. In other words, every positive integer can be written as a product of prime powers, which is unique up to the sequence of the factors:

$$n = \prod_p p^{\nu_p(n)},$$

where $\nu_p(n)$ denotes the multiplicity with which a prime p divides an integer n. Note that only finitely many of the exponents differ from zero, and the empty product is by definition equal to one. By the unique prime factorization, the ring of integers $\mathbb{Z}$ is an example of a **unique factorization domain**. It is clear that the same holds for the quotient field $\mathbb{Q}$ too.

Taking into account the unique prime factorization we can define the **greatest common divisor** and the **least common multiple** of two integers m and n by

$$\gcd(m, n) - \prod_p p^{\min\{\nu_p(m), \nu_p(n)\}} \quad \text{and} \quad \operatorname{lcm}[m, n] = \prod_p p^{\max\{\nu_p(m), \nu_p(n)\}},$$

respectively. However, for computations this representation is not very convenient since factorization takes time. The gcd and the lcm of more than two integers is defined recursively.

There are infinitely many primes. If $2, p_1, \ldots, p_m$ are prime, then the integer

$$2 \cdot p_1 \cdot \ldots \cdot p_m + 1$$

must have a prime divisor different from $2, p_1, \ldots, p_m$ (otherwise it would be a divisor of 1). This ingenious argument is from Euclid (an analytic proof by Euler was given in Section 4.1). It is natural to ask how the infinitude of primes is distributed among the integers. Let $\pi(x)$ count the number of primes $p \le x$. The **prime number theorem** with the strongest existing remainder term due to Korobov and Vinogradov (independently) states

Theorem A.2. *There exists a positive constant c such that, as $x \to \infty$,*

$$\pi(x) = \int_2^x \frac{du}{\log u} + O\left(x \exp\left(-c \frac{(\log x)^{\frac{3}{5}}}{(\log \log x)^{\frac{1}{5}}} \right) \right).$$

This shows that, in first approximation, the number of primes less than or equal to x is asymptotically $\frac{x}{\log x}$ (see Exercise 4.2). Thus, the set of prime numbers is a rather *thin* set within the set of positive integers.

A.3. Riemann's hypothesis

We have already mentioned some of the basic facts on the Riemann zeta-function in Section 4.1. The error term in the prime number theorem is intimately related to the distribution of the zeros of the zeta-function. In fact,

$$(A.2) \quad \zeta(s) \ne 0 \quad \text{for} \quad \operatorname{Re} s > \theta \quad \Longleftrightarrow \quad \pi(x) - \int_2^x \frac{du}{\log u} \ll x^{\theta + \varepsilon}$$

for any $\theta \in [\frac{1}{2}, 1)$ and any positive ε. The famous yet unproved **Riemann hypothesis** states that

$$\zeta(s) \ne 0 \quad \text{for} \quad \operatorname{Re} s > \frac{1}{2}$$

or, equivalently, that all nontrivial (non-real) zeros lie on the so-called **critical line** $\operatorname{Re} s = \frac{1}{2}$. Riemann's hypothesis is number 8 on Hilbert's famous list of 23 problems. This far-reaching conjecture is yet unsolved and is one of the seven millennium problems.*

In support of his conjecture, Riemann calculated some zeros; the first one with positive imaginary part is $\varrho = \frac{1}{2} + i 14.134 \ldots$. It is known that infinitely many zeros lie on the critical line. In view of (A.2) Riemann's hypothesis implies that the error term in the prime number theorem is as small as possible. If so, in some sense, the prime numbers are as uniformly distributed as possible.

Surprisingly, the Riemann hypothesis is related to the deviation of the Farey sequence $\mathcal{F}_n$ from being distributed equidistantly. Denote the jth

---

*For more information see http://www.claymath.org/millennium/.

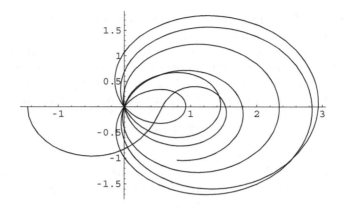

FIGURE A.1. The first nontrivial zeros of $\zeta(\frac{1}{2}+it)$ for $0 \leq t \leq 40$.

element of $\mathcal{F}_n \setminus \{0\}$ by f_j (in ascending order). Then, by results from Franel and Landau, Riemann's hypothesis is true if and only if

$$\sum_{j=1}^{F_n} \left| f_j - \frac{j}{F_n} \right| \ll n^{\frac{1}{2}+\varepsilon},$$

where $F_n := \sharp\mathcal{F}_n - 1$. For this and more information on zeta see [**86**] and [**162**].

A.4. Modular arithmetic

Let m be a positive integer. We say that the integers a and b are **congruent modulo** m if m divides $b - a$, and we write

$$b \equiv a \quad \text{mod } m.$$

The **residue class** a mod m is the set of integers b which are $\equiv a$ mod m. Obviously, there are m different residue classes modulo m, and we may represent them by $0, 1, \ldots, m - 1$ mod m. We denote the set of all these residue classes by $\mathbb{Z}/m\mathbb{Z}$. Then, carrying addition and multiplication from $\mathbb{Z}$ to $\mathbb{Z}/m\mathbb{Z}$ by

$$(a \bmod m) + (b \bmod m) = (a + b) \bmod m,$$
$$(a \bmod m) \cdot (b \bmod m) = (a \cdot b) \bmod m,$$

the set $\mathbb{Z}/m\mathbb{Z}$ becomes a ring, the **ring of residue classes modulo** m.

Theorem A.3. *The ring $\mathbb{Z}/m\mathbb{Z}$ is a field if and only if m is prime.*

If m is not prime, then $\mathbb{Z}/m\mathbb{Z}$ contains **divisors of zero**, i.e., there are non-zero integers a, b such that $ab \equiv 0$ mod m.

We call a residue class a mod m **prime** if a and m are coprime. In view of the Euclidean algorithm a residue class a mod m is prime if and only if it is invertible in $\mathbb{Z}/m\mathbb{Z}$. The set

$$(\mathbb{Z}/m\mathbb{Z})^* = \{a \bmod m : \gcd(a, m) = 1\},$$

called **group of prime residue classes modulo** m, is the multiplicative group of $\mathbb{Z}/m\mathbb{Z}$; its group order is $\varphi(m)$, where $\varphi(m)$ is Euler's totient (see Section 2.7). In the sequel we write simply a instead of a mod m.

Dirichlet's prime number theorem for arithmetic progressions states that the prime numbers are equally distributed in the prime residue classes, i.e., the number of primes $p \leq x$ which are $\equiv a$ mod m is asymptotically

$$\frac{1}{\varphi(m)}\pi(x),$$

provided a is coprime with m.

The **order** of an element $a \in (\mathbb{Z}/m\mathbb{Z})^*$ is the smallest positive integer k such that $a^k \equiv 1$ mod m. **Fermat's little theorem** states that, for prime p and an integer a not divisible by p,

$$a^{p-1} \equiv 1 \bmod p.$$

However, the converse implication does not hold since, for example,

$$a^{561-1} \equiv 1 \bmod 561 \qquad \text{for all} \quad a \quad \text{with} \quad (a, 561) = 1,$$

but $561 = 3 \cdot 11 \cdot 17$. Another example is the taxicab number 1729. Such numbers are called **Carmichael numbers** and, unfortunately, there are infinitely many of them; otherwise Fermat's little theorem could be used as a very fast primality test. Nevertheless, in practice the converse of Fermat's little theorem is often applied to generate numbers which are *very likely* prime.

Note that for prime p the group $(\mathbb{Z}/p\mathbb{Z})^*$ is cyclic; the generating element is called **primitive root** mod p (in fact there are $\varphi(p-1)$ many such primitive roots).

There is a slightly stronger formulation of Fermat's little theorem due to Euler (which we also refer to as Fermat's little theorem):

Theorem A.4. *For a coprime with m,*

$$a^{\varphi(m)} \equiv 1 \bmod m.$$

This is nothing else but Lagrange's theorem (A.1) applied to the multiplicative group modulo m.

A very useful tool in elementary number theory is the Chinese remainder theorem. It shows that one can reconstruct integers in a certain range from their residues modulo a set of coprime moduli.

Theorem A.5. *Let $m_1, \ldots, m_r$ be positive integers, pairwise coprime, and let $a_1, \ldots, a_r$ be arbitrary integers. Then there exists an integer x such that*

$$x \equiv a_j \bmod m_j \qquad \text{for} \quad 1 \leq j \leq r;$$

moreover, x is uniquely determined mod $m_1 \cdot \ldots \cdot m_r$.

In view of Fermat's little theorem A.4

$$x \equiv \sum_{j=1}^{r} a_j \left(\frac{m_1 \cdot \ldots \cdot m_r}{m_j}\right)^{\varphi(m_j)} \bmod m_1 \cdot \ldots \cdot m_r$$

solves the system of congruences above; the uniqueness is a bit harder to establish.

A.5. Quadratic residues

Let p be an odd prime. A non-zero residue class $a \bmod p$ is said to be a **quadratic residue modulo** p if the quadratic congruence

$$X^2 \equiv a \bmod p$$

has a solution; otherwise a is called a **quadratic non-residue**. We define the **Legendre symbol** by

$$\left(\frac{a}{p}\right) = \begin{cases} +1 & \text{if } a \text{ is a quadratic residue } \bmod p, \\ -1 & \text{otherwise.} \end{cases}$$

It is easily seen that half of the $p-1$ non-zero residue classes modulo p are quadratic residues and the other half are non-residue classes.

Conjectured by Euler and Lagrange and proved by Gauss, the **quadratic reciprocity law** plays a central role in the theory of quadratic residues.

Theorem A.6. *For two different odd primes* p, q,

$$\left(\frac{q}{p}\right)\left(\frac{p}{q}\right) = (-1)^{\frac{p-1}{2}\cdot\frac{q-1}{2}}.$$

Moreover,

$$\left(\frac{-1}{p}\right) = (-1)^{\frac{p-1}{2}} \qquad and \qquad \left(\frac{2}{p}\right) = (-1)^{\frac{p^2-1}{8}}.$$

Due to Gauss the first of the statements is often called *theoreum aureum*; the second and third formula are the supplementary laws.

The Legendre symbol $(\frac{\cdot}{p})$ is multiplicative and p-periodic, and so Theorem A.6 allows one to compute quickly whether a given integer is a quadratic residue modulo some given prime or not. For instance,

$$\left(\frac{7}{19}\right) = -\left(\frac{19}{7}\right) = -\left(\frac{21-2}{7}\right) = -\left(\frac{-1}{7}\right)\left(\frac{2}{7}\right) = +1,$$

and so 7 is a quadratic residue $\bmod 19$. Conversely, one can compute for which primes p a given integer is a quadratic residue. For example, by the quadratic reciprocity law, 7 is a quadratic residue $\bmod p$ if and only if

$$\left(\frac{p}{7}\right) = (-1)^{\frac{p-1}{2}}.$$

The quadratic residues $\bmod 7$ are $1, 2, 4$, and the non-residues are $3, 5, 6$. Hence it follows that 7 is a quadratic residue modulo p if and only if

$$p \equiv 1, 3, 9, 19, 25, 27 \bmod 28.$$

(This is of interest for Exercise 14.6.)

A.6. Polynomials

Let $\mathcal{R}$ be a ring. A **polynomial** in one variable X over $\mathcal{R}$ is given by

$$P(X) = a_d X^d + a_{d-1} X^{d-1} + \ldots + a_1 X + a_0,$$

where the coefficients a_j lie in $\mathcal{R}$. If a_d is not zero, then we say that a_d is the **leading term** and that P has **degree** d, denoted by $\deg P = d$. If the leading term is equal to 1, P is called **monic**. The set of all polynomials in one variable with coefficients in $\mathcal{R}$ forms a ring, the **polynomial ring** over $\mathcal{R}$, and is denoted by $\mathcal{R}[X]$. Addition and multiplication is defined by superposition. Many properties of the ring of coefficients $\mathcal{R}$ can be transported to the polynomial ring $\mathcal{R}[X]$.

A polynomial $P \in \mathcal{R}[X]$ is said to be **reducible** if it splits into a product of non-constant polynomials (within $\mathcal{R}[X]$); otherwise the polynomial is called **irreducible**. If the ring $\mathcal{R}$ is a unique factorization domain, then an analogue of division with remainder holds also for polynomials in $\mathcal{R}[X]$: for $P_1, P_2 \in \mathcal{R}[X]$ with $P_2(X) \not\equiv 0$, there exist polynomials $Q, R \in \mathcal{R}[X]$ such that

$$P_1(X) = Q(X)P_2(X) + R(X) , \qquad \text{where} \quad \deg R < \deg P_2;$$

note that Q and R are unique and the size of the remainder is measured by the degree. Consequently, one can define the greatest common divisor of polynomials with integer coefficients in, more or less, the same way as for integers.

The **fundamental theorem of algebra** states that every non-constant polynomial P with complex coefficients has a root or, equivalently, that every non-constant P splits over $\mathbb{C}$ into linear factors. A field $\mathbb{K}$ is said to be **algebraically closed** if it contains all roots of any polynomial in $\mathbb{K}[X]$. Therefore, an alternative formulation of the fundamental theorem of algebra is

Theorem A.7. $\mathbb{C}$ *is algebraically closed.*

Now we consider multivariate polynomials. In some sense, the polynomial ring over $\mathbb{Z}$ has a similar arithmetic as the ring of integers $\mathbb{Z}$. The role of the prime numbers is played by irreducible polynomials. An important result concerning factorization of polynomials over $\mathbb{Z}$ is **Gauss' lemma**:

Theorem A.8. *For $P_1, P_2 \in \mathbb{Z}[X_1, \ldots, X_n]$ let $\gamma(P_i)$ denote the greatest common divisor of the coefficients of P_i. Then*

$$\gamma(P_1 P_2) = \gamma(P_1)\gamma(P_2).$$

It follows that every polynomial which can be factored over $\mathbb{Q}$ can also be factored over $\mathbb{Z}$ and vice versa. Without loss of generality, let P be a polynomial with coprime integer coefficients. If now $P = P_1 \cdot P_2$, where the P_i are polynomials with coefficients in $\mathbb{Q}$, then we can find rational numbers a_1, a_2 such that $a_1 P_1$ and $a_2 P_2$ are polynomials with integral coefficients and $\gamma(a_1 P_1) = \gamma(a_2 P_2) = 1$. From Gauss' lemma we obtain $\gamma(a_1 a_2 P) = 1$. Since $\gamma(P) = 1$ by assumption, it follows that $a_1 a_2 = \pm 1$. Choosing the sign of a_1 we get $P = a_1 P_1 \cdot a_2 P_2$. The converse implication follows in a similar manner.

A polynomial $P(X_1, \ldots, X_n)$ is said to be **symmetric** if for any permutation σ of $\{1, \ldots, n\}$

$$P(X_{\sigma(1)}, \ldots, X_{\sigma(n)}) = P(X_1, \ldots, X_n).$$

The **elementary symmetric polynomials** $s_1, \ldots s_n$ in n variables $x_1, \ldots, x_n$ appear as coefficients in the monic polynomial of degree n and roots $x_1, \ldots, x_n$:

$$(X - x_1) \cdot \ldots \cdot (X - x_n) = X^n - s_1 X^{n-1} \pm \ldots + (-1)^n s_n.$$

It is easily seen that the s_j are homogeneous and, indeed, symmetric polynomials. For instance,

$$s_1 = x_1 + \ldots + x_n \qquad \text{and} \qquad s_n = x_1 \cdot \ldots \cdot x_n.$$

The **fundamental theorem of symmetric polynomials** claims that every symmetric polynomial can be written as a polynomial in the elementary symmetric polynomials.

Theorem A.9. *Let* $P \in \mathbb{Q}[x_1, \ldots, x_n]$ *be symmetric. Then there exists a polynomial* $Q \in \mathbb{Q}[X_1, \ldots, X_n]$ *such that*

$$P(x_1, \ldots, x_n) = Q(s_1, \ldots, s_n).$$

The constructive proof goes by induction on n and also provides an estimate for the degree of Q (see e.g. [**13**] or [**99**]).

A.7. Algebraic number fields

In Section 9.1 we have already introduced the notion of an algebraic number. Recall that a complex number α is said to be algebraic over $\mathbb{Q}$ if there exists a non-zero polynomial $P(X)$ with integer coefficients such that $P(\alpha) = 0$; the polynomial with coprime coefficients and least degree having this property is called the **minimal polynomial** of α, and we denote it by $P_\alpha(X)$. The degree of the minimal polynomial is said to be the **degree** of α; for short $d := \deg \alpha = \deg P_\alpha$. For algebraic α, the set

$$\mathbb{Q}(\alpha) := \mathbb{Q}[X]/P_\alpha(X)$$

is a finite algebraic extension of the field of rational numbers, the **algebraic number field** associated to α. A more convenient form is

$$\mathbb{Q}(\alpha) \cong \left\{ a_0 + a_1 \alpha + \ldots + a_{d-1} \alpha^{d-1} : a_j \in \mathbb{Q} \right\}.$$

The **degree** of the field extension $\mathbb{Q}(\alpha)/\mathbb{Q}$ is equal to the dimension of the field $\mathbb{Q}(\alpha)$ as a $\mathbb{Q}$-vector space; we write $d = [\mathbb{Q}(\alpha) : \mathbb{Q}(\alpha)]$. Note that any number in $\mathbb{Q}(\alpha)$ is algebraic of degree $\leq d$.

The zeros $\alpha'_1, \ldots, \alpha'_d$ of the minimal polynomial $P_\alpha(X)$ are the **conjugates** of α (in fact, they are the images of α under the field automorphisms) and have degree equal to $d = \deg \alpha$. The product of all conjugates is the **norm** of α:

$$N(\alpha) := \prod_{j=1}^{d} \alpha'_j.$$

This quantity is, up to the sign, equal to the constant term $P_\alpha(0) \in \mathbb{Z}$ in the minimal polynomial. The norm provides a measure for the size of algebraic numbers.

An algebraic number is said to be an **algebraic integer** if its minimal polynomial is monic. This extends the notion of integers from $\mathbb{Q}$ to number fields. In fact, one can show that an algebraic integer, which is rational, is a rational integer. The set of all algebraic integers in a number field forms a ring, the so-called **ring of integers**. Unfortunately, these rings in general do not have a unique prime factorization. For example, the identity

$$2 \cdot 3 = (1 - \sqrt{-5}) \cdot (1 + \sqrt{-5})$$

gives two distinct factorizations of 6 in the ring $\mathbb{Z}[\sqrt{-5}]$ into irreducible factors; one can overcome this problem by introducing *prime ideals* but this is another story.

In the case $\deg \alpha = 2$ we call $\mathbb{Q}(\alpha)$ a **quadratic** number field. It is easily seen that there always exists a squarefree $d \in \mathbb{Z}$ such that $\mathbb{Q}(\alpha) = \mathbb{Q}(\sqrt{d})$. We say that $\mathbb{Q}(\sqrt{d})$ is a **real** or **imaginary** quadratic number field according to $\sqrt{d} \in \mathbb{R}$ or not. One can show that the ring of integers equals

$$\mathbb{Z}[\sqrt{d}] \qquad \text{or} \qquad \mathbb{Z}\left[\frac{1 + \sqrt{d}}{2}\right]$$

according to $d \equiv 2, 3 \bmod 4$ or $d \equiv 1 \bmod 4$. For example, the golden ratio $G = \frac{1+\sqrt{5}}{2}$ is an algebraic integer in the number field $\mathbb{Q}(\sqrt{5})$ and $\mathbb{Z}[G]$ is the ring of integers in $\mathbb{Q}(\sqrt{5})$.

The conjugates of algebraic integers are of the form $\mathcal{X} \pm \mathcal{Y}\sqrt{d}$, where $\mathcal{X}, \mathcal{Y}$ are rational with denominators ≤ 2. Consequently, the norm of an element $\mathcal{X} + \mathcal{Y}\sqrt{d}$ is given by

(A.3) $$N(\mathcal{X} + \mathcal{Y}\sqrt{d}) = \mathcal{X}^2 - d\mathcal{Y}^2.$$

Thus, the question whether or not a rational number q is the norm of some algebraic integer is related to the solvability of the equation

$$aX^2 + bY^2 + cZ^2 = 0 \qquad \text{with} \quad d = -\frac{b}{a}, \; q = -\frac{c}{a},$$

where a is a non-zero rational number. By the theorem of Hasse–Minkowski in the form of Theorem 14.3, the answer to this problem is encoded in the local representations of zero of the underlying ternary quadratic form.

For an understanding of the structure of number fields it is important to study its integers and, in particular, its units. An algebraic integer is said to be a **unit** if the absolute value of its norm is equal to one. By (A.3), the units of a real quadratic number field are of the form

$$\varepsilon = x + y\sqrt{d} \qquad \text{with} \quad N(\varepsilon) = x^2 - dy^2 = 1$$

if $d \equiv 2, 3 \bmod 4$, resp.

$$\varepsilon = x + \frac{1 + y\sqrt{d}}{2} \qquad \text{with} \quad N(\varepsilon) = x^2 + xy + \frac{1-d}{4}y^2 = \pm 1$$

if $d \equiv 1 \bmod 4$, where $x, y \in \mathbb{Z}$. Both norm unit equations are included under the form

$$X^2 - dY^2 = \pm 4.$$

A real quadratic number field has an infinitude of units; in fact, this follows from Theorem 6.2 if $d \equiv 2, 3 \bmod 4$ (the other case can be treated in a similar way). An imaginary quadratic number field has only finitely many units, each of them being a root of unity (for more details see [40] or [76]). In both cases, real quadratic or imaginary quadratic, the units form a cyclic group. A generator of this group is called **fundamental unit**.

Thus, finding the minimal solution of the Pell equation is equivalent to computing the fundamental unit ε of a quadratic number field $\mathbb{Q}(\sqrt{d})$. This is an important problem in computational algebraic number theory. Hallgren [75] obtained an algorithm based on quantum computation for finding the fundamental unit of $\mathbb{Q}(\sqrt{d})$ (i.e., the minimal solution of Pell's equation) in polynomial time; for a deterministic approach see Lenstra [103].

A.8. Kummer's work on Fermat's last theorem

The most remarkable approach toward a solution of Fermat's last theorem in the nineteenth century was taken by Kummer. His research led to the foundation of algebraic number theory. In fact, Kummer discovered a close relation with questions concerning the arithmetic of fields of the form $\mathbb{Q}(\zeta_n)$, where ζ_n is a root of unity $\zeta_n = \exp(\frac{2\pi i}{n})$ (or any other root of degree n over $\mathbb{Q}$); such a field is called a **cyclotomic field**. His initial idea was to factor the Fermat equation over $\mathbb{Q}(\zeta_n)$:

$$Z^n = X^n + Y^n = (X + Y)(X + \zeta_n Y) \cdot \ldots \cdot (X + \zeta_n^{n-1} Y).$$

Since $\mathbb{Q} \subset \mathbb{Q}(\zeta_n)$, for a proof of Fermat's last theorem it suffices to show that there are no non-trivial solutions in the cyclotomic field $\mathbb{Q}(\zeta_n)$. But what is the advantage of enlarging the field of possible solutions? Recall the proofs of Theorems 1.1 and 1.2, the solution of Fermat's last theorem for the exponents $n = 2$ and 4. The crucial step was a factorization of the Fermat equation which implied that all factors are powers. Kummer's factorization leads to a similar start for the general case; however, one cannot conclude that the factors on the right-hand side above are powers since, in general, there is no unique prime factorization in number fields.

An important notion in Kummer's work is the one of regular primes. A prime p is said to be **regular** if p does not divide the class number h_p of $\mathbb{Q}(\zeta_p)$, where the **class number** h_p is defined as the order of the group of non-zero fractional ideals of $\mathbb{Q}(\zeta_p)$; we do not give a definition of the notion of fractional ideals but note that, in some sense, the class number measures how far the ring of integers is from having unique prime factorization. On behalf of deep investigations on the arithmetic of cyclotomic fields Kummer proved in 1847 that Fermat's last theorem is true for regular prime exponents p. Since a prime p divides h_p if and only if p divides the numerator of some Bernoulli number B_{2k} for $k \leq \frac{p-3}{2}$ (introduced in Section 4.2), this gives a criterion for checking whether a given prime is regular or not. For instance, 13 does not divide the numerators of the Bernoulli numbers

$$B_0 = 1, \ B_2 = \frac{1}{6}, \ B_4 = -\frac{1}{30}, \ B_6 = \frac{1}{42}, \ B_8 = -\frac{1}{30}, \ B_{10} = \frac{5}{66}.$$

Consequently Fermat's last theorem is true for the exponent $p = 13$. This criterion is based on Kummer's congruences (12.11) for Bernoulli numbers. The first prime which is not regular is $p = 37$. It is known that infinitely many primes are not regular but it is still unknown whether there are infinitely many regular primes or not.

Bibliography

[1] N.H. ABEL, Sur l'intégration de la formule différentielle $\varrho\,dx/\sqrt{R}$, R ét ϱ etant des fonctions entières, in: *Oeuvres Complètes de Niels Henrik Abel*, L. Sylow and S. Lie (eds.), Christiania, t. 1 (1881), 104-144

[2] M. AGRAWAL, N. KAYAL, N. SAXENA, Primes is in P, preprint 2002, available at http://www.cse.iitk.ac.in/news/primality.html

[3] S. ALBEVERIO, A. KHRENNIKOV, P.E. KLOEDEN, Memory retrieval as a p-adic dynamical system, *BioSystems* **49** (1999), 105-115

[4] K. ALLADI, M.L. ROBINSON, Legendre polynomials and irrationality, *J. Reine Angew. Math.* **318** (1980), 137-155

[5] R. APÉRY, Irrationalité de $\zeta(2)$ et $\zeta(3)$, *Astérisque* **61** (1979), 11-13

[6] S. ASTELS, *Cantor sets and numbers with restricted partial quotients*, Ph.D. thesis, University of Waterloo, 1999

[7] R.M. AVANZI, U.M. ZANNIER, Genus one curves defined by separarted variable polynomials and a polynomial Pell equation, *Acta Arith.* **99** (2001), 227-256

[8] D. BAILEY, P. BORWEIN, S. PLOUFFE, On the rapid computation of various polylogarithmic constants, *Math. Comp.* **66** (1997), 903-913

[9] A. BAKER, Approximations to the logarithms of certain algebraic numbers, *Acta Arith.* **10** (1964), 315-323

[10] A. BAKER, *Transcendental number theory*, Cambridge University Press, Cambridge, 1975

[11] A. BAKER, G. WÜSTHOLZ, Number theory, transcendence and diophantine geometry in the next millenium, in *Mathematics: frontiers and perspectives*, V. Arnold et al. (eds.), Am. Math. Soc., Providence, RI, 2000, 1-12

[12] I.N. BAKER, On a class of meromorphic functions, *Proc. Am. Math. Soc.* **17** (1966), 819-822

[13] E.J. BARBEAU, *Polynomials*, Springer, New York, NY, 1989

[14] K. BARNER, How old did Fermat become?, *NTM (N.S.)* **9** (2001), 209-228

[15] D. BARSKY, B. BENZAGHOU, Nombres de Bell et somme de factorielles, *J. Théo. Nombres Bordeaux* **16** (2004), 1-18

[16] M.A. BENNETT, B.M.M. DE WEGER, On the diophantine equation $|ax^n - by^n| = 1$, *Math. Comp.* **67** (1998), 413-438

[17] D.J. Bernstein, Sharper ABC-based bounds for congruent polynomials, *J. Théo. Nombres Bordeaux* (to appear)

[18] L. BERGGREN, J. BORWEIN, P. BORWEIN, π: *a scource book*, Springer 1997

[19] F. BEUKERS, A note on the irrationality of $\zeta(2)$ and $\zeta(3)$, *Bull. London Math. Soc.* **11** (1979), 268–272

[20] F. BEUKERS, Lattice reduction, in: *Some tapas of computer algebra*, A.M. Cohen et al. (eds.), Springer, Berlin, 1999

[21] F. BEUKERS, J.P. BÉZIVIN, P. ROBBA, An alternative proof of the Lindemann-Weierstrass theorem, *Am. Math. Mon.* **97** (1990), 193-197

[22] J.P. BÉZIVIN, P. ROBBA, A new p-adic method for proving irrationality and transcendence results, *Ann. Math.* **129** (1989), 151-160

[23] H. BOHR, E. LANDAU, Über das Verhalten von $\zeta(s)$ und $\zeta^{(k)}(s)$ in der Nähe der Geraden $\sigma = 1$, *Nachr. Ges. Wiss. Göttingen Math. Phys. Kl.* (1910), 303-330

[24] E. BOMBIERI, The Mordell conjecture revisited, *Ann. Sc. Norm. Super. Pisa*, Cl. Sci. IV. **17** (1990), 615-640

[25] E. BOMBIERI, Roth's theorem and the *abc*-conjecture, preprint ETH Zürich, 1994

[26] E. BOMBIERI, Forty years of effective results in diophantine theory, in: *A panorama of number theory or the view from Baker's garden*, Proceedings from the "Conference on Number Theory and Diophantine Geometry at the Gateway to the Millennium," Zürich 1999, G. Wüstholz (ed.), Cambridge University Press, Cambrdige, 2002, 194-213

[27] E. BOMBIERI, A.J. VAN DER POORTEN, Continued fractions of algebraic numbers, in: *Computational algebra and number theory*, W. Bosma and A.J. van der Poorten (eds.), Kluwer, Dordrecht, 1995, 137-152

[28] J. BORWEIN, D. BRADLEY, Empirically determined Apéry-like formulae for $\zeta(4n+3)$, *Exp. Math.* **6** (1997), 181-194

[29] D. BRESSOUD, S. WAGON, *Computational number theory*, Key College Publishing, Emeryville, CA, 2000, in cooperation with Springer

[30] C. BREZINSKI, *History of continued fractions and Padé approximants*, Springer, Berlin, 1991

[31] J. BRILLHART, M.A. MORRISON, A method of factoring and the factorization of F_7, *Math. Comp.* **29** (1975), 183-205

[32] R.B. BURCKEL, Vol.1, *An introduction to classical complex analysis*, Academic Press, New York–San Francisco, 1979

[33] E.B. BURGER, *Exploring the number jungle: a journey into Diophantine Analysis*, Am. Math. Soc., Providence, RI, 2000

[34] N. CALKIN, H.S. WILF, Recounting the rationals, *Am. Math. Mon.* **107** (2000), 360-363

[35] J.W.S. CASSELS, *An introduction to diophantine approximation*, Cambridge University Press, Cambridge, 1957

[36] J.W.S. CASSELS, *An introduction to the geometry of numbers*, Springer, Berlin, 1959

[37] J.W.S. CASSELS, *Local fields*, Cambridge University Press, Cambridge, 1986

[38] R. CHAPMAN, A proof of Hadjicosta's conjecture, preprint arXiv:math.NT/0405478 v2 (15 Jun 2004) available at http://front.math.ucdavis.edu/

[39] T. COCHRANE, R.E. DRESSLER, Gaps between integers with the same prime factorization, *Math. Comp.* **68** (1999), 395-401

[40] H. COHEN, *A course in computational algebraic number theory*, Springer, Berlin, 1993

[41] J.H. CONWAY, N.J.A. SLOANE, *Sphere packings, lattices and groups*, Springer, New York, NY, 1999, 3rd ed.

[42] R. CRANDALL, C. POMERANCE, *Prime numbers. A computational perspective*, Springer, Berlin, 2001

[43] H. DARMON, A. GRANVILLE, On the equations $z^m = F(x, y)$ and $Ax^p + By^q = Cz^r$, *Bull. London Math. Soc.* **27** (1995), 513-543

[44] H. DAVENPORT, On $f^3(t) - g^2(t)$, *K. Norske Vid. Selsk. Forrh. (Trondheim)*, **38** (1965), 86-87

[45] M. DAVIS, H. PUTNAM, J. ROBINSON, The decision problem for exponential diophantine equations, *Ann. Math.* **74** (1961), 425-436

[46] W. DIFFIE, M. HELLMAN, New directions in cryptography, *IEEE Trans. Inform. Theory* **22** (1976), 644-654

[47] T. DOKCHITSER, *LLL & ABC*, *J. Number Theory* **107** (2004), 161-167

[48] A. DUBICKAS, J. STEUDING, The polynomial Pell equation, *Elem. Math.* **59** (2004), 133-143

[49] A. DUJELLA, Continued fractions and RSA with small secret exponent, preprint arXiv:math.CR/0402052, available at http://front.math.ucdavis.edu/

[50] E. DUNNE, M. MCCONNELL, Pianos and continued fractions, *Math. Mag.* **72** (1999), 104-115

[51] H.-D. EBBINGHAUS ET AL., *Numbers*, Springer, New York–Heidelberg–Berlin, 1991

[52] H.M. EDWARDS, *Fermat's last theorem*, Springer, New York–Heidelberg–Berlin, 1977

[53] N.D. ELKIES, *ABC* implies Mordell, *Intern. Math. Res. Not.* **7** (1991), 99-109

[54] J.S. ELLENBERG, Congruence *ABC* implies *ABC*, *Indagationes Math.* **11** (2000), 197-200

[55] C. ELSNER, J.W. SANDER, J. STEUDING, Kettenbrüche als Summen ebensolcher, *Math. Slov.* **51** (2001), 281-293

[56] M. EINSIEDLER, A. KATOK, E. LINDENSTRAUSS, Invariant measures and the set of exceptions to Littlewood's conjecture, preprint 2003, available at http://www.math.princeton.edu/simmeinsied/research/

[57] P. ERDÖS, Representation of real numbers as sums and products of Liouville numbers, *Michigan Math. J.* **9** (1962), 59-60

[58] P. ERDÖS, How many pairs of products of consecutive integers have the same prime factors?, *Am. Math. Mon.* **87** (1980), 391-392

[59] J.-H. EVERTSE, The subspace theorem of W.M. Schmidt, in: *Diophantine approximations and abelian varieties*, Introductory lectures, B. Edixhoven, J.-H. Evertse (eds.), Lecture Notes in Mathematics 1566, Springer, Berlin, 1993, 31-50

[60] G. FALTINGS, Diophantine approximations on abelian varieties, *Ann. Math.* **133** (1991), 549-576

[61] P. FLAJOLET, B.VALLÉE, I. VARDI, Continued fractions from Euclid to the present day, preprint 2000, http://www.lix.polytechnique.fr/Labo/Ilan.Vardi/publications.html

[62] L.R. FORD, Fractions, *Am. Math. Mon.* **45** (1938), 586-601

[63] M. VAN FRANKENHUYSEN, The abc-conjecture implies Roth's theorem and Mordell's conjecture, *Math. Contemp.* **16** (1999), 45-72

[64] G.A. FREIMAN, *Diophantine approximations and the geometry of numbers (Markov's problem)*, Kalinin. Gosudarstv. Univ., Kalinin 1975 [Russian]

[65] C.F. GAUSS, *Disquisitiones arithmeticae*, Yale University Press, New Haven 1966 [translated by A.A. Clarke]

[66] D. GOLDFELD, Modular forms, elliptic curves and the abc-conjecture, in: *A panorama of number theory or the view from Baker's garden*, Proceedings from the "Conference on Number Theory and Diophantine Geometry at the Gateway to the Millennium," Zürich 1999, G. Wüstholz (ed.), Cambridge University Press, Cambridge, 2002, 128-147

[67] J. GOLLNICK, H. SCHEID, J. ZÖLLNER, Rekursive Erzeugung der primitiven pythagoreischen Tripel, *Math. Sem.* **39** (1992), 85-88

[68] F.Q. GOUVÊA, *p-adic numbers*, Springer, Berlin, 1993

[69] A. GRANVILLE, H. STARK, abc implies no "Siegel zeros" for *L*-functions of characters with negative discriminant, *Invent. Math.* **139** (2000), 509-523

[70] A. GRANVILLE, T.J. TUCKER, It's as easy as abc, *Not. Am. Math. Soc.* (2002), 1224-1231

[71] A. GRYTCZUK, F. LUCA, M. WÓJTOWICZ, The negative Pell equation and Pythagorean triples, *Proc. Jpn. Acad. Ser. A Math. Sci.* **76** (2000), 91-94

[72] T.C. HALES, Cannonballs and honeycombs, *Not. Am. Math. Soc.* **47** (2000), 440–449

[73] M. JR. HALL, On the sum and product of continued fractions, *Ann. Math.* **28** (1947), 966-993

[74] M. JR. HALL, The diophantine equation $x^3 - y^2 = k$, in *Computers in number theory*, A.O.L. Atkin and B.J. Birch (eds.), Academic Press, London 1971, 173-198

[75] S. HALLGREN, Polynomial-time quantum algorithms for Pell's equation and the principal ideal problem, preprint 2002, available at http://www.cs.caltech.edu/~hallgren/

[76] G.H. HARDY, E.M. WRIGHT, *An introduction to the theory of numbers*, Clarendon Press, Oxford, 1979, 5th ed.

[77] H. HASSE, Über die Darstellbarkeit von Zahlen durch quadratische Formen im Körper der rationalen Zahlen, *J. Reine Angew. Math.* **152** (1923), 129-148

[78] F. HAZAMA, Pell equations for polynomials, *Indagationes Math.* **8** (1997), 387-397

[79] H. HEILBRONN, On the average length of a class of finite continued fractions, in *Number Theory and Analysis (Papers in Honor of Edmund Landau)*, Plenum, New York 1969, 87–96

[80] K. HENSEL, Über eine neue Begründung der Theorie der algebraischen Zahlen, *Jahresber. Dtsch. Math. Ver.* **6** (1897), 83-88

[81] M. HINDRY, J.H. SILVERMAN, *Diophantine geometry. An introduction*, Springer, New York, NY, 2000

[82] M.M. HJORTNAES, Overforing av rekken $\sum_{k=1}^{\infty}(1/k)^3$ til et bestemt integral, in *Proc. 12th Cong. Scand. Maths* (at Lund 1953), Lund 1954

[83] E. HLAWKA, *Theorie der Gleichverteilung*, BIB, Mannheim, 1979

[84] M. VAN HOEIJ, Factoring polynomials and the knapsack problem, *J. Number Theor.* **95** (2002), 167-189

[85] L.K. HUA, *Introduction to number theory*, Springer, Berlin–Heidelberg–New York, 1982

[86] M.N. HUXLEY, *The distribution of prime numbers*, Oxford, Clarendon Press, 1972

[87] M.N. HUXLEY, Exponential sums and lattice points. III, *Proc. London Math. Soc.* **87** (2003), 591-609

[88] H. IWANIEC, E. KOWALSKI, *Analytic number theory*, Am. Math. Soc., Providence, RI, 2004

[89] J.P. JONES, Y.V. MATIYASEVICH, Proof of recursive unsolvability of Hilbert's tenth problem, *Am. Math. Mon.* **98** (1991), 689-709

[90] A. KHINTCHINE, *Continued fractions*, Noordhoff Ltd., Groningen 1963, English translation of the 3rd Russian edition

[91] N. KOBLITZ, *p-adic numbers, p-adic analysis, and zeta-functions*, Springer, New York–Heidelberg–Berlin, 1984, 2nd ed.

[92] N. KOBLITZ, *A course in number theory and cryptography*, Springer, New York, NY, 1994, 2nd ed.

[93] D. KÖNIG, A. SZÜCS, Mouvement d'un point abandonné à l'intérieur d'un cube, *Palermo Rend.* **36** (1913), 79-90 [Hungarian]

[94] M. KRAITCHIK, Recherches sur la Théorie des Nombres, *Tome II Factorization*, Paris 1929, 1-7

[95] T. KUBOTA, H.W. LEOPOLDT, Eine p-adische Theorie der Zetawerte I. Einführung der p-adischen Dirichletschen L-Funktionen, *J. Reine Angew. Math.* **214/215** (1964), 328-339

[96] L. KUIPERS, H. NIEDERREITER, *Uniform distribution of sequences*, John Wiley & Sons, New York, NY, 1974

[97] J.C. LAGARIAS, The $3x+1$ problem and its generalizations, *Am. Math. Mon.* **92** (1985), 3–23

[98] S. LANG, *Introduction to diophantine approximations*, Springer, New York, 1995, 2nd ed.

[99] S. LANG, *Algebra*, Springer, New York, NY, 2002, 3rd ed.

[100] M. LANGEVIN, Sur quelques conséquences de la conjecture abc en arithmétique et en logique, *Rocky Mt. J. Math.* **26** (1996), 1031-1042

[101] D.H. LEHMER, R.E. POWERS, On factoring large numbers, *Bull. Am. Math. Soc.* **37** (1931), 770-776

[102] A.K. LENSTRA, H.W. LENSTRA, L. LOVÁSZ, Factoring polynomials with rational coefficients, *Math. Ann.* **261** (1986), 515-534

[103] H.W. LENSTRA, Solving the Pell equation, *Not. Am. Math. Soc.* **49** (2002), 182-192

[104] L. LOVÁSZ, *An algorithmic theory for numbers, graphs and convexity*, CBMS-NSF Regional Conference Series in Applied Mathematics, Philadelphia 1986

[105] K. MAHLER, *Lectures on transcendental numbers*, Lecture Notes in Mathematics 546, Springer, Berlin–Heidelberg–New York, 1976

[106] YU. I. MANIN, *A course in mathematical logic*, Springer, New York–Berlin–Heidelberg, 1977

[107] A.A. MARKOFF, Sur les formes binaires indefinies I+II, *Math. Ann.* **15** (1879), 281-309; **17** (1880), 379-400

[108] R.C. MASON, *Diophantine equations over function fields*, LMS Lecture Notes **96**, Cambridge University Press, Cambridge, 1984

[109] D.W. MASSER, Note on a conjecture of Szpiro, *Astérisque* **183** (1990), 19-23

[110] Y. MATIYASEVICH, The Diophantineness of enumerable sets, *Doklady Akad. Nauk SSSR* **191** (1970), 279-282 [Russian]

[111] A. MCBRIDE, Remarks on Pell's equation and square root algorithms, *Math. Gazette* **83** (1999), 47-52

[112] J. MCLAUGHLIN, Polynomial solutions of Pell's equation and fundamental units in real quadratic fields, *J. London Math. Soc.* **67** (2003), 16-28

[113] A.J. MENEZES, P.C. VAN OORSCHOT, S.A. VANSTONE, *Handbook of applied cryptography*, CRC Press, Boca Raton, FL, 1997

[114] M. MIGNOTTE, An inequality about factors of polynomials, *Math. Comp.* **28** (1974), 1153-1157

[115] P. MIHĂILESCU, Primary cyclotomic units and a proof for Catalan's conjecture, *J. Reine Angew. Math.* **572** (2004), 167-195

[116] H. MINKOWSKI, Über die Bedingungen, unter welchen zwei quadratische Formen mit rationalen Koeffizienten ineinander transformiert werden können, *J. Reine Angew. Math.* **106** (1890), 5-26

[117] H. MINKOWSKI, *Geometrie der Zahlen*, Teubner, Leipzig, 1896

[118] R.A. MOLLIN, *Quadratics*, CRC Press, Boca Raton, FL, 1996

[119] L.J. MORDELL, On the rational solutions of the indeterminate equation of third and fourth degrees, *Proc. Cambridge Philos. Soc.* **21** (1922), 179-192

[120] L.J. MORDELL, On the magnitude of the integer solutions of the equation $ax^2 + by^2 + cz^2 = 0$, *J. Number Theor.* **1** (1969), 1-3

[121] M.B. NATHANSON, Polynomial Pell equations, *Proc. Am. Math. Soc.* **56** (1976), 89-92

[122] Y.V. NESTERENKO, On the algebraic independence of numbers, in: *A panorama of number theory or the view from Baker's garden*, Proceedings from the "Conference on Number Theory and Diophantine Geometry at the Gateway to the Millennium," Zürich 1999, G. Wüstholz (ed.), Cambridge University Press, Cambridge, 2002, 148–167

[123] J. NEUKIRCH, *Algebraic number theory*, Springer, Berlin 1999

[124] I. NIVEN, A simple proof that π is irrational, *Bull. Am. Math. Soc.* **53** (1947), 509

[125] J. OESTERLÉ, Nouvelles approches du 'theoreme' de Fermat, *Astérisque* **161/162** (1988), 165-186

[126] M. OVERHOLT, The diophantine equation $n! + 1 = m^2$, *Bull. London Math. Soc.* **25** (1993), 104

[127] O. PERRON, *Die Lehre von den Kettenbrüchen*, vol. I, Teubner, Stuttgart, 1954, 3rd ed.

[128] B. POONEN, Computing rational points on curves, in *Surveys in number theory; papers from the Millennial conference on number theory*, University of Illinois, Urbana-Champaign, M.A. Bennet et al. (eds.), AK Peters, Natick, MA, 2003, 149-172

[129] A. VAN DER POORTEN, A proof that Euler missed, *Math. Intelligencer* **1** (1979), 195-203

[130] G. RHIN, C. VIOLA, The group structure for $\zeta(3)$, *Acta Arith.* **97** (2001), 269-293

[131] P. RIBENBOIM, *The book of prime number records*, Springer, New York–Heidelberg–Berlin, 1989, 2nd ed.

[132] P. RIBENBOIM, *Fermat's last theorem for amateurs*, Springer, New York, NY, 1999

[133] G.J. RIEGER, On the modular figure and Ford-spheres, *Math. Nachr.* **186** (1997), 225-242

[134] H. RIESEL, *Prime numbers and computer methods for factorization*, Birkhäuser, Boston–Basel–Stuttgart, 1985

[135] T. RIVOAL, La fonction zeta de Riemann prend une infinité de valeurs irrationnelles aux entiers impairs, *C.R. Acad. Sci. Paris Sér. I Math.* **331** (2000), 267-270

[136] R. RIVEST, A. SHAMIR, L. ADLEMAN, A method for obtaining digital signatures and public-key cryptosystems, *Comm. ACM* **21** (1978), 120-126

[137] A. ROBERT, *A course in p-adic analysis*, Springer, New York, NY, 2000

[138] K.F. ROTH, Rational approximations to algebraic numbers, *Mathematika* **2** (1955), 1-20

[139] K.F. ROTH, Rational approximations to algebraic numbers, in *Proceedings of the ICM 1958*, Edinburgh, Cambridge University Press, Cambridge 1960, 203-210

[140] N. SCHAPPACHER, Wer war Diophant?, *Math. Semesterber.* **45** (1998), 141-156

[141] H.P. SCHLICKEWEI, The subspace theorem and applications, *Doc. Math. J. DMV*, Extra volume ICM 1998, Vol. 2, 197-206

[142] W.M. SCHMIDT, Norm form equations, *Ann. Math.* **96** (1972), 526-551

[143] W.M. SCHMIDT, *Diophantine approximation*, Lecture Notes in Mathematics 785, Springer, Berlin–Heidelberg–New York, 1980

[144] M.R. SCHROEDER, *Number theory in science and communications*, Springer, Berlin, 1997, 3rd ed.

[145] J.A. JR. SEEBACH, L.A. STEEN, *Counterexamples in topology*, Springer, New York–Heidelberg–Berlin, 1978, 2nd ed.

[146] J.P. SERRE, *A course in Arithmetic*, Springer, New York–Heidelberg–Berlin, 1973

[147] J.O. SHALLIT, Simple continued fractions for some irrational numbers, *J. Number Theor.* **11** (1979), 209-217

[148] H.N. SHAPIRO, G.H. SPENCER, Extension of a theorem of Mason, *Comm. Pure Appl. Math.* **47** (1994), 711-718

[149] A.B. SHIDLOVSKII, *Transcendental numbers*, de Gruyter, Berlin 1989

[150] P. SHOR, Quantum computing, *Doc. Math. J. DMV*, Extra volume ICM 1998, Vol. 1, 467-486

[151] C.L. SIEGEL, *Transcendental numbers*, Annals of Mathematics studies, no. 16, Princeton University Press, Princeton, NJ, 1949

[152] J.H. SILVERMAN, J. TATE, *Rational points on elliptic curves*, Springer, New York, NY, 1992

[153] S. SINGH, *Fermat's last theorem*, Fourth Estate, London 1997

[154] S. SINGH, *The Code book*, Fourth Estate, London 1999

[155] D. SINGMASTER, The legal values of Pi, *Math. Intelligencer* **7** (1985), 69-72

[156] R. ŠLEŽEVICIENĖ, J. STEUDING, Factoring with continued fractions, weighted mediants and the Pell equation, *Proc. Sci. Semin. Fac. Phys. Math. Šiauliai Univ.* **6** (2003), 120-130

[157] J. STEUDING, Extremal values of Dirichlet L-functions in the half plane of absolute convergence, *J. Théo. Nombres Bordeaux* **16** (2004), 221-232

[158] C.L. STEWART, K. YU, On the abc conjecture, II, *Duke J.* **108** (2001), 169-181

[159] W.W. STOTHERS, Polynomial identities and hauptmoduln, *Q. J. Math.* **32** (1981), 349-370

[160] A. THUE, Über Annäherungsswerte algebraischer Zahlen, *J. Reine Angew. Math.* **135** (1909), 284-305

[161] R. TIJDEMAN, Roth's theorem, in: *Diophantine approximations and abelian varieties*, Introductory lectures, B. Edixhoven, J.-H. Evertse (eds.), Lecture Notes in Mathematics 1566, Springer, Berlin, 1993, 21-30

[162] E.C. TITCHMARSH, *The theory of the Riemann zeta-function*, Clarendon Press, Oxford, 1986, 2nd ed. (rev. by D.R. Heath-Brown)

[163] I. VARDI, Archimedes' cattle problem, *Am. Math. Mon.* **105** (1998), 305-319

[164] R.C. VAUGHAN, *The Hardy-Littlewood method*, Cambridge University Press 1981

[165] F. VIVALDI, *Experimential mathematics with Maple*, CRC Press, Boca Raton, FL, 2001

[166] P. VOJTA, *Diophantine approximations and value distribution theory*, Lecture Notes in Mathematics 1239, Springer, Berlin, 1987

[167] M. WALDSCHMIDT, Open diophantine problems, *Moscow Math. J.* **4** (2004), 245-305

[168] J.T.-Y. WANG, An effective Roth's theorem for function fields, *Rocky Mountain J. Math.* **26** (1996), 1225–1234

[169] L.C. WASHINGTON, *Elliptic curves*, CRC Press, Boca Raton, FL, 2003

[170] W.A. WEBB, H. YOKOTA, Polynomial Pell's equation, *Proc. Am. Math. Soc.* **131** (2002), 993-1006

[171] B.M.M. DE WEGER, A curious property of the eleventh Fibonacci number, *Rocky Mountain J. Math.* **25** (1995), 977-994

[172] A. WEIL, *Number theory. An approach through history. From Hammurapi to Legendre*, Birkhäuser, Boston–Basel–Stuttgart, 1983

[173] M.J. WIENER, Crypt analysis of short RSA secret exponents, *IEEE Trans. Inform. Theory* **36** (1990), 553-558

[174] A. WILES, Modular elliptic curves and Fermat's last theorem, *Ann. Math.* **141** (1995), 443-551

[175] H. ZASSENHAUS, On Hensel factorization, *J. Number Theor.* **1** (1969), 291-311

[176] W. ZUDILIN, One of the numbers $\zeta(5), \zeta(7), \zeta(9), \zeta(11)$ is irrational, *Uspekhi Mat. Nauk [Russian Math. Surveys]* **56** (2001), 149-150

Index